Praise for **The Upside of Down**:

"Anyone who doubts the seriousness of the human predicament should read Thomas Homer-Dixon's brilliant *The Upside of Down*. Anyone who understands the seriousness should also read it."
—Paul R. Ehrlich, President,
Center for Conservation Biology at Stanford University

"This book is an impressive achievement . . . thoughtful and thought-provoking. . . . For those who want a clear and accessible overview of this catastrophist debate, and one with a Canadian flavour . . . this is a useful place to start."
—*The Globe and Mail*

"[*The Upside of Down*] introduces general readers to . . . the links between population growth, environmental degradation, and global security. . . . Like any good futurist, Homer-Dixon repeatedly reminds the reader that the future is never truly written."
—*Toronto Star*

"*The Upside of Down* is a potent distillation of information from a range of disciplines that is both stimulating and profoundly disturbing."
—*The StarPhoenix* (Saskatoon)

"An extraordinarily important book. If you read only one book this year about the end of civilization, let this be it."
—*Quill & Quire*

"There isn't any book out there quite like it. . . . [Thomas Homer-Dixon] is fast emerging as an especially eloquent voice among the English-speaking world's most insightful thinkers on the great challenges of our time. . . . Well-written, engaging, and lively, from the beginning to the end."
—*Georgia Straight*

"[T]he details of *The Upside of Down*—an important and sometimes hopeful book—offer a . . . compelling picture of human ecology in the broadest imaginable sense."
—TheTyee.ca

"Thomas Homer-Dixon is the giant-killer of overwhelming issues."
—*Toronto Star*

"For over a decade, Thomas Homer-Dixon has provided that rare thing: a bridge between leading-edge research and the lay reader. Now, addressing the greatest problems of our time, he points us towards a path forward."
—Robert D. Kaplan, Senior Fellow,
Foreign Policy Research Institute

THE
UPSIDE OF
DOWN

Catastrophe, Creativity, and the Renewal of Civilization

Thomas Homer-Dixon

Vintage Canada

VINTAGE CANADA EDITION, 2007

Published in Canada by Vintage Canada, a division of Random House of Canada Limited, Toronto, in 2007. Originally published in hardcover in Canada by Alfred A. Knopf Canada, a division of Random House of Canada Limited, Toronto, in 2006. Distributed by Random House of Canada Limited, Toronto.

Vintage Canada and colophon are registered trademarks of Random House of Canada Limited.

www.randomhouse.ca

LIBRARY AND ARCHIVES CANADA CATALOGUING IN PUBLICATION

Homer-Dixon, Thomas, 1956–
      The upside of down : catastrophe, creativity and the renewal of civilization / Thomas Homer-Dixon.

Includes bibliographical references and index.
ISBN 978-0-676-97723-3

      1. Regression (Civilization). 2. Social change. 3. Problem solving.
4. Regression (Civilization)—History. 5. Social change—History.
6. Problem solving—History. I. Title.

CB161.H636 2007   901   C2007-900748-1

Text design: CS Richardson
Printed and bound in the United States of America

10 9 8 7 6 5 4 3 2 1

For Sarah
Per annos amor

CONTENTS

FIRESTORM

San Francisco, Thursday, April 19, 1906

THE WIND had shifted. Now the inferno turned its attention west-ward. Block by block, it savaged some of the city's finest houses. As the mayor, chief of police, and members of the municipal council retreated from building to building before the flames, they decided the city would make one last stand.

The final line of defense, they announced, would be Van Ness Avenue—a broad residential boulevard bisecting San Francisco from north to south. The street lay directly in the fire's path: if they could use it as a firebreak, they might be able to halt the advance. But if this last effort failed, what remained of the city would surely be lost.

Early the previous day, an enormous earthquake had shattered the city's core, snapping cast-iron water mains like twigs, toppling thou-sands of chimneys, and upending coal-burning stoves and boilers. Electrical utility poles fell over, bringing down live wires in showers of sparks. Gas lines ruptured. Kerosene and oil poured out of burst fuel tanks. In seconds, sparks and fuel combined, and dozens of fires exploded across the city. Then, energized by the wood in the city's buildings, small fires coalesced into mighty firestorms. Even when fire-fighters could maneuver around the piles of earthquake debris in the streets, they found no water in the hydrants.

By noon on the 19th, the fire had destroyed almost ten square kilo-meters of the city east of Van Ness Avenue. The financial district, Market Street, and the district south of Market were smoking ruins,

Chinatown was ablaze, and the docks, ferry terminal, and Telegraph Hill were under siege. The U.S. army had tried to deprive the fire of easily combustible material by blowing up hundreds of undamaged buildings in front of the flames. But so far their efforts had been futile, and supplies of dynamite were almost gone.

Orders went out to concentrate all soldiers, police, workers, and fire engines for the climactic fight along a sixteen-block section of Van Ness. They would raze the houses along the east side of the boulevard. So the last pounds of dynamite were brought on wagons from the Presidio and Alcatraz, placed in the buildings' basements, and connected to fuses. Police and volunteers rushed from house to house to evacuate residents. And because there wasn't enough dynamite, the army wheeled field cannon into position along the west side of Van Ness. The guns' muzzles pointed across the street.

The fire crested Nob Hill a few blocks to the east, enveloped the brand-new Fairmont Hotel, and now surged down California, Sacramento, and Washington Streets toward Van Ness. A broad wall of flames and smoke closed in on the defenders.

At 4 p.m., the cannon opened fire on the elegant mansions lining the east side of the street. "The sight was one of stupendous and appalling havoc," wrote a correspondent for *The New York Times*, "as the cannons were trained on the palaces and the shot tore into the walls and toppled the buildings in ruins." Simultaneously fuses were lit, and as the dynamite exploded, "the dwellings of millionaires were lifted into the air by the power of the blast and dropped to the earth a mass of dust and debris."

For hours, above the roar of the approaching flames, the air shook with the steady concussion of exploding artillery shells and dynamite. When the fire reached Van Ness, it seemed it would breach the defensive line. "The fire spread across the broad thoroughfare," wrote the *Times* correspondent, "and the entire western addition, which contains the homes of San Francisco's wealthier class, seemed doomed." But when the smoke cleared the next morning, the defenders found to their joy that their strategy had been largely successful. The flames had jumped the street in only a few places.

By then, though, it was obvious that much of the city had been obliterated. Hundreds of thousands of people had no shelter or food, and authorities feared famine and epidemic. Looking across San Francisco's smoldering hulk that day, no one could have imagined that such appalling destruction would also produce some good; that it would not only lead to a rejuvinated city but also trigger a wave of events that would sweep around the world and, years hence, help create the Federal Reserve System of the United States—the country's bank of last resort, an essential defense against financial panic, and one of the most important new institutions of the twentieth century.[1]

Rome, Tuesday, May 13, 2003

The late-afternoon Italian sun is low in the sky, but still hot. In the shade of a tree, I'm enjoying a blessed moment of tranquillity. I'm perched atop the stubby base of a pillar among the ruins of the Forum in Rome—the center of political, religious, and public life through much of Ancient Rome's history.

I look northwest toward the Arch of Septimius Severus and the Temple of Saturn. In front of me is a broad expanse of wild grass that shimmers green and gold. Here and there stand pitted and cracked imperial columns, crumpled brick arches, and bits and pieces of travertine steps— broken remnants of Rome's triumph and power.

A breeze ruffles the grass.

I'm visiting Rome on a journey to better understand the complex problems we face—problems like energy shortages, climate change, disease, and economic crisis. And while it may be a moment of tranquillity for me, in mid-May 2003 much of the world is in turmoil. In the past weeks, a string of suicide bombings ripped through Israel. Terrorists attacked Western targets in Riyadh and Casablanca. Indonesia launched a war against rebels in the province of Aceh. The United Nations warned of a new genocide in northeastern Congo. A virulent form of pneumonia, SARS, caused near panic from Beijing to my home city of Toronto. And the United States and its allies achieved a lopsided military triumph over Saddam Hussein's regime in Iraq. The victory seemed to confirm the United States as an imperial power, certainly the greatest since the Roman empire, and perhaps—in terms of the gulf between America and its nearest competitors in military and economic strength—the greatest of all time.

All this turbulence makes it seem as if nothing is dependable. Shocks and surprises seem to rush toward us faster than ever before. As I sat among the Forum's scattered ruins, trying to imagine what the place must have

The Roman Forum

looked like two thousand years ago, I asked myself, Did the Romans ever have the same feelings? Were their certainties ever challenged, and did events ever seem out of control? I wondered whether the pressured, chaotic circumstances of today's world are in any way like those that existed when the western Roman empire crumbled in the fifth century. How could anything that seemed so permanent and consequential, as Rome must have seemed in its heyday, be reduced to these scraps of rubble? Of course in the centuries since Rome's fall, countless others have asked the same kind of questions, but I suspected we'd learn something new by asking them again now, in light of new studies of why societies sometimes collapse.[2]

Isn't everyone intrigued by the idea of the fall of Rome? As a boy, I was fascinated by it. I marveled at Rome's feats of conquest and engineering. They were the stuff of wonder. Rome's legions subjugated Europe and North Africa and reached deep into western Asia, while its engineers built roads, aqueducts, temples, baths, and amphitheaters across the empire's landscape. But what really drew me to the story was what it revealed about our human frailty. There was something both spectacular and eerie about this civilization that so dominated much of the world—and then almost completely disappeared. Rome's vast influence on Western cultures endures, but we can see today only scattered fragments of its incalculable physical effort. For a ten-year-old on the cusp of adolescence, this tale was mysterious and subtly frightening. It hinted that—in the sweep of time—all our striving and building and all our passion about issues of the day are almost wholly inconsequential; that when viewed across thousands of years, even our most prodigious achievements will seem ephemeral.

At the very least, Rome's story reveals that civilizations, including our own, can change catastrophically. It also suggests the dark possibility that our human projects are so evanescent that they're essentially meaningless.

Most sensible adults avoid such thoughts. Instead, we invest enormous energy in our families, friends, jobs, and day-to-day activities. And we yearn to leave some enduring evidence of our brief moment on Earth, some lasting sign of our individual or collective being. So we construct a building, perhaps, or found a company, write a book, or raise a family.

We seldom acknowledge this deep desire for meaning and longevity, but it's surely one source of our endless fascination with Rome's fall: if we could just understand Rome's fatal weakness, maybe our societies could avoid a similar fate and preserve their accomplishments forever.

Of course, an infinite number of factors—most of them unknown and some unknowable—affect how our societies develop, and we can only rarely influence even those few factors we know about. So rather than resisting change, our societies must learn to adapt to the twists and turns of circumstance. This means we must sometimes give up our accomplishments. If we try to keep things largely the way they are, our societies will become progressively more complex and rigid and, in turn, progressively less creative and able to cope with sudden crises and shocks. Their collapse—when it eventually does happen—could then be so destructive that there would be little of the prior order left behind. And there would be little left to seed the vital process of renewal that should follow.

Here we have ancient Rome's real lesson. Most of us who recall a bit of history think that constant barbarian invasions caused the western empire to disintegrate, but actually these invasions were only the most immediate cause. In the background were more powerful long-term forces, especially the rising complexity of all parts of Roman society—including its bureaucracy, military forces, cities, economy, and laws—as the empire tried to maintain itself. To support this greater complexity, the empire needed more and more energy, and eventually it couldn't find enough. Indeed, its increasingly desperate efforts to get energy only made its bureaucracies and laws more elaborate and sclerotic and its taxes more onerous. In time, the burden on the empire's peasants became too great, while rising complexity strangled the empire's ability to renew itself. The collapse that followed was dramatic: populations of cities and towns fell sharply, interregional trade dwindled, banditry and piracy soared, construction of monumental buildings and large-scale infrastructure stopped, and virtually all institutions—from governments to armies—became vastly simpler in their operation and organization.[3]

In this book I'll argue that our circumstances today are surprisingly like Rome's in key ways. Our societies are also becoming steadily more complex and often more rigid. This is happening partly because we're trying to manage—often with limited success—stresses building inside our societies, including stresses arising from our gargantuan appetite for energy to run our factories, heat our homes, and fuel our cars. Eventually, as occurred in Rome, the stresses may become too extreme, and our societies too inflexible to respond, and some kind of economic or political breakdown will occur.

I'm not alone in this view. These days, lots of people have the intuition that the world is going haywire and an extraordinary crisis is coming. Some people, particularly those of a religious disposition, think we're entering end times. Parallels between ancient Rome and the modern world are common; and fiction, religious preaching, and even scientific analyses abound with apocalyptic images of doom. Much of this stuff is nonsensical, which makes it easy for our "experts" to dismiss it with a patronizing wave of the hand. But I think that non-experts' intuition is actually largely right. Some kind of real trouble does lie ahead.

That trouble doesn't have to be calamitous in its ultimate results, though. If we're smart and a bit lucky, we have a good chance of avoiding a terrible outcome. In fact, just as happened after the great San Francisco fire—when a new and more resiliant city rose from the ashes and America's banking system was made far more resiliant too—catastrophe could create a space for creativity that helps us build a better world for our children, our grandchildren, and ourselves.

The White Wall

It's late at night, and you're driving fast along a country road in a dense fog. Your headlights are strong, but you still can't see much. The beams reflect off the fog, creating a wall of white light that seems to hang motionless in front of your windshield.

It's a strange feeling: you know you're moving fast—your foot is pressing hard on the accelerator, the engine is roaring, and directly in front of your hood you can glimpse the pavement rushing toward you—but otherwise there's little evidence of motion, because no matter how hard you scan the white wall for details you can see only faint hints of what's coming your way.

Yet you feel calm and confident—secure, even. After all, this is sleepy farming country, and while you know you're driving a little too fast for the conditions and have never been down this road before, your map tells you it's straight and flat and that there are no side roads from which other cars can emerge. And because it's nighttime, there's little chance of oncoming traffic. Anyway, in spite of the hazards, you want to get where you're going as quickly as you can.

Sensible behavior? Most would say not. Countless things could go wrong. Perhaps the map is wrong, and there's a sharp curve ahead; perhaps a deer will jump into the road suddenly or a stranded motorist step out to flag you down.

Driving fast in the fog is, of course, not sensible. But it's exactly what we're doing today.

Think of the road as the line of time, stretching endlessly from the past behind us and into the future in front of us. We live in an instant of fast-forward motion between what has been and what's to come. We don't give the past much thought, although we might occasionally glimpse bits and pieces—like Ancient Rome—shrouded in fog in our rearview mirror and reflecting our taillights' rosy glow. The decades that roll out in front of us are obscure, try as we may to make out their features. The car's headlights are like the best experts and forecasting technologies we can muster, but they penetrate only a short distance into the haze.

Even so, and despite those who sense calamity ahead, many people believe they have a pretty good idea where we're going. They're sure things will work out fine, because they're basically optimistic about the future and our ability to deal with its surprises. They see an ever-improving future of broadened and deepened global capitalism, expanded democracy and respect for human rights, and marvelous advances in science and technology—all changes that will strengthen our progress toward greater happiness and well-being. So they ignore the fog and keep their foot pressed on the accelerator pedal.

Many other people are just passengers on this wild ride. They'd like to slow down, but they're too scared to make any real changes because they don't know what would happen if they did.

What will our future really hold? We all need to recognize that the road ahead isn't going to be straight, flat, or unobstructed. We can't possibly know when sharp corners, other vehicles, or unexpected junctions will appear, but we can be sure that there are a huge number of them ahead and that we're still driving far too fast.

In the middle of the first decade of the twenty-first century, some of us have a feeling of dread. We see headlines about avian flu, impending oil shortages, and horrible terrorism in distant places. We realize that humankind is doing more things, faster, across a greater space than ever before, and that this is producing changes of a size and speed never before seen. Globalization erases our jobs, new technologies inundate our lives with information, waves of migrants push at our borders, and pollution destabilizes our climate. Stupendous changes are converging simultaneously on our societies, on our leaders, and on each one of us—leading many people to feel that things are out of control, and we're going to crash.

And we might well ask, What kind of trouble is our civilization likely to encounter ahead? How can we cope, and how might we take advantage of opportunities that arise for civilization's renewal?

TECTONIC STRESSES

T ENS OF THOUSANDS of people are walking toward me.

 It is 6 p.m. on Thursday, August 14, 2003—three months, almost to the day, after my tranquil afternoon in Rome's Forum. I'm standing on Yonge Street, Toronto's central artery, looking down a slight grade to the skyscrapers at the city's center. For several kilometers, as far as I can see down the road, the sidewalks are choked with people trudging north.

 Two hours earlier, the power failed across an immense wedge of eastern North America extending from New York to Detroit to Toronto. It's the biggest blackout ever on the continent. Subway trains shuddered to a halt, traffic lights went dead, and all surface transport was snarled in gridlock. Unable to get home the usual way, people are leaving the urban centre on foot.

 I talk to a few of them. Some are frustrated and annoyed, but none seems angry. Some find the whole thing a novelty—a fun interruption of a hot summer afternoon's routine. Others generously pitch in to direct traffic or guide pedestrians across chaotic streets.

 But everyone seems puzzled and at least a little disconcerted. What happened? Was it terrorism? How long will it last? And how will we get home?

 The view down Yonge Street reminds me of something I'd already seen, but at first I'm not sure what. Then I realize that it resembles a grim day only two years before, when the world had gaped in horror as people

September 11, 2001, won't be the last time we walk out of our cities.

fled on foot from southern Manhattan, the smoke of the collapsed Twin Towers billowing into the sky behind them. Unlike 9/11, the great 2003 blackout didn't claim thousands of lives or trigger a war. But it echoed that earlier catastrophe. Both events were complete surprises that materialized suddenly out of a complex world we only remotely understand. Both had effects that were greatly amplified by the intricate networks that tightly connect us together and that move people, money, information, materials, and energy. And both starkly reminded us how vulnerable we've become to the abrupt failure of critical technological, economic, and social systems.

When the power went off in August 2003, all air conditioners, elevators, subways, and traffic signals failed—but that wasn't surprising. What did surprise many people, though, was the simultaneous failure of portable phones, automatic tellers, debit card machines, electronic hotel-room doors, electric garage doors, and almost all clocks. Most disconcerting of all was the loss of the constant flow of information that's become a drug in our lives, as people were cut off from television, e-mail, and—worst of all—the Web. No one could tell what was going on. It was as if darkness had fallen in mid-afternoon. People clustered around cars that boomed out reports from radio stations running on backup power.

In the sudden gridlock downtown, cars weren't much good for getting around, but at least they had batteries, so their radios worked.

Most of us in cities are now so specialized in our skills and so utterly dependent on complex technologies that we're quickly in desperate straits when things really go wrong.[1] When we can't drive, catch a cab, or take the subway, we have to fall back on such age-old methods as walking to meet our immediate needs.

When, next, will we see people walking out of our cities—in the darkness of a mid-afternoon?

Maybe not long from now, because the possibility of abrupt break-down in our vital social and technological systems is rising, and perhaps rising fast. Breakdown is often like an earthquake: it's caused by the slow accumulation of deep and largely unseen pressures beneath the surface of our day-to-day affairs. At some point these pressures release their accumulated energy with catastrophic effect, creating shock waves that pulverize our habitual and often rigid ways of doing things. Events like last century's Great Depression and two World Wars were good examples of this kind of buildup and sudden release of pressure.

Five *tectonic stresses* are accumulating deep underneath the surface of our societies, as I'll show in the next chapters. They are

- population stress arising from differences in the population growth rates between rich and poor societies, and from the spiraling growth of megacities in poor countries;

- energy stress—above all from the increasing scarcity of conventional oil;

- environmental stress from worsening damage to our land, water, forests, and fisheries;

- climate stress from changes in the makeup of our atmosphere;

- and, finally, economic stress resulting from instabilities in the global economic system and ever-widening income gaps between rich and poor people.[2]

Of the five, energy stress plays a particularly central role. I discovered in investigating the story of ancient Rome that energy is society's critical *master resource*: when it's scarce and costly, everything we try to do, including growing our food, obtaining other resources like fresh water, transmitting and processing information, and defending ourselves, becomes far harder.

Most of the five stresses spring from our troubled relationship with nature. Indeed, one of my most important points in this book is that we can't ignore nature any longer, because it affects every aspect of our well-being and even determines our survival.[3] Yet today, despite a growing intuitive public awareness of this fact, most politicians, corporate leaders, social scientists, and commentators in Western societies give nature little attention. They push it to the sidelines of public discussion, focusing instead on the headline issues that regularly hijack social, economic, and political debate. And they tend to dismiss people who concern themselves with nature as, at best, softheaded do-gooders or, at worst, eco-freak fanatics.

Most such opinion leaders imply we don't need to worry because we human beings are biologically exceptional, unlike any other species on Earth, with brains that endow us with immense ingenuity to solve our problems. And they imply that modern Western societies are historically exceptional, because no other societies in the past had our science, markets, and democracy. Today, our science gives us the knowledge, our markets give us the incentives, and our democracy gives us the social resources to solve any demographic, health, energy, or environmental crisis that might come our way.[4]

Yes, we do have exceptional brains, and Western societies are certainly among the most creative and adaptive in human history. But there are times when our problems are too hard for our brains, or when science, markets, and democracy can't generate solutions when and where they're needed.[5] And such opinion leaders conveniently overlook the fact that every great civilization believes itself to be exceptional—right up to the time it collapses.[6] Instead, unrealistically optimistic, they promote their Panglossian view almost as if it were a religion—an absolutist creed that leaves no room for uncertainty and that we're supposed to accept as a matter of faith.

Sure enough, this creed now permeates our common language and thought, and many of us truly believe we can free ourselves of the physical constraints that have otherwise governed human beings throughout history. Our recent experience has also encouraged this complacency. For a few remarkable decades—decades when energy seemed in endless supply, when our antibiotics seemed to have conquered infectious disease, when we traveled to the moon, and when the productivity of capitalist economies appeared to know no bounds—we could fool ourselves that the physical facts of life no longer applied.

But now Earth's glaciers and icecaps are disappearing, while mammoth hurricanes pound the United States, Australia, and Japan—signs that nature is reasserting its authority. The twenty-first century will, in fact, be the Age of Nature. We'll learn, probably the hard way, that nature matters: we're not separate from it, we're dependent on it, and when there's trouble in nature, there's trouble in society.

Multipliers

These stresses are of concern enough. But two other factors are likely to give them extra force. I call these *multipliers*, because they combine with the five stresses to make breakdown more likely, widespread, and severe. The first multiplier is the rising speed and global connectivity of our activities, technologies, and societies. The second is the escalating power of small groups to destroy things and people.

Humankind has been crisscrossing the globe for millennia, and we've been trading large quantities of raw materials and manufactured goods around the world for many centuries. But only in the past hundred years or so, while our population has quadrupled, have we created tightly interlinked economic, technological, and social systems—from industrial agriculture to financial markets—that penetrate virtually every corner of the planet. The kiwifruit on your breakfast plate comes from New Zealand, the plate itself comes from Malaysia, while the tantalum metal in the cell phone beside your plate comes from the jungles of eastern Congo. The globe, says the eminent historian Eric Hobsbawm, is now "a single operational unit."[7] And only in the past few decades has our impact on the natural environment become truly planetary: we're now a physical force on the scale of nature itself, disrupting the deepest processes of natural systems like Earth's climate, and massively changing global cycles of carbon, nitrogen, phosphorus, and sulfur.

This is the real face of globalization—a phenomenon that many people talk about but few really understand. It's not just a process of growing economic interdependence among countries. That's something that's been underway for hundreds of years.[8] Globalization is really a much broader and, in many ways, more recent phenomenon: an almost vertical rise in the scope, connectedness, and speed of *all* humankind's activities and impacts. It's as much about the spread of new diseases like AIDS and avian flu from one continent to another, the infestation of the Great Lakes by foreign mollusks, and the arrival of shiploads of poor migrants on our shores as it is about trade negotiations, farm subsidies, and currency convertibility.

The change has brought huge benefits. More trade in goods and services often boosts wealth for all involved: better movement of capital can aid investment and development, and mobilized global opinion brings attention to distant human-rights and environmental problems. Greater connectivity between people and a higher speed of interaction—caused mainly by lightning-fast information technology—let people far and wide combine their ideas, talents, and resources in ways that may expand everyone's prosperity.

But globalization has also created huge challenges. Greater connectivity and speed, for instance, allow what would once have been merely local shocks and disruptions to cascade outward as never before, sometimes affecting the whole planet. Just as the 2003 blackout ramified across eastern North America from its starting point in Ohio, so, earlier that year, did severe acute respiratory syndrome (SARS) emerge in southern China and explode into dozens of countries from Vietnam to Canada.

Greater connectivity and speed are especially worrisome in light of the spread of "lethal technologies" that have sharply raised the destructive power of angry and violent people. In a globalized world, an attack in one place can have instant repercussions everywhere. Lethal technologies don't have to be exotic or rare, like biochemical, nuclear, or radiological weapons. Technologies that provide impressive killing power to fanatics, insurgents, and criminal gangs are already widely available: conventional assault rifles, rocket-propelled grenades, and plastic explosives—staggeringly abundant and traded in vast quantities legally and illegally around the planet—are contributing to havoc from Chechnya to Congo and Iraq. Violent groups have also been learning how to convert civilian technologies into appalling weapons—as the

Terrorists need less than a ten-thousandth of the world's highly enriched uranium (HEU) to build a crude atomic bomb.
(Each dot above represents one hundred kilograms of HEU; the large rectangle of dots represents the amount of HEU in the world.)

Al Qaeda terrorists did so horrifically when they used passenger airliners as guided missiles.

But it's the exotic technologies—the weapons of mass destruction—that keep experts awake at night. If terrorists obtained barely one hundred kilograms of highly enriched uranium—less than one ten-thousandth of the world's stockpile, much of which is stored in insecure facilities in the former Soviet Union—they could easily build an atomic bomb that could flatten the core of any of our great cities.[9] London or New York, Paris or Washington, Moscow or Delhi, Tel Aviv or Riyadh—these metropolises are all in countries whose policies evoke hatred from fanatically violent groups, and any could be obliterated in an instant. Never before has it been possible for small groups to destroy entire cities, and this one fact by itself will ensure that our future is entirely different from our past.

Synchronous Failure

The stresses and multipliers are a lethal mixture that sharply boosts the risk of collapse of the political, social, and economic order in individual countries and globally—an outcome I call *synchronous failure*. This would be destructive—not creative—catastrophe. It would affect large regions and even sweep around the globe, in the process deeply damaging the human prospect. Recovery and renewal would be slow, perhaps even impossible.

It's the convergence of stresses that's especially treacherous and makes synchronous failure a possibility as never before.[10] In coming years, our societies won't face one or two major challenges at once, as usually happened in the past. Instead, they'll face an alarming variety of problems—likely including oil shortages, climate change, economic instability, and mega-terrorism—all at the same time.[11]

Scholars have found that bloody social revolutions occur only when many pressures simultaneously batter a society that has weak political, economic, and civic institutions.[12] These were the conditions in France in the late eighteenth century, Russia in the early twentieth century, and Iran in the late 1970s. And in many ways the same conditions are developing today for societies around the world and even for global order as a whole.

We don't usually think in terms of convergence, because we tend to "silo" our problems. We look at our challenges in isolation, so we don't see the whole picture. But when several stresses come together at the same time, they can produce an impact far greater than their individual impacts. When sparks combined with fuel immediately after the great 1906 earthquake, San Francisco exploded in flames. Today, around the world, we see similar explosive combinations of factors. For instance, just as shrinking global oil supplies are becoming ever more concentrated in some of the planet's most dangerous and politically unstable regions, more countries desperately need cheap energy to maintain their consumption-driven growth—a situation that raises the likelihood of wars over oil in places like the Persian Gulf. And just as gaps between rich and poor people are widening fast within and among our societies, new technology has put staggeringly destructive power in the hands of people who could be enraged by those gaps.

Convergence is treacherous, too, because it could lead directly to synchronous failure, if several stresses were to climax together in a way that overloads our societies' ability to cope. What happens, for example, if together or in quick succession the world has to deal with a sudden shift in climate that sharply cuts food production in Europe and Asia, a severe oil price increase that sends economies tumbling around the world, and a string of major terrorist attacks on several Western capital cities? Such a convergence would be a body blow to global order, and might even send reeling the world's richest and most powerful societies. Global financial institutions and political stability could begin to break down.

We can't estimate the exact likelihood of any one of these events, but we can say with reasonable confidence that their individual probabilities are rising. The probability that they'll happen together is, of course, much lower, but it's surely rising too.[13] And I've described only one scenario of converging stresses. If some form of synchronous failure does occur, it's likely to be in a way that we've never anticipated, because the range of permutations is almost infinite. We shouldn't be surprised by surprise.

Skeptics dismiss this kind of argument as alarmist, as a shopworn revelation of a coming apocalypse.[14] At most, they say, these are remote worst-case outcomes, so we shouldn't give them much attention. In answer I acknowledge that we can't know for sure what our future holds—what's beyond that white wall of fog. But we can still say confidently that we're sliding toward a planetary emergency; that the risk of major social breakdown in general—the result of something like synchronous failure specifically—is growing.[15]

And this is precisely the kind of outcome that disaster planners and insurance companies think about all the time: a perhaps low-probability event that would nevertheless exact a colossal toll if it happened.[16] We spend enormous amounts of money and time trying to make such outcomes even less likely—for instance, by building dikes and breakwaters along our coastlines to protect against the every-hundred-years storm or by reinforcing our buildings and bridges to withstand the every-hundred-years earthquake. Synchronous failure would be the same kind of disaster, except on a far greater scale, and it's something we must do our very best to avoid.

Beyond Management

The question is, how?

We typically respond to unfolding threats with a two-stage strategy: first denial, then reluctant management. If we can get away with denying or ignoring a problem—like the increasing risk of oil shortages or the international economy's chronic instability—we do so. We tell ourselves that the challenges aren't that serious and then simply continue with business as usual. Sometimes, lo and behold, benign neglect is the best strategy, and we muddle through successfully.

But it isn't the best strategy now because the potential costs if we're wrong are too high. In fact, we'd hotly criticize any family that ran its affairs the way we're running ours. If a family—especially a family with children—lived in a dangerous place—say, in a house on a floodplain, while massive storms brewed in nearby hills—and if they ignored the warning signs and continued as if nothing was the matter, we'd consider them irresponsible, the parents in particular. And the same criticism holds for us today if we deny the seriousness of our global situation: fundamentally, we're shirking our responsibility to our children and grandchildren who'll live in the world we're creating today.

When denial no longer works, perhaps because the signs that something's wrong have become too obvious to ignore, we may do our best to manage the challenge. We'll analyze the data, forecast the future, and lay out detailed policies to reduce the problem's seriousness and adapt

"Let's change 'brink of chaos' to 'Everything is wonderful.'"

to its consequences. Today, most experts who take our global problems seriously advocate a management response.

While this is better than simply denying our problems exist, it often doesn't help much. Any management policies that really address the underlying causes of our hardest problems usually require big changes in the existing economic and political order. After all, that order is often a central reason why our problems are so bad. But big changes always run headlong into staunch opposition from powerful and entrenched interest groups—like companies, unions, government bureaucracies, and associations of financial investors—that benefit from the status quo. So they're hardly ever carried out.

A good example is the history of the North American electrical system leading up to the 2003 blackout. The blackout wasn't, of course, the first event of its kind. In 1965, a failure left thirty million powerless from Ontario to New Jersey. Soon after, researchers began to better understand the complex behavior of the electrical grid—the continent-wide network of electricity-generating stations, high-voltage transmission lines, control centers, and substations that supplies Canadian and American consumers with power—and the peril of grid breakdown. In 1982, two of America's most thoughtful energy experts, Amory and Hunter Lovins, warned that the United States "has gradually built up an energy system prone to sudden, massive failures with catastrophic consequences."[17] Military, government, academic, and industrial experts reviewed the Lovinses' research, but willful denial, technological obstacles, and obstruction by powerful corporate and political interests blocked fundamental reform of the continent's electricity system. When asked a year before the 2003 blackout if things had improved over the previous two decades, Amory Lovins said, "I'm surprised the lights are still on."[18]

From the point of view of those with a vested interest in the status quo, efforts to manage our problems can actually be a useful diversion: such efforts provide a focus for research, discussion, and countless meetings for academics, politicians, consultants, and NGOs, while in practice nothing really changes. The Kyoto climate-change negotiations kept thousands of scientists and other experts busy for years (ironically generating vast amounts of carbon dioxide as they traveled from meeting to meeting) while providing cover for politicians who wanted to say they were doing something about global warming.

Because it's hard to challenge the arrangements that benefit vested interests, when we try to manage serious threats to our well-being we usually create new organizations, institutions, and procedures rather than reforming those that already exist. We might, for example, create another office in the government's bureaucracy to monitor the flow of nuclear materials that could fall into terrorists' hands, or we might sign a treaty like the Kyoto Protocol that says we're going to cut carbon dioxide emissions. Too often, though, this strategy simply adds another layer of complexity on top of an already cumbersome and dysfunctional management system. So, over time, our mechanisms for dealing with a more volatile world become more rigid and susceptible to catastrophic failure when exposed to severe stress.

What, then, should we do? We could adopt a more radical response to our converging challenges: while a management approach is sometimes useful, we could also recognize that sometimes it won't work, and that when it doesn't work, as time passes, breakdown becomes increasingly likely. When breakdown happens, our challenge will be to keep it from becoming synchronous failure, while at the same time exploiting the opportunities it provides to promote deep reform.

We can help keep future breakdown constrained—that is, not too severe—by making our technological, economic, and social systems more *resilient* to unexpected shocks. For example, to lessen the risk of cascading failure of our energy system—failure that spreads through the system just the way a row of dominoes falls—we can make much greater use of decentralized, local energy generation and alternative energy sources (like small- and medium-scale solar, wind, and geothermal power) so that individual users are more independent of the grid. We might lose some economic efficiency, and our economy's total output of wealth might be smaller, but we'd benefit from a more stable and resilient energy system—and that benefit could far outweigh the cost.

I'll outline later other such resilience-enhancing strategies. If widely adopted, they would profoundly alter the course of our societies. In truth, by shifting us away from a monomaniacal focus on greater economic productivity, efficiency, and growth, they would represent a wholesale challenge to current economic orthodoxy.

We can also get ready in advance to turn to our advantage any breakdown that does occur. Breakdown happens—in our personal lives as well as in our societies. If seldom desirable in itself, it's nonetheless rarely the end of the world, and much good can come of it. We can boost the chances that it will lead to renewal by being well prepared, nimble, and smart and by learning to recognize its many warning signs.

To help us recognize the signs and prepare for breakdown is a central purpose of this book. In the following pages, I don't provide a checklist of technical and institutional solutions we might apply to manage the world's tectonic stresses. Instead my aim here is to begin a conversation about why breakdown of some kind is becoming more likely, how we can keep it from being so severe that it's debilitating, and what we can do to exploit the opportunities it presents when it happens. If breakdown is to have an upside—and I fervently believe that it can—we have to work together to develop a wide range of scenarios and explore what we can do individually and together in each situation. You can join me in this conversation at www.theupsideofdown.com.

From Crash to Creativity

At some time or other in our lives, most of us have been humbled by a professional or personal crisis—say the bankruptcy of a business, the loss of a job, or the death of a loved one. In response we've examined our basic assumptions, gathered together our remaining resources, and rebuilt our lives—surprisingly, often in a new and better ways.

Surprisingly too, there's no term in English for this commonplace occurrence of renewal through breakdown. So I found a label for it. I call it *catagenesis*, a word that combines the prefix *cata*, which means "down" in ancient Greek, with the root *genesis*, which means "birth." The word is used in some scientific fields—for instance, ecologists use catagenesis to refer to the evolution of a species toward a simpler, less-specialized form.[19] In my use of the term here, I retain the idea of a collapse or breakdown to a simpler form, but I especially emphasize the "genesis"—the birth of something new, unexpected, and potentially good. In my view of it, whether the breakdown in question is psychological, technological, economic, political, or ecological—or some combination of these forms—catagenesis is, in essence, the everyday reinvention of our future.

I developed this idea of catagenesis after much study of how some systems adapt very well to changes in their surrounding environments. All systems—whether a windup clock, the Earth's climate, or a country's government—are made up of interacting components that stay together, as a set, over time.[20] But not all systems adapt well to new challenges or stresses. I learned that that those that do adapt well are generally called "complex adaptive systems," and they include things like tropical forests, private corporations, human societies, and even individual people. Each one of us is actually a complex adaptive system.

But what, exactly, makes a system complex? Partly it's the fact that it has lots of bits and pieces—in the case of a society, a lot of people, organizations, machines, and flows of material and energy. But that's not the only factor. If it were, a complex system would merely be complicated. Complex systems have other characteristics that we'll discover later.[21] At this point let's just say that they generally have a wider range of potential behaviors than simple systems. So machines like windup clocks or car engines aren't complex. They may be extremely complicated—they may have thousands of parts—but all their parts work together to produce a system with a relatively narrow and predictable range of behaviors.

Also, we can take machines apart to find out why they behave the way they do. We can, for instance, dismantle a windup clock to discover its various cogwheels, bushings, and springs; and then, by examining each of these parts and how they fit together, we can figure out how the clock works. Its behavior is the direct result of the characteristics of its component parts, and if the clock doesn't work, or if it does something weird—such as go backward—we can attribute its unfortunate conduct to the failure of particular parts.

Complex systems, on the other hand, have properties and behaviors that can't be attributed to any particular part but only to the system as a whole. A stock market is a complex system, and its overall behavior—whether it's a bull market going up or a bear market going down, for example—is the result of the buying and selling of thousands, maybe even millions, of individual investors. A person is a complex system too. Let's take an average male adult—call him John. There are aspects of John's physiology, personality, and actions that we can't understand no matter how well we understand his discrete bits and pieces, like his spleen, his right big toe, or even his brain's frontal lobes. Like all complex systems, John has *emergent* properties: he is more than, and different from, the sum of his parts. Once all those parts are linked together and operating in their right places, we get characteristics and behaviors—perhaps his body's ability to regulate its temperature or his whimsical fascination with butterflies—that we couldn't have anticipated or understood beforehand, even with complete knowledge of all his separate parts.

Now recent research—which we'll get to know in later chapters—shows that some kinds of complex systems adapt to their changing environment by going through a four-stage cycle of growth, breakdown, reorganization, and renewal (the last three of these stages are what I call catagenesis).[22] There's an important caveat to this general idea of a four-stage cycle, though: while breakdown is essential to long-run adaptation and renewal, it must not be too severe. In other words, breakdown must be constrained—just as the great San Francisco fire was constrained when it was stopped at Van Ness Avenue—for catagenesis to happen.[23]

Of course, even constrained breakdown, when it affects our communities, companies, and societies, can hurt many people, sometimes very badly. But it can also shatter the forces standing in the way of change and the deeply entrenched and too-comfortable mindsets that keep people from seeing exciting possibilities for renewal. It can, in short, be a source of immense creativity—a shock that opens up political, social, and psychological space for fresh ideas, actions, institutions, and technologies that weren't possible before. In capitalist economies, this often happens when companies fail or go bankrupt. There are many examples in history too. The implosion of Soviet power in Eastern Europe in the late 1980s and early 1990s created stunning opportunities (some exploited, some not) for fundamental renewal of the region's political and economic systems. Longer ago, the profound shock of the Great Depression in the 1930s allowed President Franklin D. Roosevelt to enact vital reforms in the U.S. economy.

But beware. Breakdown can also usher in a period of great danger—of turmoil, confusion, frustration, and anger—a period when demagogues can rush into the breach and turn one group against another with ferocious violence. While the Great Depression gave Roosevelt the impetus and opportunity to reform American capitalism, it also gave Hitler the chance to establish one of history's most evil regimes.

In times of upheaval, wrote the great Irish poet W. B. Yeats in "The Second Coming," "the best lack all conviction, while the worst are full of passionate intensity." When social breakdown happens, as it will in coming years, we can be sure that the worst will be full of passionate intensity. We must be equally sure that the best will have the conviction, the knowledge, and the resources to prevail.

Thresholds

Complex systems have a number of other essential features that affect how they respond to stress and also whether we can predict their future behavior.

Sometimes, for instance, small changes in a complex system produce huge effects, while large changes make little difference at all.[24] In other words, cause and effect aren't proportional to each other. Specialists call this *nonlinear* behavior, and we encounter it all the time in our daily lives—even in relatively simple systems. A warming of one degree in temperature in our kitchen's freezer may be imperceptible to touch, but it can thaw all our food. A light switch doesn't budge with a gentle push, but apply slightly more pressure and it suddenly flips from off to on.

In the case of a complex system, nonlinear behavior can happen as disturbances or changes in the system, each one relatively small by itself, accumulate. Outwardly, everything seems to be normal: the system doesn't generate any surprises. At some point, though, the behavior of the whole system suddenly shifts to a radically new mode. This kind of behavior is often called a *threshold effect*, because the shift occurs when a critical threshold—usually unseen and often unexpected—is crossed. (In our everyday conversation, when we say something was "the straw that broke the camel's back," we're saying it caused a threshold effect.)

Threshold effects can be good or bad for us, depending on the circumstances and one's point of view. The end of apartheid in South Africa and the collapse of the Grand Banks cod fishery are both great examples of threshold effects, but the former was a positive development for many people and the latter wasn't. The international economy often exhibits threshold effects. The 1997–98 Asian financial crisis was a sobering case. A devaluation of the Thai bhat, a minor currency, launched a financial crisis that ricocheted through the international economy for months, cost trillions of dollars in lost economic output, and threw tens of millions of people out of work. One day the Asian economy was booming; the next it was in a nosedive.

We often see beneficial threshold effects in the evolution of technologies: when just the right confluence of enabling factors occurs, technological progress surges. For instance, once lots of people were using the Internet and once an effective browser had been invented, the World Wide Web spread around the planet like wildfire—with the number of Web servers (the powerful computers that host Web pages on the Internet) soaring from a few thousand in the mid-1980s to over two million by 1994 and almost 400 million now.[25]

The behavior of a complex system with these features is highly *contingent*—how it behaves at any given time, and how it evolves over time, depends on a host of factors, large and small, knowable and unknowable. I've come to think of such systems as encountering many junctions as they move through time—just like the junction that Robert Frost's traveler encounters in his famous poem "The Road Not Taken."

> Two roads diverged in a yellow wood,
> And sorry I could not travel both
> And be one traveler, long I stood
> And looked down one as far as I could
> To where it bent in the undergrowth

Frost's traveler, like all human beings, is a complex system. The route we take through the woods depends on the number of junctions we encounter, on the number of paths available at each junction, and on countless subtle, imponderable things that influence us to choose one path over another at each junction. We can't hope to forecast our ultimate route. And the same is true of any other complex system. The further we try to predict into the future, the more bewildering the task of predicting the system's route becomes.

Once a complex system goes down a particular path, it can't easily jump from one path to another or retrace its steps to try a different path. Frost seemed to understand this inescapable feature of our world. After having made a choice to follow one path, his traveler laments,

> Oh, I kept the first for another day!
> Yet knowing how way leads on to way,
> I doubted if I should ever come back.

Specialists have a term to describe this characteristic of complex systems: *path dependent*. Where the system is at any particular time depends on the tortuous, circuitous route by which it got there—"how way leads on to way," as Frost marvelously puts it. A complex system's history turns out to be crucially important because it profoundly shapes what the system becomes, and it can't be rewritten or repealed. Frost finishes,

> I shall be telling this with a sigh
> Somewhere ages and ages hence:
> Two roads diverged in a wood, and I—
> I took the one less traveled by,
> And that has made all the difference.

Many people focus on the decision made in the poem's penultimate line, taking it as an affirmation of why it's important to be different from the crowd. But I believe Frost's real message is in the last line, and it's much more disturbing and, in the end, poignant. He is telling us that a choice that appears insignificant can "make all the difference," and that there may be no going back.

When small things can make a big difference, and when it's impossible to know which small things matter and which don't, predicting the

The traveler confronts path dependency.

future becomes formidably difficult. This is especially true of human affairs.[26] Even more than the behavior of other complex systems, human social, economic, and political behavior is often extremely sensitive to serendipity, to fad and the whims of leadership, and to sudden technological, economic, political, and environmental developments. Also, we can't know exactly when or how any complex system that crucially affects our lives will cross a critical threshold and flip to a new mode of behavior. Thirty years ago, who anticipated the implosion of Soviet communism, the widespread adoption of personal computers, the emergence of AIDS, or the opening-up of a gaping hole in the stratospheric ozone layer over the Antarctic? Or, for that matter, on September 10, 2001, who among us predicted that terrorists would fly planes into the World Trade Center?

Try it. Try to come up with a plausible scenario for what the world will look like in, say, 2025, or even 2015. After even a moment's reflection, you'll realize that the range of possibilities is almost infinite and that given the blazing rate of change in today's world, there's something profoundly unknowable about the future, even a future that could arrive within a decade or two.[27]

Because we don't have anything firm to guide us, when we try to predict what our world will be like, we tend to think that things will continue the way they are going now. If a technology like the computer microchip has steadily improved in one direction—in the microchip's case, by doubling its power every eighteen months—we tend to assume that the trend will continue in the same direction in the future. We also fall back on our underlying personal temperaments. Our natural optimism or pessimism powerfully affects whether we believe technological, social, and environmental trends will bring us happiness or grief. And in our data-saturated lives, it's easy to find evidence that confirms our biases.

Few people actually recognize how bad we are at predicting the future. Recently, though, I discovered an exception in an obscure collection of essays written at the time of the 1893 Chicago World Exposition. Various famous Americans were asked to describe what American society would be like a century hence—in 1993. In today's light, many of their predictions seem downright bizarre, and almost all are infused with the exuberant American optimism of the late nineteenth century—only two decades before the twentieth century's calamities began to unfold. To be fair, though, envisioning life a hundred years in the future is astonishingly hard, as is thinking free of our culture's dominant sentiments.

But one comment especially caught my eye. In his short essay, "The Future Is a Fancyland Palace," James William Sullivan, a prominent newspaper editor and a follower of the American economist and reformer Henry George, offered more insight than all the book's other prognostications. And in his poetic yet blunt acknowledgment of our inability to see the future, Sullivan was far more sensible than most of our "experts" today:

> I find that I am unable to prophesy. The future is a fancyland palace whose portals I cannot enter. Moving toward it from Here, I am charmed with its brilliant façade. What sculptured splendors—porticoes, pillars, statues, windows! What is within? As I advance, however,

the airy structure recedes. I cannot push beyond its threshold; its doors never open; on their other side are silence and mystery.[28]

The Prospective Mind

On the other side may be silence and mystery, but here's one thing we do know about the future: surprise, instability, and extraordinary change will be regular features of our lives. Some events, such as 9/11, will even transform our outlook forever. They'll be like massive social earthquakes, rupturing the order of things—the routines and regularities we rely on for a sense of safety and a sense of who we are and where we are going. Our surroundings won't ever look the same again. The reliable landmarks of life will become strange and distorted—recognizable, yet at the same time weirdly unrecognizable.

If we're going to choose a good route through this turbulent future, we need to change our conventional ways of thinking and speaking. Too often today we talk about our world as if it's a machine that we can precisely manipulate. We talk as if we can understand and master everything around us, keeping what we want and discarding what we don't want. This attitude is deeply dangerous. The surest way for us to crash disastrously is to believe that we know and can master it all, because then we'll lose our capacity for self-criticism and self-reflection. We'll no longer see the signals around us that tell us things are going wrong and that we should adjust our course.

We need, instead, to adopt an attitude toward the world, ourselves within it, and our future that's grounded in the knowledge that constant change and surprise are now inevitable. The new attitude—which involves having a *prospective mind*—aggressively engages with this new world of uncertainty and risk. A prospective mind recognizes how little we understand, and how we control even less.

There's no delusional optimism here. The prospective mind knows that severe pressures are building around the planet. But neither is this viewpoint relentlessly pessimistic. The coming decades will be perilous, but we shouldn't enter them with fear. Human beings are first and foremost problem solvers, and the prospective mind tries to anticipate harmful outcomes in the future by better understanding the pressures affecting our world and how they might act, singly or together, to cause our undoing. It also knows, though, that the future is opaque. We can't really see beyond the white wall because as prognosticators we come up against two formidable obstacles: the higly nonlinear systems that surround us and the biases of our temperament.

Still, we can create a rough image of the future. It's not really a prediction. Instead, it's a bit like a French Impressionist painting that when viewed as a whole is a vivid, cohesive image, meaningful and rich with movement and feeling, but when examined closely consists of discrete brushstrokes and dollops of color. Our image of the future might be crude, but it can still be grounded in sensible judgments about the deep trends and forces affecting us and about the boundary between what's plausible and what's wholly unlikely.

The prospective mind then looks for ways to prevent or forestall horrible outcomes, not just through managing things—an approach that's often ineffective and sometimes even counterproductive—but also by imagining and implementing more radical and far-reaching solutions. It recognizes that we're unlikely to prevent all forms of breakdown and that sometimes breakdown can open up opportunities for deep and beneficial progress—for catagenesis—if men and women of courage and good sense are prepared to act. Most fundamentally, the prospective mind seeks to make our societies—and each one of us—more resilient to external shock and more supple in response to rapid change. At the end of the day, the western Roman empire wasn't supple, and if our fate is be different from Rome's in a world of relentless change and surprise, we must constantly reinvent our societies, ourselves, and our future.

We're entering a crucial time in our history. In coming decades we'll come upon one critical junction after another in rapid succession. The choices we make and the paths we choose at each junction will be irreversible. The stakes are as high as they can get. But as we rush forward into the fog, very few of us actually have our hands on the steering wheel. Most of us are just passengers in the front seat. Sometimes we stare—wide-eyed with anxiety—through the windshield, and other times we just sink back into our seat in denial—denial of our speed, of the dangers ahead, and of our lack of control.

It's time we turned passengers into drivers.

A KEYSTONE IN TIME

ANCIENT ROME was built out of concrete, brick, and quarried rock. Today, you can find elegantly carved rock everywhere in the Forum, most obviously in the grand ruins of the site's many temples and triumphal arches. But chunks of lovely fluted column and fragments of carved plinths, capitals, entablatures, and lintels are also strewn everywhere, often half buried in the dirt and heaped in rough piles.

Few visitors pause to think about the enormous labor it took to cut these rocks out of the ground, carve them, and carry them to Rome from distant quarries, or about the ingenuity needed to accomplish such a feat. Yet these scattered, broken stones often have fascinating stories to tell.

So it was with one piece of rock that caught my attention. It wasn't in the Forum, but a five-minute walk away, to the southeast, in the Colosseum—the greatest of all Roman amphitheaters, built to gratify the Roman lust for gladiatorial combat and wild beast shows. I spied this piece of rock as I passed underneath an arch to climb from the Colosseum's street-level entrance to the main promenade inside. I looked up briefly and noticed the arch's keystone. For some reason—I don't really know why; maybe it was something about the keystone's sheer unremarkableness—I stopped to look more closely. I saw that the arch itself was huge—over seven meters high and four meters wide—but in the context of the amphitheater's great structure it didn't stand out. It was only one of eighty of the same size and construction that formed the bottom tier of the building's immense first interior wall. But looking

The keystone in the Colosseum

more closely I realized that while seemingly insignificant, the arch and its keystone were actually a brilliant feat of engineering.

The Roman arch is a type of structure that engineers designate a "voussoir arch" because its wedge-shaped stones are called voussoirs. It was an engineering breakthrough. Prior to its invention, someone building a bridge across a river, say, would usually span the horizontal distance using a long length of rigid material, like rock or timber. But this type of structure is vulnerable to collapse in the middle of its span. Worse, it's bedeviled by a basic contradiction: if the span's substance is thickened to make the bridge stronger, the span itself becomes heavier. Eventually, it becomes so heavy that the bridge is weakened, not strengthened. The voussoir arch circumvents this contradiction. It gains its strength not just from its constituent materials but also from its design: because it's an arch, the structure's load is distributed downward into the pillars on each side, and because its voussoirs are wedge shaped, up to a point their weight and that of any material they carry actually strengthens the structure by driving the voussoirs together more tightly.

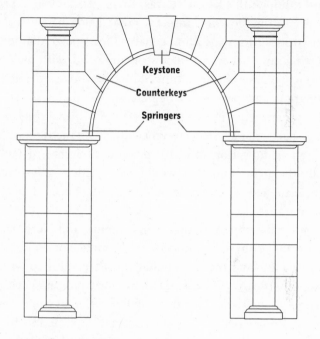

The design of the voussoir arch gives it strength.

The Romans didn't discover this idea; they probably adopted it from the Greeks and Etruscans.[1] They did master the technology and worked out the surprisingly intricate engineering principles that govern the voussoir arch's proper construction. Then they spread the technology far and wide, using it to build bridges, aqueducts, temples, domes, and amphitheaters throughout their empire. Today, a large proportion of the residue of ancient Rome—the most durable bits, in fact, like the occasional segments of aqueduct that can be spied in the French, Italian, or Spanish countryside—incorporate voussoir arches.

The keystone lodged above my head was massive: about two meters deep, one meter high, one and a half meters wide at the top of its wedge, and three-quarters of a meter wide at the bottom. It was made of travertine, a kind of limestone that has been quarried for millennia near the present-day town of Tivoli (known to the Romans as Tibur), about thirty kilometers to the east. Tivoli travertine is a cream color, streaked with darker indentations and small holes that make it look a bit like white cork. But here, over time, age, pollution, and years of inattention had allowed a thick layer of grime to coat the rock's white surface, turning it a muddy brown.

Scruffy though this keystone was, I was struck by how precisely it had been cut: one couldn't insert a razor blade between it and the adjacent stones. Romans didn't have high-speed diamond-edged rock-cutting saws—they cut their stone by hand, with iron picks and chisels, one hammer blow at a time. The labor involved, and the sheer determination and patience, boggled my mind. I was struck most, though, by something more essential: stone is heavy stuff, and I estimated this particular keystone weighed about 5.7 metric tons.[2] I started to think about how Roman engineers and laborers had put it there, imagining the giant crane they would have needed—constructed in an A-frame, festooned with wooden pulleys—with dozens of men straining on ropes at its base.

The powerful emperors Augustus, Caligula, and Nero had all considered erecting a huge amphitheater in the middle of Rome. But Nero's death in 68 CE—and the brutal civil war that followed—spurred his successor, the emperor Vespasian, to build the Colosseum as a gift to the city's people to show that his rule would be different from Nero's vainglorious and cruel reign. Vespasian began construction between 72 and 75 CE, likely financing the scheme using spoils from his crushing victory over the Jews and the sacking of Jerusalem a few years before. His son Titus inaugurated the building in 80 CE with one hundred days of games in which some ten thousand beasts were killed.

In as little as five years, the Colosseum's architects, engineers, and laborers had erected a building with the dimensions and seating capacity of New York's Yankee Stadium. It was, the architectural historian Rabun Taylor writes, "the most complex structure ever successfully completed in antiquity." Not only could it support the combined weight and simultaneous swaying and stomping of over fifty thousand spectators, but its architecture also provided for the smooth circulation of these spectators (the stadium could be emptied in as little as fifteen minutes), while an elaborate arrangement of pipes brought perfumed water to fountains throughout the building and carried away waste from lavatories. "It is surpassed in daring, originality, and beauty by other buildings," Taylor continues, "but as a monument to architectural process it stands alone."[3]

Public buildings like the Colosseum were designed to astonish, awe, and even humble Rome's citizens with the sheer audacity of their engineering. They were the physical expressions of an ideology of power. "They spoke of strength, control, and stability," says Taylor. "The intent was to induce participatory pride and willing submission and allegiance to the emperor." In fact, the Colosseum's most spectacular views, available only from the uppermost seating farthest from the arena, were reserved for the common person, whose allegiance was most uncertain and most in need of reinforcement.[4]

The Romans began this monument to their power by draining and excavating a lake that had been on Nero's palace grounds. They then laid an elliptical ring of cement and crushed stone—thirteen meters deep, fifty meters wide, and over five hundred meters in circumference—to serve as a foundation, reinforcing each side with brick walls three meters thick. On top of this foundation they placed a floor of travertine almost a meter thick, and then anchored the base blocks for the building's main pillars to the floor with molten metal.[5] On these blocks, the Romans erected the columns and arches of the amphitheater's eighty radial walls that fanned outward from the central field like the spokes of the wheel—each intersecting the three immense outer walls that girdled the building.

The final building's total mass was simply staggering. The builders extracted, moved, shaped, mixed (in the case of concrete), and assembled

The Colosseum and its foundation originally contained about a million metric tons of rock, concrete, and brick.

about a million metric tons of raw material, including 295,000 tons of travertine, 653,000 tons of concrete, 54,000 tons of tufa (a soft volcanic rock abundant along the west coast of Italy), 58,000 tons of clay brick, 6,000 tons of marble, and 300 tons of metal to connect the major stones.[6]

That scruffy keystone in the Colosseum could tell a story, I thought, about Rome's huge energy requirements and about how those requirements shaped the empire's evolution. Empires run on energy. A central task of any empire is to produce, transport, and focus enough energy to maintain and extend its economic and political power. Acquiring and protecting the sources of this energy, the routes along which it's carried, and the people and organizations responsible for generating and transporting it becomes a key job of an empire's security and military forces.

Like any ancient society, Rome was powered in essence by the sun's energy—absorbed by plants in Roman fields and converted into food.[7] As I inspected the keystone, I asked myself what area of farmland was needed to grow the calories that powered the muscles that put just this stone in place? And how much was needed to power the construction of the whole Colosseum? This was not an idle question, I believed, but tied to Rome's rise and fall.

The stone could also tell us something, I suspected, about the similarities and differences between Rome's empire and our own of more recent times—including today's American imperium.

The Thermodynamics of Empire

Energy is the lifeblood of all societies. Just as we can understand fundamental things about the human body by tracing its flow of blood, we can understand much about a society's activities by tracing its flows of energy.

Through the ages, the ways in which human beings have produced and used energy have sporadically advanced, and each advance has defined a new stage in technology and the civilization it sustains.[8] Ancient Romans, for instance, not only lacked rock-cutting machines, they also had no dump trucks or hoists powered by electricity, gasoline, or diesel fuel. Almost all the work they applied to the job of cutting, moving, and lifting stone and other materials was done by human and animal muscles; the energy fueling those muscles came of course from food, and the food was mainly grain and hay, such as wheat, barley, millet, rye, spelt, and alfalfa grown by Roman farmers (most Romans ate only very small amounts of meat).[9]

We readily agree that energy is vital, but it's harder to grasp its role in our affairs.[10] When we think of energy, we usually think of the gasoline that fuels our cars, the electricity that lights our homes, and maybe the natural gas and coal that we burn in our power plants. In other words, we usually think of energy as fuel, and we tend to think that it's useful because of the immediate services this fuel provides—services like transport, light, and heat.

Of course, these are critical services, but energy's role in our lives is actually much more fundamental, essential, and subtle. We extract energy from our environment to create order out of disorder and complexity out of simplicity.[11] We often use this order and complexity, in turn, to help us solve the problems we face—for instance, to shelter ourselves from our harsh environment and to protect ourselves from attack. Put simply, societies with access to lots of energy are generally more adaptive, resilient, and better at solving problems.

To understand the link between energy and complexity, we need to explore some of the scientific principles that govern energy's behavior. Although people usually think of energy as a tangible thing, like gasoline, it's actually not a material substance at all. It's really a property of things. A handful of wheat kernels, a lead-acid battery, a stream of photons of light, or a rushing river all have the property of energy. This property can move from one place to another, and when it moves we can sometimes use the resulting flow of energy to do what physicists call "work"—that is, to change things in our physical world. For example, the energy in a rushing river can do work for us when it moves to another system like a water mill or turbine. The energy in a lead-acid battery does work only when it flows to a light bulb or an electric motor.

Some sources of energy are more useful for doing work than others, provided that we know what to do with the energy available in them. The *high-quality* energy available from gasoline, for instance, is better for doing work than the *low-quality* energy of the diffuse heat present in the ground under our feet, which isn't much good for anything, except perhaps heating other things, like our buildings and homes.[12] So sources of high-quality energy, such as oil, natural gas, and rushing rivers, are extraordinarily valuable, because we can use them to provide many different types of service.[13]

There are, fundamentally, only two forms of energy: kinetic energy, which is the energy of material in motion, and potential energy, which is energy that's trapped in something. We can convert potential energy to kinetic energy and vice versa. When we burn gasoline to power a car engine, we're converting the potential energy in the gasoline's chemical bonds to the kinetic energy of the engine's motion. On the other hand, when we use a river's rushing water to turn an electric turbine and then use the turbine's electricity to charge a battery, we're transferring the river's kinetic energy to the turbine to do work, and then we're using this work to create potential energy in the battery.

Heat is a special type of kinetic energy. It's stored in a material like the walls of our house or the air around us by the vibrations and other motion of the atoms and molecules making up that material. We often use heat as an intermediary form of energy to run our machines. For example, when gasoline is ignited in the cylinders of a car engine, high-temperature gases are created, and the heat of these gases provides the kinetic energy that drives the engine's pistons. Heat also drives our jet and rocket engines. Understanding the nature of heat can tell us important things about the flow of energy in natural and human-made systems, such as how much work a given system can do.

This is the province of the sub-field of physics called thermodynamics. The nineteenth-century discovery of the laws of thermodynamics—among the most important in all science—was a stunning breakthrough. These laws tell us two essential things about our natural world. The first says that energy can't be created or destroyed: the total energy of any system and its surroundings (generously defined) stays constant, whether the system is a mechanical device like a car engine, a biological system like a human body, or a social system like ancient Rome's.[14] The second law says that during the normal operation of most systems, energy degrades in quality: high-quality energy degrades to progressively lower-quality energy, with the end result being simply low-grade heat.[15] It's as if energy always flows down a slope from forms that we can use to do lots of work—like hoisting rocks in the Colosseum—to forms that aren't very useful to us at all. And every time we use energy to do work, we further degrade its quality. As a system's energy degrades, physicists say its "entropy"—often described as its disorder or randomness—increases.

To see all of this better, imagine you're sealed inside a big box with a newspaper, a chair, and a battery wired to a light bulb. Let's assume that the box is isolated from the rest of the world, so energy can't flow into or out of it. Inside the box, high-quality chemical energy is concentrated in the battery; it's also concentrated in you in the form of the sugars and proteins that you've obtained from food and that power your body. Physicists say that such energy is "coherent" and "ordered." Sitting inside the box, you can read the newspaper because the battery converts its chemical energy into electrical energy that the bulb then transforms into light energy. That light—speeding away from the bulb—hits the newspaper, walls, and other things in the box where it's absorbed and degraded into low-quality heat. The light's energy is essentially spread around the box as heat, slightly increasing the incoherent and disordered vibrations of the molecules making up the things in the box. But the chemical energy in the battery is finite, so eventually the battery is exhausted, and the light goes out.

The light won't last forever in a closed system.

You (the person in the box) are much like the battery: you convert your chemical energy into other forms of energy, such as kinetic energy as you move and lift things. If there's a bicycle generator in the box, you can hop on board and use it to drive the light and finish reading your newspaper, warming up the box even more as the heat from your exertion is carried away from your body by your sweat. But just like the battery, you too will run down and eventually die. All the high-quality energy in the battery and your body will be then degraded, and the box itself will wind down—just the way a mechanical clock winds down—to a black three-dimensional space of uniform temperature.

We'll see later that societies that don't have access to enough high-quality energy are likely to disintegrate. The laws of thermodynamics tell us that despite the fact that energy can't be created or destroyed, it does inevitably degrade, which makes it progressively less useful for work. And if it's less useful for work, it's less useful for maintaining a society's complexity and resilience.

Our light-in-a-box illustration is entirely imaginary and artificial. It's a "closed" system, which means it's sealed off from the outside world. Scientists use such imaginary systems to tease out implications of their theories. But in the real world almost all physical, biological, technological, and social systems are "open." They interact with their surroundings. Most important, these systems often extract high-quality energy from their environment to do work or to reduce disorder at their core, and they expel waste heat and material back into that environment. A city like ancient Rome, for example, imported from its hinterland timber, fresh water, and energy in the form of food and released back into its surroundings heat, sewage, and garbage (at its peak population, Rome probably produced around one million cubic meters of human waste each year).[16] A modern car engine takes in fuel to do the work of moving the car, and it expels heat and exhaust gases into the atmosphere. A steel mill takes in raw materials like iron ore and high-quality energy (in the form of coking coal, for instance) to create coherent and ordered materials like steel rods; in the process, the mill discards heat, carbon dioxide, and pollution into nearby air and water.

At first glance, open systems appear to violate the classical thermodynamic principle that disorder, randomness, and entropy always increase. After all, the mill produces low-entropy steel rods. And over time Rome's internal arrangements became more ordered and complex, as its various social and technological parts became more diverse, specialized, and interdependent.

The principle that a system's entropy must always increase, scientists eventually realized, applies only when its boundaries are defined to encompass virtually all its interactions with its surrounding environment. The system of a steel mill then includes the entire technological infrastructure that produces the iron ore and coke it uses as well as the atmosphere and waterways into which it expels its pollutants. And the system of ancient Rome included the solar energy provided by the sun and the city's entire natural hinterland of land, forest, water, and air. Within this generous boundary, the average quality of the system's energy always declines, and entropy always increases.

All the same, there can be parts of the broader system that have a very high degree of order: Rome was a zone of low entropy within its larger system. In fact, things like cities, ecosystems, and even our human bodies can create order and complexity spontaneously, decreasing their entropy even further in the process.[17] Cities build elaborate transportation, water, and energy infrastructures; ecosystems become more biologically diverse as new species evolve; and human embryos develop into people, with all their complex organs and structure. How do such amazing things happen?

Scientists still aren't sure. But they now know that systems like cities, ecosystems, or human bodies are, as they say, "far from thermodynamic equilibrium."[18] They can spontaneously create order inside themselves. But maintaining this order is a bit like holding a marble on the side of a bowl with your finger: the marble wants to sit at the bottom of the bowl—that's its equilibrium point; so holding the marble on the side takes a constant input of energy. Similarly, cities, ecosystems, and human bodies must have a constant input of high-quality energy to maintain their complexity and order—their position far from thermodynamic equilibrium—in the face of nature's relentless tendency toward degradation and disorder. And, as the system gets larger and more complex, more and more energy is needed to keep it operating.

All these ideas can help us grasp why the Roman empire fell and, ultimately, discern the fate of our own societies. The Romans employed astonishing technological prowess to construct buildings like the Colosseum. Less obviously but just as critically, they needed considerable social prowess to assemble themselves into work units, coordinate the efforts of these units, encourage specialization skills, and provide themselves with public services like governance, tax collection, and security. Codified laws regulated everything from money and debt to property rights, corporate organization, guilds, and the employment of laborers and slaves.[19]

Complex social organization doesn't appear out of thin air. Courts must be staffed, functionaries paid, and armies fed and supplied with weapons. More fundamentally, to create and sustain organizations, rules, and laws, people must move around, communicate, discuss, argue, and negotiate with one another; educate and train one another; and record—in some stable medium like rock, parchment, or CD ROM— basic rules, contracts, and bargains. All these activities once again require high-quality energy.

So the Romans used farms to capture the sunlight falling on wide swathes of land around the Mediterranean basin. Some of the farms existed before the Romans arrived. Especially in the eastern Mediterranean—for instance, in modern-day Lebanon and Syria—the expanding empire often simply took over existing cities and their food-production and tax systems, while in northwestern Europe, in places like the Rhône valley, new land was sometimes converted into farms. But wherever the farms were located, they played a role in the Roman energy economy similar to that of solar battery chargers: they converted sunlight into a form of high-quality potential energy, especially fodder and grain, that was storable and transportable.

The Romans then focused this energy—they used their food bat-teries, so to speak—to create a productive, resilient, and phenomenally complex system of public buildings, manufacturing facilities, housing, roads, aqueducts, and social organization. And here's the punch line: recent research (which we'll come to later) shows that the Roman empire was eventually unable to generate enough high-quality energy to support its technical and social complexity. This shortfall—more than proximate events like incompetent emperors and invasion by Visigoths—was the fundamental cause of Rome's fall. In other words, the empire's loss of internal order, coherence, and complexity was, in significant part, a thermodynamic crisis. The empire tipped into irreversible decline pre-cisely because it couldn't feed its energy hunger.[20]

This was Rome's fate. Will it be ours as well? A closer look at energy's role in ancient Roman society will help us find out.

A Stone's Journey

When I returned to Canada from Rome a few weeks after my visit there, I set out to see if I could follow my particular keystone's journey from its beginnings to its final placement in the great arch. I felt that focusing on that single stone could help me understand energy's role not only in the overall effort to build the Colosseum—and, more generally, in the life of Rome—but also in the empire's ultimate decline.

I began by turning to my research assistant at the University of Toronto, Karen Frecker. A painter, pianist, and electric utility analyst with a truly remarkable mind, Karen is skilled at applying technical analysis to social questions. She set to work enthusiastically to find out how much farmland the Romans needed to build the Colosseum. She divided the project into two parts—demand and supply.

First, she studied the details of the Colosseum's construction; calculated the total mass of the different types of materials used in the building; estimated how much work—or how much energy—was required to cut, move, and place all those materials; and estimated how many people and oxen were required to do that work. We called this the "demand side" of the problem. Second, she calculated how much agricultural land was needed to grow the food to feed these people and oxen. This was the "supply side" of the problem.

It sounded simple at first, but all these calculations turned out to be vastly more involved than we'd expected—and much more interesting.[21] We made a series of assumptions about Rome's technology, workforce, and agricultural practices and about the physics of Roman transportation and construction. To ensure these assumptions were sound we consulted ancient texts, literature on Rome, treatises on mechanics, and experts on the Colosseum's architecture and engineering.

Early on, we learned about a crucial difference between Roman and modern buildings. Today's buildings are constructed mostly of materials fabricated by men and women using huge quantities of energy—steel, aluminum, glass, plastics, brick, and composites. But Rome's buildings were made mainly of rock that nature had provided. In those days, the materials that took the most energy to gather and shape for human use were metals—the lead used widely in water pipes, the silver in coins, and the iron and copper in armor, swords, and other weapons, as well as the iron used to hold stones together in many buildings. Tiles and bricks also required a lot of energy: after being fashioned from clay, straw, and small amounts of high-quality pozzolana sand, they were dried in the sun and baked in wood-fired kilns at temperatures reaching hundreds of degrees Celsius.[22]

As for the energy spent to cut, move, and place the Colosseum's rock, my keystone told much of the story. Some 1,927 years before I spied it in its arch, workers at a Tibur quarry chiseled cracks into a natural bed of travertine. Then they drove wedges into the cracks, splitting off a rough six-ton block.[23] The block was rolled on logs and hoisted with pulleys onto an oxcart pulled by two oxen. It then began its two-day, thirty-kilometer journey to Rome along a specially built road. According to ancient records, every day of the Colosseum's construction, two hundred carts of stone destined for the building entered the city.[24] Unlike my travertine keystone, though, the tufa used in the building was found at sites within a few kilometers of Rome, while the bricks and concrete were also likely produced from materials close at hand. Indeed, much of the aggregate used as the binding agent in the Colosseum's concrete probably came from the rubble of Nero's nearby demolished palace.[25]

Once the rough six-ton block arrived at the construction site, masons chiseled the rock into the precise wedge designed to fit between the voussoirs—called counter-keys—that would be its neighbors. Workers used a giant wooden crane, fitted with a block and tackle, to lift the counter-keys into the wood scaffold they had built to support the arch; then the keystone itself was hoisted and maneuvered into the space between them and joined with iron clamps. Now the arch had structural integrity, and the scaffolding could be removed.[26]

The A-frame cranes used for lifting the building's rock were massive, sometimes twenty meters high. Some of these cranes were located around the inside of the Colosseum leaning outward, with stay cables made of hemp stretching behind them hundreds of meters to their anchors at the center of the arena. Others were positioned on the outside leaning inward, with their cables splaying from the building. All these cranes could tilt vertically, but none could move horizontally, so the builders had to release the stays and disassemble and assemble the cranes repeatedly.[27] Most employed ropes with pulleys and hand-cranked winches, but treadmills powered the largest cranes of all: several workers, probably slaves, got inside a device that looked like a giant hamster wheel—up to eight meters in diameter and called, in Latin, *majus tympanum*. By clambering up the curved wall of the treadmill's interior, they automatically turned a shaft that acted as the crane's winch. Such devices could lift rocks weighing tens of tons. The Romans, as one scholar puts it, "considered it a point of honor . . . to work with blocks of enormous size simply out of the desire for technical achievement."[28]

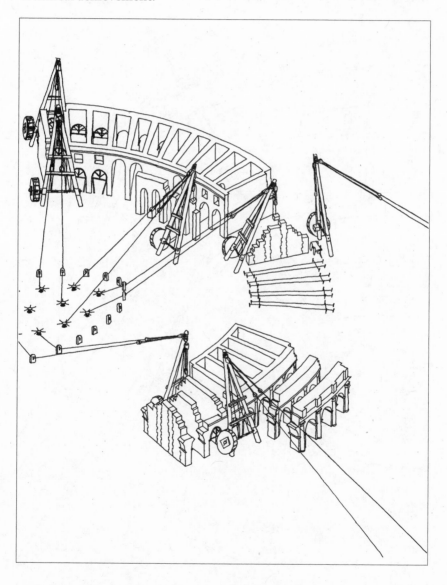

A-frame cranes lifted the rock for Level 1 of the Colosseum.

As the first floor was completed, the Romans rolled their cranes up the Colosseum's sloped seating area so they could hoist the stone elements for the higher floors. Many of the walls of the upper levels were made of concrete faced with brick; this material was lighter than the first floor's travertine and tufa, and it didn't have to bear as much weight. Wet concrete was probably mixed at ground level and lifted by

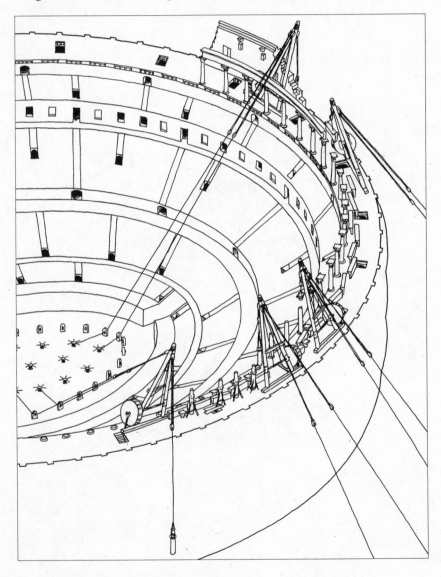

Cranes were rolled up the sloped seating area to construct Level 4.

ropes tied to buckets; or workers carried it, along with loads of brick, on their backs up a winding route of stairways and scaffolding to the ongoing work on the upper levels. When construction finally reached the fourth and highest level, nearly fifty meters above the ground, the builders installed a row of columns—eighty in all—along the inside edge of the gallery. Each column shaft alone weighed nine tons, and together they held aloft a sturdy entablature and roof of travertine designed to shelter the spectators seated in the amphitheater's upper reaches.[29]

Colosseum Calories

Karen discovered ingenious ways to estimate the total mass of various materials in the Colosseum and the distances they traveled to the building.[30] Then she plugged these figures into standard equations used to calculate the amount of energy needed to carry, push, pull, and hoist a particular mass a certain distance. This gave her an estimate—grounded in hard physics—of the minimum amount of human energy needed to *move* the building's materials. But this was only a start, because it left out most of the work done to construct the Colosseum. Among other things, oxen were used for most of the transportation of materials from quarries to the site.[31] Also, the Romans excavated and filled the foundation, drove oxcarts to and from the quarry, shaped the rocks at the construction site, built and removed the scaffolding, and mixed and laid concrete. All these activities were administered, and everyone involved was housed, clothed, and fed. In fact, over a dozen workers' guilds participated in this astonishing endeavor, including construction workers, bronze workers, blacksmiths, carpenters, porters, brick makers, marble workers, pavement layers, and masons.[32]

So Karen calculated the work required for each one of these sub-tasks.[33] As our project expanded (and as we increasingly questioned our sanity), she crunched through endless calculations and added hundreds of numbers to her lengthening spreadsheets.

Despite our attention to detail, we still had to leave many things out. For instance, our Colosseum was just a shell: it didn't have any of the real building's elaborate external and internal decorations, statuary, and water fountains—there were probably over 150 fountains in the building—or even internal piping for its lavatories.[34] Also, because we weren't able to find accurate data for the work involved in assembling, disassembling, and moving the cranes, we had to omit these important tasks from our estimate. Finally, the Colosseum project was, in thermodynamic terms, a classic open system. For example, the Romans had to harvest and transport timber for scaffolding and cranes, and they needed enormous amounts of firewood to slake lime, melt metal, and fire brick. Their wood-harvesting operation must have extended far into the city's hinterland and involved not only woodcutters but also countless teams of oxen and drivers, road builders, and a host of people supplying tools and food to these workers and animals. To make our task tractable, however, we decided that all this work fell outside the boundary of the Colosseum system; we focused instead simply on the caloric needs of the people and animals directly involved in the building's construction.

Even then, the final tally was staggering: erecting the Colosseum required more than 44 billion kilocalories of energy.[35] Over 34 billion of these kilocalories went to feed the 1,806 oxen engaged mainly in transporting materials. More than 10 billion kilocalories powered the skilled and unskilled human laborers, which translates into 2,135 laborers working 220 days a year for five years.[36]

Some of our results on the demand side of the problem surprised us. We found that the scholarly literature's standard estimate of the amount of concrete that Romans used—some six thousand metric tons—was far off the mark; the actual amount in the building alone is nearly thirty times larger. In fact, the Colosseum is a bit like an iceberg: most of its volume is underground, and almost 90 percent of this underground structure is concrete.[37] Indeed, the visible travertine, tufa, and brick make up barely a third of its volume. We were surprised, too, to find that three-quarters of the total energy expenditure went to oxen. Supporting the workers when they weren't working also took a lot of energy—after all, they had to be kept alive over holidays and rest days, and ancient Rome had many holidays.

Then there was the most surprising fact of all. When I first spied the keystone, all I could think of was the monumental effort needed to move rock and other heavy materials around the construction site and to hoist multi-ton stones and columns up to fifty meters in the air. But when we'd completed our demand-side analysis, we discovered that all these activities used only a small fraction—about 4 percent—of the total energy needed to build the Colosseum.

This was an eye-opener. It reminded me that human societies expend a great deal of energy in non-obvious ways and places—for example, to excavate and build the Colosseum's foundation. Yet most of the time we don't think about unseen but essential activities. And this would become a key discovery in my larger investigation: to understand energy's role in our societies' ability to adapt and survive, we need to develop the everyday habit of recognizing energy in all its uses and consequences.

Energy Return on Investment

Once we'd calculated the total number of calories needed to build the Colosseum, we could turn our attention to the supply question. How much cropland did the Romans need to grow the food that provided the calories for the animals and men who did all this work?

The Roman diet consisted of a mixture of grain (especially wheat), fruit (including olives and figs), legumes, small amounts of meat, vegetables, and wine.[38] Oxen were fed hay, legumes, millet, clover and other grasses, tree foliage, wheat chaff, and bean husks.[39] But to make our calculation manageable, Karen assumed Romans produced only two crops: wheat for humans and hay for oxen, with alfalfa the main type of hay.[40] Although Romans did allow their oxen to graze on pastureland, it seems that working oxen were fed mainly cropped fodder. As Cato the Censor marvelously observed in his second-century BCE treatise *De agricultura*, oxen "should not be put out to graze except when they are not worked; for when they eat green stuff they expect it all the time, and it is then necessary to muzzle them while they plough."[41]

A lot of Roman grain came from northern Africa, Egypt, Spain, Sardinia, and Sicily.[42] The regions of Etruria and Campania—just north and south of Rome along Italy's western coast—were also very fertile and probably provided some food for the surrounding area, including Rome. So grain came by ship across the Mediterranean and along the coast of the Italian peninsula.[43] Arriving at the port of Ostia, located at the mouth of the Tiber, the grain was offloaded, weighed, sorted, and reloaded on barges that were pulled by men and oxen the twenty kilometers up the river to Rome.[44] Records indicate that during the reign of Augustus, the city had to store annually half a million cubic meters

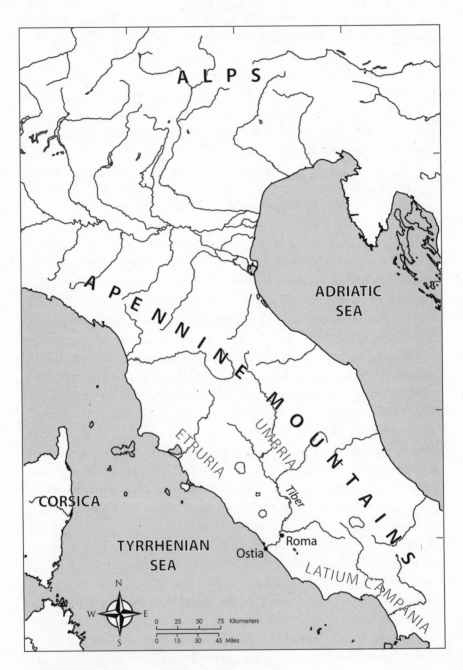

Ancient Rome and its surrounding regions

of grain from Egypt and North Africa alone, enough to fill a cube almost as long as a football field on each side.[45]

In Roman times an average hectare of good cropland in Etruria produced 1,158 kilograms of wheat and 2,600 kilograms of dry alfalfa.[46] But farmers always set aside some harvested seed for the next year's planting, and a significant portion of any harvest was also lost to rot and vermin during storage and transport.[47] Taking these factors into account, we estimated that a hectare of Roman agricultural land could produce wheat containing about 2.4 million kilocalories or dry alfalfa containing about 4 million kilocalories of usable energy.[48] Yet we'd still neglected a key complication: many people were involved in growing and harvesting the wheat and alfalfa, and all of them had to be fed. In other words, Roman farmers had to burn calories to produce calories. So before arriving at a final figure for the total land needed to build the Colosseum, we had to determine how much labor was needed to farm a hectare of land.

It turns out that we can evaluate any project intended to generate energy, including farming, in the same way that we'd evaluate a financial investment: we compare the size of the investment with the size of the return on that investment. In the case of energy projects, energy experts call this ratio the EROI, which stands for energy return on investment.[49] We calculate the EROI by dividing the amount of energy a project produces by the amount it consumes. For a modern coal mine, for instance, we divide the useful energy in the coal that the mine produces by the total of all the energy needed to dig the coal from the ground and prepare it for burning, including the energy in the diesel fuel that powers the jackhammers, excavation shovels, and off-road dump trucks and the energy in the electricity that powers the machines that crush and sort the coal.

If you're interested in the role of energy in human society—including our own societies, as we'll see—keep an eye on the EROI for different sources of energy, because it's one of the most useful statistics to compare the relative value of the sources. "An EROI of much greater than 1 to 1 is needed to run a society," note the system ecologist Charles Hall and his colleagues, "because energy is also required to make the machines that use the energy, [and to] feed, house, train and provide health care for necessary workers and so on."[50] In agrarian societies like ancient Rome's, where people did most of the farm work, food production also had to have an EROI of well over 1 to 1 for the society to support any kind of social and technological complexity. Indeed, in any complex society, those of us who aren't farmers are essentially parasites on those of us who grow the sources of energy—the grain, vegetables, fruit, and meat—that keep all our bodies running.

To see the truth of these assertions, imagine an agrarian society in which human beings do all the work, including all farm work—where there are, in other words, no oxen or other draft animals—and which has an EROI of only 1 to 1. In this kind of society, the farmers themselves would consume all the energy generated by farming: everyone would have to work exclusively to feed himself or herself, so there wouldn't be surplus energy to support people doing other things—even to build houses for the farmers, make their farm tools, or prepare their food. If this society's EROI fell below 1 to 1, the farmers wouldn't be able to generate enough energy even for themselves, and they'd slowly starve to death.

Two thousand years ago, Roman farms were often organized as large plantations—or *latifundia*—worked by slaves.[51] Farming technologies were remarkably similar to those used today in many rural zones of Asia and Africa. Oxen pulled iron-tipped plows, and farmers sowed seeds by hand, fertilized their fields with manure, and separated grain from chaff using winnowing baskets. The agricultural historian M. S. Spurr has concluded that Romans needed about fifty-eight days of work each year to farm a hectare of wheat and about forty days to farm a hectare of hay.[52] Since one day of labor burns about 3,000 kilocalories, farm workers therefore had to spend about 174,000 kilocalories to grow and harvest a hectare of wheat and about 120,000 kilocalories for a hectare of alfalfa. But this wasn't the end of the story, of course, because farm laborers had to be fed during the portions of the year when they weren't working in the field. Taking all these factors into account, we estimated that the Roman EROI for wheat was around 12 and for alfalfa around 27.[53] In other words, for each kilogram of wheat that the Romans invested in farming, they got about twelve kilograms of wheat in return.

Karen had finally arrived at the last steps in the process of estimating the total land area needed to produce the energy to build the Colosseum. Now she could determine the crucial figures for the surplus energy generated per hectare: 2.2 million kilocalories for wheat and 3.8 million kilocalories for alfalfa. She then divided these figures into the number of calories needed by Colosseum laborers and oxen.

The results were impressive: to build the Colosseum the Romans had to dedicate, every year for five years, at least 19.8 square kilometers to grow wheat and 35.3 square kilometers to grow alfalfa. That's a total of 55 square kilometers of land—or almost the area of the island of Manhattan.[54] And to capture the solar energy needed to extract, move, carve, and hoist the single keystone that I spied on my visit to the Colosseum, they needed nearly 1,300 square meters of farmland, or about three times the area of a modern city lot.[55]

The Exigencies of Energy

A crude but useful measure of a society's complexity is its level of urbanization. By this measure, for an agrarian society the Roman empire was extraordinarily complex. The size of the city of Rome and the level of urbanization in key regions like the Italian peninsula and Egypt were probably unmatched until the industrial era.[56] To sustain this urbanization, complexity, and order, the Romans had to invest immense effort to generate and transport energy.[57] And this may have been the empire's fatal weakness.

At the height of the empire in the first and second centuries CE, the population of Rome itself probably approached 1 million and may even have reached 1.5 million.[58] The entire empire's population is far harder to estimate because solid data are almost nonexistent.[59] On balance, evidence suggests that it likely peaked in the neighborhood of 60 million toward the end of the second century. Of this population, between 15 and 20 percent lived in a network of several thousand cities.[60] Most of these cities were small, with 10,000 to 15,000 people, while a handful like Alexandria, Antioch, and Carthage had a few hundred thousand. Urbanization was likely over 30 percent in Egypt, while it might have been over 20 percent in the Italian peninsula, given the massive size of Rome itself.[61]

No European city again approached 1 million people until early nineteenth-century London, and as recently as the late eighteenth century the populations of many Western societies were still around 10 percent urban.[62] The historical demographer E. A. Wrigley writes,

The ultimate reason for the comparatively low levels of urbanization in pre-industrial societies is not far to seek. To live, one must eat.

Only if levels of output per head in agriculture rise to the point
where one man on the land can feed ten, twenty or even fifty off the
land can a very high degree of urbanization be reached. [In] many
pre-industrial societies levels of output per head were so modest, and
so variable from year to year with the vagaries of the harvest, that
there was no margin available to support any considerable proportion of the population outside agriculture.[63]

The city of Rome, where at one point hundreds of thousands of
people survived on the emperor's grain handouts, required at least 8,800
square kilometers of agricultural land to grow enough wheat to feed
itself, an area not much smaller than the entire country of Lebanon
today. The population of the Roman empire in total required more than
530,000 square kilometers, or an area equivalent to modern-day France.

Around the Mediterranean basin and deep inland along the fertile
valleys of great rivers like the Rhône and the Nile, Roman engineers
built irrigation systems, networks of waterways for transportation, and
grain, oil, and wine storage facilities.[64] Boatbuilders constructed an
armada of grain ships.[65] Meanwhile, officials surveyed the countryside,
classified its land by quality and use, and established and enforced property rights—all practices that allowed them to identify, and bring under
the authority of Roman law, taxable cropland and individuals.[66] Indeed,
the late empire's most important tax, which probably generated at least
90 percent of its revenue, was levied on agriculture.[67]

What does this initial analysis of Roman energy tell us about
humanity's predicament at the beginning of the twenty-first century? It
highlights, I believe, two critical lessons.

First, when it comes to the exigencies of energy, our rich, high-tech
Western societies aren't any different from poor developing societies
or, for that matter, from ancient Rome. All our societies require enormous flows of high-quality energy just to sustain, let alone raise, their
complexity and order (to keep themselves, in the clumsy terminology
of physics, far from thermodynamic equilibrium). Without constant
inputs of high-quality energy, complex societies aren't resilient to
external shock. In fact, they almost certainly can't endure. These ever-
present dangers drive societies to relentlessly search for energy sources
with the highest possible return on investment (EROI). They also
drive societies to aggressively control and organize the territories that

supply their energy and to extend their interests, engagements, and often their political and economic domination far beyond their current borders—as we see today with American involvement in Iraq and the Persian Gulf.

The second lesson is less obvious but more important: after a certain point in time, without dramatic new technologies for finding and using energy, a society's return on its investments to produce energy—its EROI—starts to decline. The Roman empire was locked into a food-based energy system. As the empire expanded and matured; as it exploited, and in some cases exhausted, the Mediterranean region's best cropland and then moved on to cultivate poorer lands; and as its grain supply lines snaked farther and farther from its major cities, it had to work harder and harder to produce each additional ton of grain.

Today humankind is facing the same trend with many of its vital energy sources, like conventional oil, natural gas, and hydropower. We've already found and tapped the biggest and most accessible oil and gas fields, and we've already exploited the best hydropower sites, as we'll see in chapter 4. Now, as we're drilling deeper and going farther abroad for our oil and gas, and as we're turning to alternatives like tar sands and nuclear power, we're finding that we are steadily spending increasing amounts of energy to get energy.

Even though today's societies confront the same energy exigencies as ancient Rome, they're different in one key respect: they're vastly more complex and ordered, and they're much further from thermodynamic equilibrium. In other words, our societies are like the marble that wants to roll back to the bottom of the bowl, and compared with ancient Rome we're holding that marble much farther up the bowl's side. Colossal flows of high-quality energy make this possible. If we can't sustain these flows, our societies will fall back toward equilibrium—which means, essentially, that their complexity will unravel. And that unraveling, should it occur, would make Rome's decline pale by comparison.

WE ARE LIKE
RUNNING WATER

THE POPULAR ACCOUNT of the Roman empire is essentially a political story—a story that focuses on the empire's rulers and their military conquests, the endless court intrigues and brutal assassinations, and the final surrender of the West's power and glory to barbarians. And, highly condensed, it goes something like this.

In the centuries following the Colosseum's completion in 80 CE, the fortunes of the empire—and of the city of Rome itself—surged, ebbed, and surged again, only to fade permanently, in the West at least, in the fifth century. A year after the great amphitheater was inaugurated, the emperor Titus died. His brother ruled harshly until murdered in 96 CE, and shortly afterward Marcus Ulpius Trajanus, a brilliant soldier and governor of Upper Germany, came to power. The eighty years that followed were a true golden age. External threats were limited, and those that arose were manageable. Each of the emperors from Trajan to the great Stoic philosopher Marcus Aurelius reigned for long periods and governed competently. Their combined reigns saw the peak of Rome's imperial achievement—a pax Romana of unprecedented stability and prosperity.[1]

Then things went horribly wrong. Marcus Aurelius's son Commodus—callow and ill suited to rule—was assassinated in 192, and in the subsequent nine decades some twenty-three emperors came to power in the city of Rome itself. "Succession by murder and civil war now became the norm," writes the military historian Edward Luttwak. Many other usurpers controlled large parts of the empire outside of Rome, often for years at a time. The empire was attacked from the north by the Franks and the Alamanni and from the east by the Sassanid dynasty of Persia. "Much that had been built and achieved since Augustus was irreparably destroyed," continues Luttwak. "Destroyed as well was an entire conception of empire."[2]

But it wasn't yet the end. In 284 CE, Diocletian rose to power, and through immense will and acumen stabilized the empire's finances, secured its frontiers, and re-established internal order, partly by dividing the empire into eastern and western sections (with Rome, Italy, and what is now western Europe in the western section). His efforts, though, only delayed the inevitable. While the eastern part continued on for another millennium, often with great success, the western part gradually succumbed to internal stresses and constant attacks by barbarians. In 476, the German chieftain Odoacer overthrew Rome's last emperor and proclaimed himself king.

To the extent that this political story says anything about the causes of the empire's decline, it emphasizes the incompetence, corruption, and decay that ultimately led to a collapse of central authority. But this is mostly a story about—and originally reported by—the empire's uppermost elites. It doesn't really tell us what happened inside Roman society— what happened, in other words, in the lives of the empire's millions of faceless and unremarked inhabitants.

Yet it's here that we can begin to see the relevance of Rome's fate to ours. We can especially begin to see how various challenges arising from the growth, decline, movement, and concentration of a society's population can shape its fate. Such challenges make up the first of the five tectonic stresses I identified in chapter 1—stresses arising from population, energy scarcity, environmental damage, climate change, and economic inequality. Changes in the size of a society's population can sometimes be a source of stress; changes in population distribution can also signal that something is gravely wrong inside the society. In Rome's case, the empire's decline was accompanied by a precipitous drop in the population of its cities. "The collapse of urban life [occurred] from one end of the Mediterranean to the other," writes the archaeologist David Whitehouse, citing recent findings from excavations. "The five most important cities of late antiquity—Rome, Carthage, Constantinople, Antioch, and Alexandria—all contracted; indeed, Carthage and Antioch virtually ceased to exist."[3] This transformation represented a colossal simplification of Mediterranean society—a widespread loss of complexity and order—that was almost certainly tied to the empire's waning capacity to extract energy from its territory.

Take the city of Rome, by far the empire's largest. Although the available data on the city's population are extremely unreliable, those estimates that we do have tell a grim story. Whereas between 200 BCE and 100 CE, the city's population apparently soared from around 300,000 to more than 1 million, by the beginning of the sixth century it had been cut in half (food imports from distant provinces stopped by 530), and by end of the sixth century it had plummeted to 100,000. After that, for more than four centuries, the slope was steadily downward to a nadir of around 15,000 inhabitants—a number that would have filled fewer than a third of the Colosseum's seats. From the turn of the millennium till the seventeenth century, Rome was little more than a small town, with the Colosseum and most of the ruins of imperial Rome relegated to its outskirts. It was only in the twentieth century that the city surpassed the population it had seen in the empire's heyday.[4]

Some of the drop in Rome's population was undoubtedly caused by the high mortality rate of its ill-starred inhabitants. They endured such an endless succession of calamities that they must have thought the gods had forsaken them. Between the end of the first century CE and the end of the sixth century, the city was devastated by fire six times; hit by nine major earthquakes and ten disastrous floods; ravaged by plague in the

Date	Population	Date	Population
800 BCE	5,000	1084	15,000
800-500	80,000	1377	17,000
400	300,000	1527	55,000
200	300,000	1550	60,000
100	800,000	1748	150,000
100 CE	1,000,000	1800	153,000
500	500,000	1870	226,000
600	100,000	1895	450,000
700	80,000	1950	1,000,000
900	35,000	1980	3,000,000

The city of Rome's population rose and fell dramatically between 200 BCE and 600 CE.

second century, three times more in the third, and yet again in the sixth; and sacked in 410, 455, 472, and 546 by assorted Goths and Vandals.[5] Invaders sometimes slaughtered large numbers of the city's inhabitants and at other times carried them off to be ransomed or sold into slavery. Such mayhem created havoc in the region's agriculture, causing famines that in turn ripened conditions for epidemics of disease. Combined with these horrible events, dwindling urban industry meant there were far fewer jobs in cities. Because it was harder to make a living there, many people were forced to leave.[6]

During the same period, the total population of the empire also seems to have fallen—although not catastrophically—from its peak of around 60 million at the end of the second century.[7] Demographers have discovered only a few fragments of census evidence. But it seems that the economy was chronically short of labor in the western empire's later days. More important, from the end of the second century onward, farmers began abandoning their land: the empire's total area of cultivated land fell, with losses of up to 50 percent in northern Africa in the fifth century. Because the empire as a whole was neither importing nor exporting food, and because food consumption was already at a very low level for most of the population, this loss of farmland indicates that the empire's overall population number was falling.[8]

The evidence we have about changes in the region's population gives us a slim but valuable insight into what was happening in the lives of the empire's average inhabitants. As cities and towns shrank, especially in the western part, economic interactions, social structures, and institutions became simpler.[9] Peasants became more tightly tied to local landlords in feudal relationships, taxes dropped, civil administration declined, the construction of monumental architecture ceased, and everyone traveled far less. Even the wealthy and powerful ventured little beyond the towns and villages near their manors.

Demographic Momentum

Today, more than one and a half millennia later, our world seems entirely different. But is it? Just as in Roman times, a society's demographic characteristics in modern times give us vital clues to its destiny. A society's degree of urbanization tells us something about its complexity and its vulnerability to shortages of energy, while its population growth or contraction tells us something about the well-being of its people and the health of its overall economy.

If we want to better understand the destiny of our global society—and the stresses that will affect us in coming decades—it's good to start with global demographics. As it turns out, the Italian peninsula today is itself the stage for a demographic drama just as gripping as the one that took place within the Roman empire over fifteen hundred years ago. In fact, Italy sits on the front line of one of the greatest demographic challenges humankind has ever seen—the astonishing gap in the population growth rates between today's rich and poor societies.

At the end of the second century, the boundaries of the Roman empire probably encircled a large fraction—maybe even a fifth—of the world's population, which scholars variously put at between 170 and 400 million people.[10] For many centuries, the world's population changed little; it probably didn't exceed 500 million until at least 1500. It reached a billion in the early nineteenth century and 2 billion somewhere around 1930. Then it started to grow very fast, adding another billion by 1960, another by 1974, another by 1987, and yet another by 1999.

Today, in the first decade of the twenty-first century, humankind is right in the middle of a historic population transition from the high birthrates and death rates that prevailed centuries ago—including during Roman times—to the low birthrates and death rates already typical of modern industrial society. Some of today's rich societies began this transition in the nineteenth century and have essentially completed it. Other, poorer societies are still in the middle of it, and a number of the poorest have barely begun. Normally, as a population enters this transition, death rates fall first, mainly because of better nutrition and sanitation. Later, birthrates fall, because of a combination of economic development, urbanization, higher female literacy, and spreading ideas about the benefits of smaller families. During the period of time between these two events—when death rates have declined but birthrates remain high—a society's population can grow very fast.

As humankind as a whole has moved through this transition, its population has grown at an unprecedented rate. In 1900, the world's population rose by about 10 million people a year; today the figure is about 76 million a year.[11] As a result, the world's population has nearly quadrupled since 1900, from 1.65 to 6.5 billion people. By the time the transition is finished, probably in the second half of this century, our population will probably be about 50 percent larger than it is now. In the next fifty years, we're going to see as much absolute growth in our population—some 2.7 billion people—as we've seen in the past thirty-five years.[12] Combined with higher standards of living, this growth will likely double or even triple our already breathtaking consumption of the world's natural resources, from oil and natural gas to water, wood, and fish.

Yet some conservative commentators say that that the human population explosion is over and that the real problem is now a global "birth dearth" or "population implosion."[13] In many rich countries—those, say, with per capita incomes at or above the Portuguese 2005 average of $19,000 a year—birthrates have fallen far below the 2.1 births for each woman needed to replace the population.[14] The decline is particularly pronounced in Europe, where Italy's fertility rate, for example, is currently 1.38. Some countries will see their populations actually shrink in coming decades. Projections suggest that by 2050 Italy's population will decline 12 percent, Japan's by the same amount, and Russia's by 22 percent or more. Such changes, of course, pose major challenges for economic and social policy.[15] Meanwhile, birthrates have dropped sharply in many poorer countries too—much faster than demographers once predicted. In Thailand, an average woman had more than six children in 1960, whereas today she has fewer than two.

But the claim that we don't have to worry any more about population growth is entirely premature. Commentators who make this claim misinterpret and take out of context recent data on fertility trends. In fact, the world's population will continue to grow rapidly even while birthrates in most countries are falling and the populations of some countries are even shrinking.

How? Some growth will occur simply because birthrates in many of the world's poorest countries, though falling, will remain well above the replacement level for decades; currently, birthrates are above five children a woman in thirty-five countries, most of which have per capita incomes below $2,000 a year. A second factor is the declining death rate in many poor countries as health improves and people live longer. This decline will slow once average life expectancy reaches sixty-five to seventy years, but many poor countries haven't reached that level yet.

There's another vitally important and often-ignored factor at work too: about 50 percent of future growth in the world's population will come from what experts call "demographic momentum." Because many poor countries' populations were growing so fast one or two decades ago, many girls are now entering their reproductive years. Even if these females replace only themselves and their mates, they'll still propel population growth. Just as the baby boomers in the West have produced an "echo" of their population boom through their children, so the large number of girls in poor countries will produce a surge of population growth over the next twenty years.[16]

Because of these three factors—greater than replacement birthrates, declining death rates, and demographic momentum—the populations of most poor countries will continue to grow for many decades, even as their birthrates fall. Bangladesh's population, for instance, will probably increase by over 70 percent by the year 2050, and Kenya's by over 140 percent. Dozens of countries—including several of great importance to world security, like Iraq, Saudi Arabia, Pakistan, and Nigeria—will see their populations double. Afghanistan's population will more than triple in size, from 29 to an astonishing 97 million—equivalent to almost one-third of the current U.S. population crammed into about one-twentieth of the land area.

And it's worth emphasizing that if a country's population is already large, even apparently low growth rates still produce huge increments in absolute numbers. India's population is growing only 1.4 percent a year, but because the country now has over a billion people, that low rate still means an additional 16 million people annually—equivalent to adding an extra Calcutta to India's population every year, or a city about the size of New York and Newark together. China's rate of growth is well under 1 percent, but the country's population still grows by 8 million a year, which is like adding a city nearly the size of Beijing.

Steady population growth in poor countries will more than counterbalance the falling population in rich countries. So global population will grow by over 700 million people—about two and a half times today's U.S. tally—in both the current decade (2000–2010) and the next one (2010–2020). Almost all of this increase will occur in poor countries, and half will come from just nine nations: India, Pakistan, Nigeria, Congo, Bangladesh, Uganda, the U.S., Ethiopia, and China.

What this all boils down to—and what we should pay good attention to—is that by 2050 the population of rich countries will be almost exactly what it is today—about 1.2 billion—while that of poor countries will have surged from about 5.3 to 7.8 billion.

No Land Is an Island

And that's a big problem, not just for the poor countries themselves but also for all humankind. The demographic imbalance between rich and poor countries is clearly already severe—and it's an imbalance that's going to get much worse and have far-reaching implications for social conflict between these countries, and inside them too.

The situation is particularly serious for Europe, as the continent's population shrinks while nearby populations explode. "Europe is not an island, surrounded by uninhabited deserts or endless oceans," writes a leading demographer, Paul Demeny. "It has neighbors that follow their own peculiar demographic logic."[17] He illustrates this point by comparing the past, present, and future populations of the European Union's twenty-five countries with the past, present, and future populations of the twenty-five neighboring countries in North Africa and West Asia. These countries, stretching from Morocco and Algeria to Pakistan and Afghanistan, are all large sources of migration to Europe.[18]

Demeny's results are astonishing: in 1950, Europe's neighbors had less than half Europe's population (163 to 350 million); by 2000, their population had almost quadrupled to surpass Europe's (587 to 451 million); and by 2050, according to UN projections, their population will be more than three times larger than Europe's (1.3 billion to 401 million).[19] By mid-century, Europe's population will have shrunk by 50 million people from today's level. In fact, it will be falling by 2 million annually, the decline will be accelerating, and half the population will be older than fifty. Meanwhile, the population of North Africa and West Asia will be growing by about 16 million a year, and almost half will be under thirty. Demeny notes, "Apart from catastrophic events of incalculable magnitude, there is no demographic scenario that could substantially modify [these] ongoing shifts in relative population sizes."[20]

The growing population imbalance between Europe and its neighbors is just one example of a worsening global imbalance. We can see this global imbalance better if we divide the world's countries (as I have above) into two rough categories—rich and poor. I'll show in later chapters that the gulf between rich and poor countries is not only remarkably wide, it's almost certainly getting wider, and there are surprisingly few countries in the middle ground between rich and poor. As population growth rates in poor countries remain far above those of rich countries, the proportion of poor people on the planet is rising fast. In 1950, there were about two poor people for every rich person on Earth; today there are about four; and in 2025, when the world's population will be about 8 billion, there will be nearly six poor people for every rich person.[21]

This trend matters much more than most of us realize, because it has grave implications for global peace.

On one side of this rich-poor divide, static or shrinking populations enjoy vast and increasing wealth, while on the other side, hundreds of millions of people are unemployed or underemployed. They are, to use the economists' euphemism, "economically surplus"; and thanks to TV, videos, and the Web, they know what life is like in rich countries. So millions in these populations, especially young men who are highly mobile, are migrating toward rich countries in search of jobs. Often aided by unscrupulous human traffickers, they cross jungles, mountain ranges, deserts, and oceans. They travel by foot, on top of dilapidated flatbed trucks, in the stinking holds of tramp freighters, and in the suffocating heat and darkness of shipping containers. "We are like running water," says Yusuf Marwan, a young Ghanaian boarding a truck in northern Niger for the dangerous trip through the Sahara to the Mediterranean. "We know our source, but we don't know where we are running to."[22]

This migration is most visible at the interfaces between rich and poor regions—along the Rio Grande between the United States and Latin America, across the Timor and Arafura Seas between Australia and Southeast Asia, and along the boundaries between Europe, North Africa, and Western Asia. Large numbers of Africans cross the Mediterranean at the Strait of Gibraltar and also at the relatively narrow stretch of water between Tunisia and the Italian islands of Pantellenia, Lampedusa, and Sicily.[23] Many of these people die horrible deaths: they drown when their rickety boats founder off the coast of Sicily and Spain; they die of thirst in the blistering heat of the southern Arizona desert; and they freeze to death after hiding in the wheel wells of intercontinental jets bound for North America.[24]

Migrants trek across Africa in overloaded trucks, hoping to get to Europe. Here a truck departs northward into the Sahara from Dirkou, Niger.

But many pay a terrible price. Above, a migrant's body washed onto a Spanish beach near the Strait of Gibraltar.

Border security agents capture tens of thousands of migrants who survive the perilous voyage. They either send them back to their home countries or hold them in squalid detention centers scattered across the landscape of rich countries, from the Spanish Canary Islands off the coast of Africa to Heathrow airport in England to barbed-wired camps in Australia's desert.

If one stops to listen to the men and women held in these camps, their stories are heartrending. Most have left behind everything and everyone they love—their homes, spouses, children, brothers, sisters, and parents. They've traveled not out of a foolhardy sense of adventure but because their lives back home offered little hope. They're escaping economic crisis, endemic unemployment, ruined natural environments, disease, and often war. Only truly dire conditions—and only the prospect of being able to send a bit of money back home to their families—would drive people to undertake a voyage promising such hardship and risk. "You know what the situation is like in Africa," says a young man from Mali sitting in a detention center in Morocco across the strait from Gibraltar. He's left behind his wife and two children. "I don't have the means to feed my family. There comes a moment when you have to take a risk to do better for them. But now we are here like cattle in this enclosure."[25]

Few citizens of rich countries see these migrants as individual people or, except for the very occasional article in a newspaper, have a chance to hear their individual stories. To many, they are a faceless foreign horde pushing into the country from every direction, threatening to disrupt the existing economic, cultural, and demographic balance.[26] Increasingly, citizens of rich countries are tempted to support the latest plans for more coastal radar, barbed-wire fences, and police to round up migrants who make it across the border. Impatience for some kind of action, indeed *any* kind of action, is growing fast in Europe, the United States, and Australia—especially among those who believe that migrants threaten their jobs—and is causing a resurgence of far-right political parties. Laws restricting immigration are proliferating too. In 2002, for example, the conservative government of Italy's then prime minister Silvio Berlusconi passed regulations making it harder for illegal immigrants to get residency permits, easier for Italian officials to deport them, and riskier for employers to hire them.[27]

Such laws are largely pointless, says Jagdish Bhagwati, an economist at Columbia University. "[Borders] are beyond control and little can be done to really cut down on immigration."[28] Sure enough, in 2005, agents of the United States Border Patrol arrested nearly 1.2 million people who had illegally entered the U.S., mostly across the Rio Grande from Mexico.[29] They came from across Latin America and as far away as Brazil, India, China, and Romania.[30] "We are basically swatting flies," said an official of the U.S. Department of Homeland Security. "Essentially, we are completely overwhelmed by the numbers. They're just running over us."[31] Said another, "The border is broken. Mexico is broken [and] the immigration policy is broken."[32]

Wealthy societies have so far experienced only a tiny fraction of the pressure they'll face in coming decades. The current stream of people now arriving from poor regions is going to become a torrent. Yet, given the complexity of the situation, we can't really know what the flow's effects will be, either for the receiving countries or for the migrants themselves.

The effects could be mainly positive, because migrants bring cultural vitality and needed labor to rich countries, while the money they send home transfers capital and purchasing power to poor countries. But depending on other factors at play, the effects could also be disastrously negative. In Europe, for instance, migrants do a large fraction of the continent's menial work—cleaning, groundskeeping, providing child care, and the like—as well as much of the work in construction, food service, and agricultural industries. Their numbers keep wages for these jobs depressed, which has long helped Europe's economy grow with low inflation, and they underwrite the enviably short workweeks and long vacations enjoyed by the continent's middle class. Yet, in return, these hardworking migrants have received a raw deal. Ghettoized in suburban slums and legally and economically marginalized by laws that explicitly discriminate against them, they're almost entirely blocked from moving upward within Europe's mainstream economy. And when they put up a ruckus, they're deported. Slowly but surely the European economy has thus developed an explosive, ethnically and racially defined class structure consisting of a privileged, largely white elite dominating a chronically excluded, largely black and brown, and often Muslim, underclass.[33]

Despite the urban riots in France in 2005 led by blacks and Arabs from the country's poor suburbs, the situation is reasonably stable—for the time being. But if Europe is hit by a powerful shock—like an economic crisis or a string of terrorist attacks tied to members of a particular ethnic group—we could easily see a racist backlash by indigenous Europeans, a wave of urban violence by excluded minorities, and even shifts, as a result, to far more authoritarian governments across the continent.[34]

Growth's Consequences

The world's demographic imbalance is only one of several population challenges facing humankind. Among other things, population growth in many poor countries is also contributing to the rapid expansion of megacities as well as a destabilizing preponderance of young people.

But let's be absolutely clear: population growth isn't uniformly a bad thing. It's often a boon to countries and their economies. The buzzing complexity that comes when many different people live in one place makes societies more diverse, creative, and interesting. Population growth can also help generate wealth. Larger populations consume more, creating scarcities of goods, services, and natural resources that spur entrepreneurs to hustle and inventors to innovate. Larger populations provide more labor to produce goods and services, bigger markets to buy those goods and services, and more heads to generate the ingenuity that drives economic growth. Bigger markets in turn allow manufacturers to specialize, which boosts their productivity and lowers costs.[35]

Whether population growth is a good or bad thing depends on a country's technology, values, economics, and politics.[36] Take, for example, the impact on the environment. A country with a population that drives cars with engines that consume a lot of fuel and emit a lot of pollution will have a bigger effect on local air quality and global climate than one with the same size population that uses cleaner engines. Likewise a country in which most people place a high value on material consumption will tend to gobble up more resources and release more waste for each person. And economic incentives are important too: a large population will do less damage to its environment if its people have secure ownership of their cropland, forest, water sources, and fisheries, because they'll be motivated to take care of these resources.[37]

When it comes to population growth's effect on poverty, the story is similar. Growth can worsen poverty if a country has poor economic policies, weak institutions, and a corrupt and incompetent bureaucracy.[38] It can, among other things, create a glut of labor that depresses wages, while encouraging people to spend their money on immediate needs rather than save it.[39] When families have many children, for instance, they often have to put food on the table first and save for their children's education sometime later. When millions of families make this kind of choice, they can depress the economy's overall savings rate, which deprives it of capital needed for investment and growth. Governments of countries with rapid population growth, especially in cities, often face a similar trade-off between immediate consumption and long-term investment: they must choose between providing basic services like housing, water, and energy and investing in the transportation infrastructure, hospitals, universities, and research facilities essential to long-term growth. And when economic growth doesn't keep up with population growth, unemployment often rises. This might be socially manageable, if the government has set up some kind of welfare or unemployment-insurance system. Otherwise, people will migrate to jobs elsewhere or become increasingly restless at home.

But as we've seen, migration is less an option than it was decades ago. When European populations grew rapidly in the nineteenth and early twentieth centuries, people without jobs or enough land to feed themselves emigrated to America, Australia, or colonies in Africa and Asia. Today, similar escape routes are unavailable. Nowhere on the planet are large tracts of good and relatively uninhabited land still open to colonization, and anti-immigrant attitudes and policies are becoming entrenched practically everywhere. Although tens of millions of people will try to move from poor to rich countries, even if some of them succeed they'll relieve only a tiny fraction of the demographic pressure building in their home countries. Almost all of poor countries' growing populations will have to be absorbed at home.

Megacities

And it's cities that will absorb the brunt of this pressure. They are growing almost twice as fast as the populations of poor countries as a whole, because of the combined effect of their inhabitants' natural reproduction and the large-scale movement of people out of the countryside in search of a better life.

In some poor countries, especially in Latin America, rural populations are now actually declining. But in most of sub-Saharan Africa and parts of

Asia and Central America, they will continue to grow for several decades. In the majority of poor countries, rural people have little chance to better their lot: often they can't sell their agricultural products for a decent price; local cropland, forests, and water supplies are badly degraded; there's little non-farm industry; and there are few ways of getting more than a rudimentary education. In these circumstances, rural population growth leaves everybody worse off, because plots of farmland and other family resources are divided into smaller and smaller parcels with each successive generation. So, hoping for a better life and often attracted by the "bright lights" of urban areas, people flood into the cities, usually to live in slums and squatter settlements surrounding the urban core.[40] Today, an astonishing 43 percent of the urban population of poor countries lives in these slums.[41]

This worldwide urban growth has slackened a bit in the past two decades, but its pace is still breathtaking.[42] Fifty years ago, about 300 million people, or 18 percent of poor countries' total population, lived in cities; in 2003, the respective figures were 2.15 billion people and 42 percent; and by 2030, almost 4 billion people representing 57 percent of poor countries' total population are expected to be urban. Also by 2030, the number of slum dwellers in poor countries is likely to double to about 2 billion people, or around 50 percent of the total urban population.[43]

Forty-three percent of the urban population of poor countries lives in slums.

Probably the most visible result of this rapid urbanization is growth in the number and size of truly immense cities—those with populations of 10 million or more that experts call megacities. In 1950, only two cities in the whole world fell into this category—New York and Tokyo. By 2003, twenty cities had more than 10 million inhabitants, with fifteen in poor countries, including Mexico City (19 million inhabitants), São Paulo (18 million), Delhi (14 million), Shanghai (13 million), Dhaka (12 million), Karachi (11 million), and Lagos (10 million).[44] Some of these megacities are growing amazingly fast. Over the next dozen years, Delhi, for example, is predicted to expand almost 3.5 percent annually, a rate that will double its population in under twenty years. Dhaka in Bangladesh is growing even faster, and Lagos in Nigeria leads the pack at 4.5 percent—a doubling time of only fifteen years.[45]

And as urban populations explode in poor countries, quality of life there plummets.[46] Governments can't provide even basic services. Water systems, roads, electrical grids, sewers, and other essential infrastructures of daily life are grossly overtaxed.[47] Untreated industrial wastes and raw sewage flow into rivers and lakes that supply drinking water. One specialist notes that most rivers and canals in the megacities of poor countries are "literally large open sewers, with the organic waste from industries, drains, sewers, and urban runoff rapidly depleting the dissolved oxygen."[48] Household garbage, hazardous industrial waste, and medical refuse pile up in the streets or accumulate in unregulated dumps, often right next to slums. Sometimes the slums are built right on top of these dumps, as they are in the Payatas dumpsite on the outskirts of Metro Manila in the Philippines. Residents make their living recycling bits of metal, plastic, and other material in the city's informal economy.[49] Meanwhile, the air above is a putrid soup of ozone, sulfur dioxide, and suspended particulates. Bad water and air, poor sanitation, and slum crowding is a toxic mix that causes appalling rates of amoebic dysentery as well as pneumonia and other infectious diseases.

In many megacities in poor countries, violence is commonplace, spurred by extreme income inequality and the weakness of the government and police. The desperately poor often live in shacks right beside wealthy enclaves cordoned off by high walls and barbed wire—a situation that encourages violent property crime, including robbery and kidnapping for ransom. The rich resort to private security guards to protect their homes, businesses, and schools (these guards often outnumber municipal police). Sprawling slums become no-go zones for any form of government authority, as gangs, ethnic factions, urban warlords, and criminal syndicates fight each other for control of territory and markets. Urban gangs are increasingly internationalized, with their members joining streams of migrants flowing across borders between poor countries and onward to rich countries, where they often become involved in urban crime.[50]

Criminal elements sometimes take over entire cities. On September 30, 2002, Rio de Janeiro's most powerful gangs—rooted in the city's hillside slums (the famous favelas), financed by huge drug profits, and armed with machine guns, firebombs, and grenades—shut down the city. Their jailed leaders were incensed that authorities had taken away their air conditioning, catered food, and cell phones. When the gang's message went out, store owners were threatened with violence and shuttered their shops. Business came to a halt, and public transport stopped when buses were torched. "Black out the South Side," ordered a gang leader, whose call from jail was intercepted.[51] "Everything has to shut down, all the commerce, everything is going to be paralyzed. We're going to show them that we've got the power and they don't." Said a prominent Brazilian newspaper columnist later, "On September 30, a border was crossed. Everything that the middle class has always observed from afar suddenly descended the mountain to take over the asphalt, and the result was fear, a generalized panic like I have never seen in this city."[52]

The prevalence of such criminal and gang violence is boosted by the age structure of a poor country's population. In many countries, rapid population growth has created a "youth bulge," with young adults between fifteen and twenty-nine years old making up a large portion of the population. Urbanized and relatively well educated young men whose aspirations for jobs and social status are frustrated may be easily recruited to organized crime or violent political action.[53] Say two experts on the subject,"Young men—out of school, out of work and charged with hatred—are the lifeblood of deadly conflict."[54] In a wide range of countries where internal instability could threaten international stability, the fifteen-to-twenty-nine age cohort accounts for nearly 40 percent or more of the adult population, including Afghanistan, India, Indonesia, Iraq, Mexico, Pakistan, the Philippines, Saudi Arabia, and South Africa. This cohort makes up nearly 50 percent or more in the Congo, Ethiopia, Haiti, Iran, Kenya, Nigeria, Syria, and Yemen.[55]

Although much violence in major cities of poor countries is criminal and gang related, it sometimes also takes the form of political protests and ethnic or communal riots. But overall in recent decades, this kind of organized large-scale violence has been rare, a surprise to many researchers, because they thought the crowding, unemployment, and income inequality of fast-growing cities would be an incendiary mix.[56]

The world's megacities will certainly be less stable in the future. Organized violence is far more likely when big cities suffer a sudden economic shock due to something like a currency devaluation, debt crisis, or sharp increase in energy prices. In other words, as long as economic conditions are reasonably good, rapid urbanization doesn't make violence more likely. But when conditions deteriorate, things can get very nasty.[57] We've seen this happen in the past: in the late 1970s and early 1980s, when governments of many poor countries tried to deal with crippling debt by imposing austerity measures, they provoked waves of protests, strikes, riots, and attacks on government buildings by the urban underclass.[58]

And just like ancient Rome—the only true megacity of European antiquity—today's megacities in poor countries are far from thermodynamic equilibrium. In fact, compared with ancient Rome, they are enormously more socially and technologically complex. This is not a bad thing in itself, of course, but it does mean that every minute of every day these cities suck from their immediate hinterlands and from regions far beyond almost incalculable quantities of high-quality energy.[59] In Delhi, for instance, a significant part of the energy used by manufacturers comes from charcoal produced from the forests in northeast India, two thousand kilometers away. And much of the city's petroleum, including the diesel that powers the trucks that carry the charcoal to Delhi, comes from Iran, Saudi Arabia, and Nigeria.

Rome, too, depended on energy from far away. A million or more people lived in a region that couldn't provide even a fraction of the energy they needed. Stores of grain often ran short through the winter, because ships couldn't sail across the Mediterranean. Sometimes the shortages led merchants to hoard stocks, sending food prices soaring and causing the poor to riot.[60] Each spring, when convoys of grain ships were sighted on the horizon, relief swept the city's inhabitants— from lowly slaves to the emperor himself.[61] The great Roman historian Tacitus, writing in the early part of the second century, commented on this profound vulnerability. "Italy relies on external resources," he lamented, "and the life of the Roman people is tossed daily on the uncertainties of sea and storm."[62]

In the same way, energy is the Achilles' heel of today's megacities. Without gigantic flows of high-quality energy—almost always coming from distant places—these megacities can't survive in their current form, and they won't stay peaceful. When these flows are constricted in the future, and when urban energy prices shoot skyward, we'll discover that the world's megacities, with their surging populations, are powder kegs.

SO LONG,
CHEAP SLAVES

I ALWAYS LIKED throwing the chain. It took a lot of practice, but if I did it right—with a precisely timed flick of my wrist—the result was magical. Sinuously, like a snake, the chain wrapped itself in multiple loops around the vertical pipe in front of me. Then the winch behind my shoulder reeled in the chain's loose end, yanking it tight. As the winch pulled hard, the chain's steel links dug into the pipe, spinning it fast and screwing it into the end of another pipe below. With a bang, the upper pipe would stop spinning, and I'd know the two were joined tightly together.

It was the mid-1970s, and I was working as a motorman on an oil rig in Alberta. Throwing the chain was part of my job.

Oil rigs are like big carpentry drills directed into the ground. A drill bit, often consisting of three conical wheels as wide as the span of a hand, is attached to the end of a ten-meter-long section of heavy steel pipe. The pipe is then pointed downward beneath the rig and spun at high speed, which turns the bit at its end and grinds a hole in the rock. A viscous fluid known as drilling mud is pumped down the pipe; it shoots through jets in the bit and returns to the surface laden with rock chips. As the bit grinds deeper into the ground, new sections of pipe are added above. After weeks or even months of drilling, the full "string" of underground pipe can stretch many kilometers down.

Every time the bit becomes dull and must be changed, the whole string must be removed—a task called "tripping." Since hours tripping

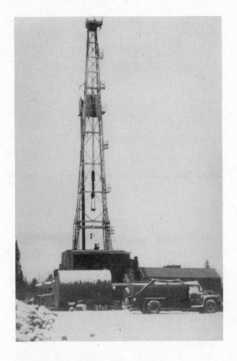

Drilling for gas at the Sikanni Chief River in northern B.C. in 1974

are hours not drilling, rig workers toil nonstop taking all the pipe out of the hole, changing the bit, and putting it back in. If the well is deep, a full trip might last a dozen hours or more, without a break. The work is hazardous, exhausting, and mesmerizingly repetitive. The men on the rig floor know the precise timing of their moves as each piece of pipe comes out of the hole or goes back in.

And it's on the trip back in that the motorman throws the chain. At least that's the way things worked in the 1970s. Unfortunately, the chains are very dangerous: sometimes they break and slash people across the face; other times they can rip off a motorman's fingers. So today most rigs use mechanical spinning devices rather than the chains I threw to tighten the sections of pipe together.

As a young man in my late teens and early twenties, I learned some of life's most important lessons in the oil patch, working across western Canada as a roughneck on oil rigs, as a laborer on turnaround crews in gas refineries, and as a welder's helper on pipeline construction teams. In these jobs, I met some of the finest men and women I've ever known. Always resourceful, usually fond of coarse language, and often highly skilled, they were—without exception—unbelievably tough. When the going got rough, and sometimes it got very rough, they didn't complain; they simply took the measure of the challenges they faced, figured out what needed to be done, and did it. Often their responses involved great financial risk to their companies and even personal risk. In their own way, they exhibited the prospective mind I described in chapter 1: they weren't surprised by surprise; they were comfortable with extraordinary uncertainty; and they intuitively understood that resilience in the face of constant change demands constant creativity—especially when it comes to finding new ways of using the materials and tools at hand to solve pressing problems.

I also learned a great deal about the energy industry in those years, from the ground up, so to speak. I came to appreciate that it takes energy to get energy—a principle formalized in the energy return on investment (EROI) concept that I described in chapter 2. And I saw that large-scale energy production often produces large-scale environmental damage, from roads that cut through fragile subarctic forest to man-made lakes overflowing with drilling mud (today, the mud is sometimes injected back underground). I became interested in the details of rig technology and subsurface geology—to the point that my endless questions sorely tried my co-workers' patience.

The author as a roughneck on the floor of the Sikanni Chief rig

Those early experiences instilled in me a lifelong interest in energy—in what it is, where it comes from, and how we use it. And as I've learned more about our global energy situation in the years since, I've realized that the growing risk of disruptions to our world energy supply combined with our societies' increased vulnerability to such disruptions are, together, a tectonic stress that threatens our future.

Energy, as we've seen, is our master resource. And it's our master resource in two senses—the profane and the sacred, one could say. In the profane or worldly sense, energy is simply a fuel. It makes things go. Without enough at the right times and places, our economies and societies would grind to a halt. Specifically, they wouldn't be able to feed themselves, supply themselves with enough clean water, or make chemicals, pharmaceuticals, clothing, shelter, and key industrial goods. They wouldn't be able to move large numbers of people and large quantities of materials and information. Without adequate energy in the future, humankind hasn't a hope of raising the standard of living of the planet's poorest 2 billion people who must try to survive on $2 a day or less.

In the sacred sense—in the sense that should really command our respect and attention—energy sustains the social order and complexity that allows us to solve our common problems and make our lives steadily better. If human beings hadn't had access to ever-larger supplies of high-quality energy, we would still be hunter-gatherers, surviving on grubs, roots, and local game. This sense of energy is larger and more informative than the notion of energy as a fuel, because it tells us why fuel is, in the end, really vital to our well-being.

Already today, energy is entangled with some of humankind's most intractable problems: its production and trade play a critical role in the skewed economic relations between rich and poor countries, its consumption causes pollution that contributes to climate change, and its uneven distribution among countries promotes war. But today's energy problems are minuscule compared with those we could face tomorrow. Our appetite for energy is rising very fast: most experts estimate that global consumption will more than double by 2050—a time span easily within the lives of today's children—and quadruple by 2100.[1] Rapidly developing poor countries, like China and India, which together make up nearly 40 percent of the world's population, are a big force behind this trend.[2]

China, for example, was a net exporter of oil as recently as 1993. Because of 8 to 9 percent annual economic growth, booming car sales, and electricity use that grows by the equivalent of Brazil's every two years, China's overall energy demand has skyrocketed. It has become the world's second-biggest consumer of petroleum products (after the United States) and a major customer for oil from the Middle East and the Sudan.[3] Between 2000 and the beginning of 2005, China's daily oil imports soared 140 percent, and in the next fifteen years the country's total energy demand is expected to double. So China is laying plans for an energy-supply network that will draw oil, natural gas, and coal from Siberia to the north, Central Asia to the west, Canada and Venezuela to the east, and Indonesia and Australia to the south—plans that are pushing it into direct conflict with other energy-hungry countries like India, Japan, and the United States.[4]

Will humankind be able to produce the energy it needs, especially as poor countries encompassing billions of people industrialize? And what problems will we create for ourselves as we try? To grasp how serious the situation is, it's best to focus our discussion on whether we're going to have enough oil.[5] Without oil, our societies, as they're currently set up, couldn't function. It provides nearly 40 percent of the world's commercial energy supply—that is, energy bought and sold in the marketplace. It's essential for farming, much manufacturing, and countless petrochemicals. And, as we're reminded every time we put gasoline in our cars, it provides nearly all our transportation fuel. Oil powers virtually all movement of people, materials, foodstuffs, and manufactured goods—inside our countries and around the world.

From Geopolitics to Geoscarcity

Human beings have known about oil for a long time. For millennia, people collected the black goo that seeped to Earth's surface and used it for a variety of purposes. Some believed it had medicinal properties, while others used it to caulk boats. But oil's value as an energy source wasn't recognized till the industrial era.

Up to the eighteenth century, most people in Europe and the United States lit their homes with wax and tallow candles and vegetable oil lamps—technologies similar to those the ancient Romans used. People then turned to whale oil and, in the first half of the nineteenth century, to town gas (a mixture of methane and hydrogen) and kerosene produced from coal.[6] But as overhunted whale stocks dwindled and as the new machines of the Industrial Revolution demanded better lubricants and unprecedented quantities of power, the stage was set for the age of oil.

In 1859, not even two lifetimes ago, Col. Edwin Drake sank the first well drilled specifically for oil at Titusville in western Pennsylvania, and by early in the twentieth century, drilling rigs had perforated Earth's crust with tens of thousands of wells, from California to the Caspian Sea.[7] Between 1950 and 2000, world energy consumption rose more than twice as fast as it had during the previous century, and oil powered most of this growth.

The oil age is only the latest stage of a larger energy transition that our species has been going through over the past two centuries. For the most part, we've shifted away from traditional sources of energy—like wood and fast-moving rivers—derived more or less directly from the sun.[8] Wood captures and stores solar energy for at most a few hundred years, and the energy in a moving river comes from the solar-powered exchange of water between the surface of Earth and the atmosphere: the sun evaporates water in lakes and the sea, it condenses and falls as rain on land at high altitudes, and then gravity makes it run in rivers back to the lakes and sea. But as our energy needs have risen, we've turned increasingly to sources derived far less immediately from sunlight, like fossil fuels that were produced when ancient organic matter was transformed by immense heat and pressure deep underground. We've also turned to sources not derived from sunlight at all, like nuclear power.

We've seen that the Roman empire relied on the sun's energy: its farms acted like battery chargers by taking solar power and converting it into storable and transportable forms of high-quality energy, like alfalfa and wheat, which in turn fueled the work of the men, women, and animals who built the empire. When it comes to fossil fuels, we can think of the entire Earth as a battery charger. This battery charger, though, operates not over the course of one agricultural growing season but over billions of years. Through these aeons, organisms have captured and stored a tiny fraction of the solar energy arriving on Earth, and the planet's gravitational energy has then compressed and metamorphosed this organic material into sources of high-quality energy like coal, oil, tar sands, oil shale, and natural gas.[9]

Of these, oil is truly special, which is why it's called black gold. It packs a huge amount of energy by volume and weight: three large spoonfuls of crude oil contain about the same amount of energy as eight hours of human manual labor, and when we fill our car with gas, we're pouring into the tank the energy equivalent of about two years of human manual labor.[10] Oil is also versatile, convenient, and still relatively cheap. No other substance or fuel comes close to matching its properties. And our species' exploitation of these properties dramatically changed the path of our economic, technological, and social evolution. Abundant oil has been a wellspring of our soaring prosperity, and it has allowed us to create today's complex globalized economy. It has also allowed the world's population to quadruple in the past century—and for people to become, on average far healthier—in large part because we can grow so much more food on our land: food output per hectare has quadrupled as we've increase energy inputs (to run tractors, power irrigation pumps, and manufacture fertilizer) more than eighty fold.[11] As the petroleum geologist Colin J. Campbell says, "It's as if each of us had a team of slaves working for us for next to nothing."[12]

But we're draining our oil battery far faster than it's being recharged. Just as we're about halfway through the boom in the human population, we're also probably about halfway through the world's available supply of oil. In fact, we can see the end of the oil age now, and—with oil prices over $70 a barrel in mid-2006 and increasingly aggressive jockeying for control of remaining reserves among companies and countries—we're likely feeling the foreshocks of the inevitable transition from oil abundance to oil scarcity.

Despite increasingly persuasive evidence that this transition is upon us, many prominent people think that such claims are utter nonsense. For instance, in 2002, Stephen Moore, then president of the influential U.S. neoconservative lobby group the Club for Growth declared, "There is a belief on the environmentalist side that we're running out of oil, that we have to conserve energy. I'm adamantly opposed to energy conservation. We're not running out. All we have to do is go out and find it and produce it."[13] In the same vein, Morris Adelman, emeritus professor of economics at the Massachusetts Institute of Technology and one of the world's foremost resource economists, has written, "Minerals are inexhaustible and will never be depleted. A stream of investment creates additions to proved *reserves*, a very large in-ground inventory, constantly being renewed as it is extracted. . . . How much was in the ground at the start and how much will be left at the end are unknown and irrelevant."[14]

There has been much concern lately about oil cost and availability, but Adelman's optimism still echoes through public-policy circles in Western countries, especially in the United States. American policy makers may worry about the scale of their country's oil imports (the U.S. now imports two-thirds of its total oil consumption, up from 28 percent twenty years ago), and they may worry about the West's dependence on oil from the Persian Gulf, but few doubt that with the application of enough new technology and capital, exploration companies will find vast new oil deposits.[15] And a powerful piece of evidence seems to support this optimism, at least at first glance: the oil price shocks in 1973–75 and 1979–81, which many pessimists thought were precursors of chronic oil scarcity, turned out to be only temporary. The higher prices brought a wave of exploration and technological development, and by the mid-1990s oil had become cheap and abundant once again.

But we've largely misunderstood the nature of those oil shocks: they had essentially geopolitical, not physical, causes. The 1973 shock, when crude oil prices rose from $4 to $12 a barrel in a few months, resulted from the deliberate action of several Arab states angered by U.S. support for Israel in the Yom Kippur War. Similarly, the 1979–81 shock, when the price went from $14 to $35 a barrel, followed the Iranian revolution, the Iran–Iraq war, and production cutbacks by the Organization of Petroleum Exporting Countries (OPEC).[16]

Oil wasn't physically scarce, at least not globally—that wasn't the problem. Instead, a large portion of world crude production had become concentrated in the hands of a small number of Persian Gulf states that had expropriated the operations of international oil companies within their territories. The oil companies were forced to go elsewhere to find oil. Because the Persian Gulf states could flood the market with cheap oil if they desired, they became what experts call "swing producers" able to determine the relative balance of supply and demand in world oil markets. And as they learned to use this enormous leverage to their advantage, world oil markets became extremely sensitive to Middle East events.

It's true that higher prices stimulated exploration, but the results were meager and had little direct effect on oil supplies.[17] The subsequent collapse of oil prices in the mid-1980s and in the 1990s (after a spike just before the 1991 Gulf War) was mainly, once again, a result of the shifting economics and geopolitics of oil: the earlier higher prices depressed oil demand for a while, countries that had previously restricted their sales returned to selling large quantities of crude, and new, non-OPEC oil fields that had been discovered earlier, like those in Alaska and the North Sea, finally came onstream. But just as the earlier higher prices didn't mean that oil was globally scarce, so this price collapse didn't mean that oil would last forever.

Oil's Peak

If the oil shocks of the 1970s and early 1980s had geopolitical causes, in coming years the story will be different: geopolitics will still play a major role, but we'll also experience a real and worsening physical scarcity of oil. This will arise from a stark imbalance between soaring demand and the inescapable reality of the planet's limited oil reserves.

I don't mean we're going to run out of oil—at least not anytime soon. Oil will be around for a long time to come. But we *are* going to run out of the cheapest oil—that is, the most accessible oil—as it becomes harder to find, costlier to produce, and more concentrated in politically volatile parts of the world.[18] It will take much more energy to get this oil in relation to the energy we get from it (the inevitable EROI, or energy return on investment). The global economy will have to adjust to permanent oil scarcity, and the transition from today's world of oil abundance to tomorrow's world of scarcity will be marked by new oil shocks, far more disruptive than those of the 1970s and '80s.

Although we don't know, of course, exactly when these shocks will hit, we're not completely in the dark about the future of the oil supply because we have a tool for prediction. In the mid-1950s, M. King Hubbert, a research geologist with Shell Oil, showed that we can roughly predict the trend of output from a specific geographic region. If unrestricted oil extraction occurs, the volume of oil pulled out of the ground tends to follow a bell curve—production starts slowly at first, builds to a peak, and then falls at about the same rate it climbed. The maximum output—represented by the curve's peak—occurs when about half the oil in the ground has been extracted.

In a sensational speech to the American Petroleum Institute in 1956, Hubbert predicted, using this technique, that oil production in the U.S. lower forty-eight states (i.e., excluding Alaska) would peak between 1965 and 1971. Most experts were outraged. They thought his forecast was nonsense and that U.S. production wouldn't fall for many decades. Indeed, Shell actually censored the written version of Hubbert's address to soften his warning.[19] Yet, sure enough, in what must be one of the most astonishingly accurate long-range forecasts in history, U.S. crude oil output crested in 1970.[20] Even with the new oil from Alaska that began to flow in the late 1970s—something Hubbert hadn't anticipated—U.S. output has never surpassed its earlier peak. By the mid-1980s, as Alaskan output itself crested, overall U.S production started an inexorable decline down the backside of Hubbert's bell curve. Today, total U.S. output of crude and natural gas liquids is 40 percent lower than it was in 1970, while energy consumption has soared nearly 30 percent. Imports of oil have become essential to bridge the ever-widening gap.[21]

If we want to use Hubbert's methods to estimate when a region's oil production will peak, we first need two other estimates: the future rate of production (in, say, millions of barrels of oil a year) and the amount of oil the region will have finally produced when, at the end of the day, all drilling and extraction are finished—what experts call the ultimately recoverable resource or URR.[22] The URR is the amount of oil we can realistically expect to get from the ground. The task of figuring out the URR is easier if we have a fairly long historical record of oil discovery in the region, because the volume of oil discovered year by year also tends

Millions of Oil-Equivalent Barrels

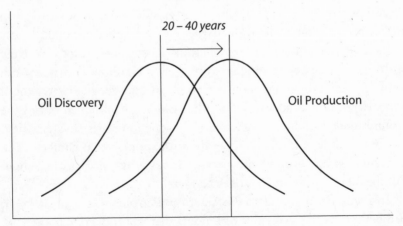

Decades can pass between oil discovery and production peaks.

to follow a bell curve.[23] The discovery curve's peak precedes the production peak, usually by a few decades. Take, for instance, the United States: it reached its discovery peak in 1930, when drillers found the East Texas field, but it reached its production peak in 1970, forty years later. Nevertheless, when Hubbert made his prediction in 1956, the rate of discovery in the U.S. had already declined enough that he could gauge the country's production trend with remarkable accuracy.[24]

Despite exploration companies' immense investments of capital and technology, oil discovery in the U.S. has declined steadily since 1930. As the petroleum geologist Colin Campbell notes, "The United States had the money to [discover more oil], it had the incentive, [and] it had the technology, so the fact that discovery reached a peak—and then declined inexorably for . . . seventy years—is not for want of trying. It was due to the physical limits of what nature gave them."[25] In oil basins around the world, petroleum geologists have seen the same sequence of a discovery peak followed by a production peak, and in about fifty countries output has crested and is now falling—including in the United Kingdom, Norway, Indonesia, Oman, and probably Russia.[26]

The story for the world as a whole, however, is more complicated. Until the early 1970s, world production followed a rising curve, much as we'd expect, given Hubbert's model. But OPEC's politically motivated output cuts in the 1970s and 1980s flattened the top of the curve before the global peak arrived (at the time, OPEC controlled about half of the world's output of oil). By the mid-1990s, the upward trend resumed. We can assume it will peak at some point, just as has happened in many individual oil-producing regions around the world. But when might the global peak occur? To answer this question we need to know something about the eventual oil yield (the URR) for the whole planet—because, as we've seen, peak will occur when about half the planet's oil has been extracted.

To estimate the planet's URR and the year of its peak oil output, Colin Campbell and his colleagues at the Association for the Study of Peak Oil and Gas have used detailed data for every oil-producing country's history of oil discovery, current oil production, cumulative oil production, estimated reserves, and prospects for finding more oil. After carefully interpreting, adjusting, and tallying all these figures, they've concluded that about 1.9 trillion barrels of conventional oil (i.e., light oil found in reasonably accessible locations) were ultimately recoverable before humans started extracting and burning it.[27] Given that humankind has consumed about 970 billion barrels, we have somewhat less than a trillion barrels left. This remainder includes, Campbell thinks, some 800 billion barrels contained in already discovered oilfields and about 140 billion barrels still to be found.[28] So, because we've burned about half of the original endowment, we must be very close to peak world production—and to the irreversible decline of oil output that will inevitably follow.

Campbell's estimate of world URR is at the low end of the range generated by oil experts, and it's contentious. In 2000, for instance, the United States Geological Survey (USGS) produced a radically different estimate—and one that's far more optimistic. It calculated that Earth's original endowment of recoverable oil was probably around 3 trillion barrels, which, if correct, would mean that we still have more than 2 trillion barrels left—double Campbell's forecast.[29] This estimate is widely cited and has been incorporated into the forecasts of the International Energy Agency in Paris, the U.S. Energy Information Administration, and even Saudi Arabia's own estimates of its reserves.[30] But close study suggests it exaggerates the amount of oil still available.[31]

In fact, we can already gauge the accuracy of the USGS forecast. It estimates global oil discovery will average about 24 billion barrels a year between 1995 and 2025. But during the first eight years of this period—that is, till 2003—actual discovery has been only 8.5 billion barrels a year, or about a third of the predicted amount.[32]

To be fair, estimating discovery trends, the size of existing reserves, and future reserve growth is less a science than a black art. There's vast room for reasonable disagreement among experts, and both Campbell and the Hubbert model have been strongly criticized.[33] Critics note that doomsayers have long predicted that world oil output would peak soon, but the much-anticipated dates have come and gone while oil output has continued to rise.[34] And some commentators have condemned Campbell and his like-minded colleagues for changing their estimates of global URR as time has gone on.[35]

Yet these criticisms suffer from their own serious weaknesses.[36] For instance, it's unreasonable to condemn Campbell and others for revising their URR estimates as they've compiled more and better data about the world's oil fields and refined their calculations. So on balance, I believe, the evidence supports the claim that the world is nearing its peak output of conventional oil.

The situation may actually be worse than the Hubbert model suggests, because the reserve estimates of oil-producing corporations are inaccurate—perhaps wildly so.[37] Reserve data are even more dubious for many oil-producing countries, because they have strong incentives to exaggerate. Belief that a country's reserves are large boosts its economic clout and creditworthiness with international banks and investors; and for countries that are members of the OPEC cartel—such as Kuwait and Saudi Arabia—estimates of large reserves ensure that the cartel will grant the country large oil-export quotas. "Countries want a higher allocation, so they tweak their numbers," says Fadel Gheit, a senior energy analyst with the investment firm Oppenheimer & Company. "Everybody lies about the reserve, so you want to make sure that you lie even more than the guy next to you."[38]

Where Are the Giants?

The oil industry has been reluctant to acknowledge the world's declining discovery trend and impending output peak. Executives tend to talk about these issues *after* they retire.[39] Once in a while, though, a senior person still in the industry's employ admits the situation is critical. A startling example appeared in the pages of the industry journal *World Energy* in 2002. Harry Longwell, then director and executive vice president of ExxonMobil, presented global oil-discovery data compiled by his company's researchers. His graph shows a clear bell-shaped discovery curve: world crude discovery peaked around 1964 and fell quickly afterward, with upward blips in the late 1970s (when Alaska's Prudhoe Bay and Europe's North Sea fields were found) and again around 2000.[40] Annual output outstripped discovery a quarter century ago, and in recent years we've extracted and consumed between four and five times the amount of oil we've found.

Most strikingly, Longwell suggested, future higher prices won't stimulate the wave of oil discovery that agencies like the USGS anticipate. "Cycles of discovery show little correlation with price over the long term."[41] Also, "it's getting harder and harder to find oil and gas. Industry has made significant new discoveries in the last few years. But they are increasingly being made at greater depths on land, in deeper water at sea, and at more substantial distances from consuming markets."[42]

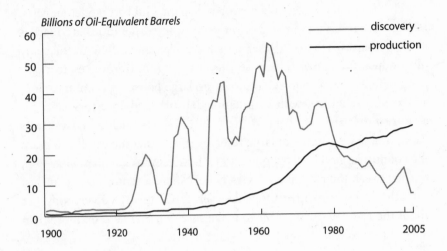

Global oil discovery peaked in the early 1960s.

When Harry Longwell says that oil and gas are getting harder to find, he means it takes more energy to find them. In other words, exploration companies are facing a declining energy return on investment (EROI). The trend is worldwide but is most pronounced in the lower forty-eight American states, earlier examined by M. King Hubbert for Shell Oil, where exploration has been going on far longer than in almost all other regions. Between the 1930s and today, oil's EROI in the U.S. fell from more than 100 to 1 to about 17 to 1, as the average depth of an oil well increased from 1,000 to 2,000 meters, and the average size of a new oil field fell from more than 20 million barrels to fewer than 1 million.[43] And because more work was needed to extract oil in the U.S., the cost of producing a barrel nearly quadrupled.[44]

But aren't there huge new deposits of oil just waiting to be discovered in distant parts of the planet like Central Asia, the Arctic, or far offshore? Probably not. Exploration companies have likely already found nearly all the biggest fields—those that experts call giants. Although the planet has about 30,000 active oil fields, just 116 giants produce half the world's conventional crude oil, and only 14 of these produce a fifth. These 14 fields are, on average, forty-four years old. As we can see in the accompanying table, over three-quarters of the world's giants were discovered prior to 1980, including the 4 largest, which together produce nearly 12 percent of global conventional output.[45] Better methods of finding

Giant Fields Production Barrels per Day	No. of Fields	Total Production 000 B/D	Era Discovered					
			Pre-1950s	1950s	1960s	1970s	1980s	1990s
1,000,000 +	4	8,000	2	1		1		
500,000 - 999.,000	10	5,900	2	3	3	1	1	
300,000 - 499,000	12	4,100	3	1	6	1	1	
200,000 - 299,000	29	6,450	8	4	6	9	1	1
100,000 - 199,000	61	7,900	5	8	13	13	11	11
Total	116	32,350	20	17	28	25	14	12

Note: Global daily conventional oil production, not including natural gas liquids, was 67 million barrels in early 2006.

Most giant oil fields were found decades ago.

and mapping deep oil deposits show that much of Earth's crust is devoid of oil, including vast tracts of the ocean floor.[46] And although new technologies—like three-dimensional imaging of oil reservoirs and steerable drilling—help oil rigs tap smaller pockets of oil tucked far underground between rock strata, we will need to find many thousands of these pockets to compensate for the decline of the handful of aging giants.

As it has become harder to find new and large oil basins, and as many existing basins outside the Middle East have begun to decline, the Persian Gulf producers (especially Saudi Arabia, Iraq, Kuwait, and the United Arab Emirates) have become correspondingly more important. All optimistic predictions of future world oil availability have these Gulf producers playing an even more dominant role in meeting our future demands.[47]

Two of the world's most important giant oil fields are, in fact, in Saudi Arabia. These two fields—named Ghawar and Safaniya—are so big that experts have labeled them supergiants. It's worth taking a moment to examine them closely, because there's a widespread perception among commentators, resource economists, and policy makers—even the U.S. Department of Energy—that Saudi Arabia has huge spare capacity. These people say that the country can meet a good fraction of the world's rising oil demand by turning on the oil spigots whenever prices soar.[48]

This isn't likely to happen. The extraordinary presence of two of the world's four supergiants in the same country has given people the misleading impression that Saudi Arabia has immense untapped potential—that it's floating on oil. But the Ghawar and Safaniya fields are old, and despite heavy exploration the country hasn't seen any major oil discoveries since the 1960s.[49] Most of the country's oil is still concentrated in these two aging fields.[50] The Ghawar field, the world's largest, came onstream fifty-five years ago, and it's still producing over 50 percent of the country's daily output—a stunning five million barrels a day. That's between 6 and 8 percent of the world's entire conventional crude output and four to five times the output of the world's next largest field.[51] But now it's filling with water because engineers are injecting millions of barrels of seawater every day to maintain the reservoir's internal pressure and to push the oil toward production wells.[52] High water concentrations are already limiting production in the south end of the field.[53]

Saudi Arabia guards its oil statistics as state secrets, but its public estimate of its reserves—at around 260 billion barrels—is almost certainly grossly inflated.[54] Also, according to some reports, the output of Saudi Arabia's mature oil fields, like Ghawar, has already peaked and is now declining at an average rate of 8 percent a year.[55] Contrary to widespread assumptions, then, the country has little spare capacity to meet our continually rising global demand for oil.[56]

If huge new discoveries aren't likely, and if many of the world's biggest fields are nearing decline or already declining, what about advanced methods for getting oil from existing fields? Yes, some of these methods will help. In the 1960s, oil-field engineers generally assumed they could "recover" only about 30 percent of a field's oil; now, by using networks of parallel horizontal wells and by injecting carbon dioxide and water into declining oil fields—as in Ghawar, where water is used to push the oil toward wells where it can be retrieved—they can often extract 50 percent or more. Oil companies are also experimenting with far more radical technologies, like using fires deep underground to decrease the oil's viscosity and injecting certain kinds of bacteria to make oil less sticky.[57] Yet these enhanced-recovery technologies generally take lots of energy—which means they lower the EROI of any additional oil produced. More important, they seem to have their greatest effect only after an oil field's output has peaked, by keeping a field's output higher for longer. When the field's final decline does occur, it's much steeper than it would have otherwise been.[58] These technologies never reverse a field's overall trend of falling output. Since 1989, for example, Alaska's oil production has fallen steadily—by over half—despite aggressive use of the best recovery technologies.

So while better oil recovery buys us a bit of time, it doesn't change the harsh reality that before long we won't be able to find enough new conventional oil to compensate for the relentless decline of many existing fields—let alone enough to meet future world demand that's expected to go up 50 percent in the next two decades.[59]

But perhaps we can tap unconventional sources of oil—like liquid fuels produced by natural gas wells, heavy oils in tar sands, or oil found by rigs drilling in very deep ocean water. Some analysts believe these resources will fill much of the gap when conventional output peaks.[60] It will certainly be important, but it's unlikely, I believe, to come online quickly enough in the volumes we need. Production of all forms of unconventional oil requires immensely challenging technologies and enormous amounts of capital—the Thunder Horse drilling platform operating in the Gulf of Mexico cost, for instance, $1 billion. And while Venezuela, Canada, Russia, and other countries have staggering reserves of heavy oil, we can't simply drill a hole in the ground and pump the stuff out. Alberta's tar sands, for example, have to be crushed and mixed with hot water and steam to separate the bitumen, which is then combined with hydrogen and cooked at high temperature to produce oil and other hydrocarbons. Natural gas fuels this process, and it takes nearly thirty cubic meters of gas to produce one barrel of oil.[61] So the EROI for tar-sands oil is very low—probably far less than 5.[62] And because North America's conventional natural gas reserves are way past peak output and declining fast, higher prices for this gas will eventually drive up the price of tar-sands oil.[63]

World oil production, then, is entangled with too many political, economic, technological, and geophysical uncertainties to allow us to predict exactly when global conventional output will peak. All the same, we can be certain the peak is coming, and we should get ready for it now, because it's likely not far away. Current high oil prices—at over $70 a barrel in mid-2006—could be telling us that peak is near, and a number of analysts like Colin Campbell have indeed predicted it will arrive in the next five years.[64] Critics, on the other hand, say that today's high prices simply reflect the past decade's low investment in exploration and refining capacity. Once new fields and refineries come online in 2007 and later, they say, oil prices will fall sharply.[65]

But given the nature of a bell-shaped production curve, even if 2 trillion barrels remain in Earth's crust—as some people claim—the peak won't be postponed for much more than a couple of decades.[66] And once we pass peak, world output could decline fast—perhaps by 3 percent or more each year.[67] In this case, the our energy situation will change radically almost overnight, because the world's economies, which are currently accustomed to a steady 2 to 3 percent annual increase in oil consumption, will have to adjust to a trend heading suddenly in the other direction—downward.

It's also possible that the peak will be followed for a time by a jagged plateau, as sharply higher energy prices dampen and perhaps even reverse economic growth, depressing demand for oil and holding down its price.[68] In this scenario, as markets and economies try to adjust, periods of oil shortage and high oil prices will alternate with periods of recession and lower prices. Meanwhile, oil companies will scour the planet for new oil fields, bring into production previously marginal fields, and use energy-hungry technologies to pull the last oil out of existing fields. They will also ramp up production of unconventional oil.

Some commentators may mistake a jagged plateau for a temporary rough patch in business as usual. But they'll be wrong: it will be a prelude to increasingly serious energy constraints, and the era of cheap oil will have come to an end.

"The World Economy Has No Plan B"

Peak oil will force us to change our entrenched energy habits. We will have to begin our much delayed, but inevitable, move away from oil to other fossil fuels, like natural gas and coal, or to green power, including solar, wind, hydrogen, and maybe even nuclear fission or fusion.

Some combination of these other energy sources could eventually take up the slack as the output of conventional oil falls. But each faces formidable obstacles, and, again, with the possible exceptions of coal and ethanol, I don't believe their energy will come online in sufficient amounts in time to meet likely demand. As Matthew Simmons, an investment banker specializing in the energy industry, puts it, "The world economy has no Plan B."[69]

Let's look briefly at some of these alternatives.[70]

Natural gas is an extremely useful fuel for heating, industrial purposes, and even transportation, so it's a good substitute for oil. But its stocks are now seriously depleted in many regions where consumption is high, like North America and Europe. In 2005, gas rigs drilled a record 27,335 new wells in the United States—up 66 percent from the number drilled in 2000—yet production was 5 percent lower, and the residential price of gas was almost two-thirds higher.[71] Unconventional gas—in deep basins offshore, in coal beds, and in low-porosity rock—is being developed, but it's very expensive and technologically difficult to tap.

A huge amount of conventional gas is still available elsewhere on the planet: on land in the Middle East, Siberia, and Central Asia and offshore in the Arctic, in the Gulf of Mexico, and along West Africa's coast. Unfortunately, most of these supplies are far from the people who could use them. Ships fitted with high-pressure storage tanks to transport liquefied natural gas (LNG)—refrigerated to minus 160 Celsius—can bring the gas from distant wells to areas in need. But the United States, Europe, and Asia will have to build dozens of new receiving terminals (there are only four in the U.S. today), and the energy needed to pressurize, refrigerate, and transport the gas greatly reduces its EROI. People are also understandably concerned that these terminals—with their berthed LNG tankers and huge fields of storage tanks—would be appealing targets for terrorists.[72] So, a combination of cost and security problems will likely keep natural gas from filling our energy gap once oil output peaks.

Coal is another possibility. Today, we burn coal mainly to produce electricity, but we can also turn it into transportation fuel. The world has immense deposits, many located in countries with big populations and heavy energy consumption—like the United States, India, and China. But getting coal out of the ground causes enormous damage: around the world, especially in countries like China, coal mining ruins vast landscapes, kills rivers and lakes with toxins, and destroys the health and lives of miners. Also, burning coal creates a lot of pollution, including sulfur dioxide, which causes acid rain, and more carbon dioxide per unit of energy than either oil or natural gas. Still, coal will almost certainly play a central role in our energy future: new types of coal gasification plants generate large amounts of electricity while producing pure streams of hydrogen and carbon dioxide. Hydrogen could be used as a transportation fuel, and the carbon dioxide could be pumped underground into abandoned oil and gas fields, unminable coalfields, and deep, briny aquifers.[73]

Recently ethanol produced from corn, other crops, and various organic wastes has been widely touted as an alternative to gasoline for transportation. United States President George W. Bush referred to ethanol in his 2006 State of the Union address, and Brazil will soon eliminate oil imports by substituting ethanol made from sugar cane in its gasoline supply. In the U.S. today, ethanol is a small percentage of transportation fuel. Some scientists estimate that more energy is used to produce this ethanol than is contained in the ethanol itself (in other words, they say that corn-derived ethanol has an EROI of less than one).[74] But new technologies are being invented that allow plant cellulose—the tough material that gives plants their structure—to be digested and converted into ethanol. These technologies could let us use a much wider range of materials—like woodchips, corn husks, wheat straw, and switchgrass, a hardy prairie plant that has a high energy content—as feedstock and might greatly improve ethanol's energy return.[75] But there are problems. Working out the technological kinks and constructing an ethanol infrastructure of farms, processing plants, and distributors will take time. Also, the more land we use to grow the feedstocks for our fuel, the less we have for growing our food. So some analysts say that a switch to ethanol would sharply raise food prices. One way or the other, though, ethanol, like coal, is likely to be a key part of our energy future.

What about solar power? Most environmentalists believe that energy from the sun is the way to go, and in some respects they're right. On a daily basis, year in and year out, the sun bathes Earth with several thousand times more radiation than all the energy humankind consumes. If we could capture even a tiny fraction of this radiation, our energy problems would be solved forever. And solar power offers the possibility of distributed, resilient, off-grid power production because individual houses and buildings can cover their roofs with solar collectors. Although it's still pretty expensive—about three to eight times more costly per kilowatt hour than coal or natural gas—the price is falling rapidly, and some analysts believe that solar will be competitive with conventional energy within a decade or two.

But it will be a long time before solar power meets a large portion of our energy needs. Many of the world's regions that consume the most energy don't have bright year-round sunlight (think of New York,

London, Berlin, Moscow, Tokyo, and Toronto in the long fall and winter seasons). More fundamentally, solar energy has a low "power density." Even in regions where sunlight is steady and bright (like the Mojave or Sahara deserts), a square meter of Earth's surface receives a maximum of only about 250 watts of solar energy when the sun shines. That's enough power density to run a household but not to run a car factory, let alone an aluminum smelter. Wind power suffers from even lower power densities. Our cities and industries, meanwhile, have very high densities of power consumption—a skyscraper can consume thousands of watts per square meter.[76] So, to supply these zones using solar or wind power, we'll need sprawling solar arrays and windmill farms many times larger in area than the zones themselves.

Nuclear fission, the technology that powers all current reactors, is another possible solution to our energy problem. Nuclear plants can be located close to populations that need energy, and new designs make the risk of a reactor meltdown very small.[77] They can also use "breeder" technology, which allows them to produce, during their normal operation, more of their own fuel by bombarding raw uranium with nuclear radiation. Because they're not constrained by fuel supply, we can build as many of these plants as we want.

But we almost certainly don't want lots of them, for security reasons. Nuclear power produces huge quantities of deadly poisonous radioactive waste. We haven't had much success figuring out where to safely store this material for the tens of thousands of years before it becomes harmless. This waste can be stolen and converted into bombs. It is extremely toxic, so terrorists can use it for radiological or "dirty" bombs; and it contains plutonium, which can be extracted to make atomic weapons. A world awash in nuclear waste would be a thoroughly terrifying place. Guarding and ensuring control of nuclear plants and their waste would require near-police-state security.

And nuclear fusion—the power that drives the sun and the stars—isn't an answer either, at least not in the foreseeable future. Fusion may be the ultimate energy source, but physicists have had huge difficulty harnessing it here on Earth. Fifty years ago they widely predicted that fusion energy would be commercially available by the end of the twentieth century. Today that goal is still fifty years away.

Finally, there has been a lot of recent excitement about hydrogen power. We can use this gas, potentially, to power our homes, factories, and appliances, and to fuel our vehicles. Burning hydrogen creates only water and heat. It's made using common chemical processes—so no heroic new technologies are required—and it can be shipped around the countryside in pipelines, just like natural gas. And best of all, it's the most abundant element in the universe, existing in huge quantities in, among other places, seawater. We'll have to overcome significant technical hurdles—for instance, scientists haven't yet invented a truly satisfactory way to store large amounts of hydrogen in a small space, like a car's fuel tank. But these difficulties are solvable in time. The real problem with hydrogen is that it's not a primary energy source—it's an energy carrier. Hydrogen may be abundant in nature, but it's not readily accessible. We can't drill for it like oil, and we can't mine it like coal. Instead, we have to *produce* it, and this takes lots of energy.

To see just how much energy, imagine we want to use hydrogen instead of oil to fuel the American transportation fleet of 230 million cars, trucks, and buses. Every day, it turns out, the country would need 230,000 tons of hydrogen—enough to fill about 13,000 Hindenburg dirigibles. To produce this hydrogen by the standard method of electrolyzing water, the country would have to double its capacity to generate electricity. And if it wanted to generate this electricity using renewable-energy technologies, it would have to cover an area the size of Massachusetts with solar panels or the size of New York State with windmills.[78]

All this means that if we focus only on the supply side of the energy problem, there are, at the moment, no really good alternatives to oil. So we need to work on the demand side of the equation too: one straightforward answer to our energy problem, and to conventional oil's impending peak, is simply conservation. We need to reduce, stop, or even reverse the growth in our overall energy demand. Those of us in rich countries can turn off our lights, heat and cool our buildings less, and drive fewer kilometers. But some people object to such strategies because they lower our standard of living, at least as it's conventionally defined. We can, instead, keep our current standard of living but use less energy by using new technologies—like advanced fluorescent light bulbs, gas-electric hybrid car engines, and high-quality public transportation—that boost our energy efficiency. These technologies give us more bang for each watt, so to speak, by lowering what specialists call the "energy intensity" of our economy—that is, the amount of energy needed to produce a dollar of economic output.

Some analysts believe that rich countries can reduce their energy intensity by 2 percent annually, more or less indefinitely.[79] Poor countries can lower their energy intensity even faster, because they use energy so inefficiently that they have huge room for gains.

Over the past few decades, rich countries have indeed made some extraordinary strides. The American steel industry, for instance, has improved its energy efficiency by an impressive 72 percent since 1955—28 percent since 1990 alone.[80] The U.S. has also steadily boosted its efficiency in glass, aluminum, concrete and fertilizer manufacturing; in refining fuel; in construction practices and materials; and in the day-to-day energy consumption of many household appliances like dishwashers and fridges.

Unfortunately, though, efficiency can't improve at such high rates forever because manufacturers and entrepreneurs exploit the easiest and cheapest ways to save energy first. After they grab this low-hanging fruit, they soon discover that each increment of greater efficiency is more technologically difficult—and more costly—than the last.

Also, although conservation is essential, it won't be enough to solve our energy problem as long as our societies and economies depend on high rates of economic growth for their well-being. Take, again, the example of the United States. Despite the country's progress in boosting its efficiency, it still hasn't reduced its total energy consumption, because economic output has grown faster than energy efficiency has improved. While U.S. energy intensity has fallen almost 2 percent annually over the past twenty years, the country's economy has grown faster, at over 3 percent annually. So although energy intensity has dropped almost 40 percent since 1980, the total U.S. appetite for energy has still gone up more than 27 percent.[81]

We're led to a striking conclusion about conservation. Even if countries reduce their energy intensity 2 percent each year—and it's unlikely they can do so indefinitely—they're not going to stabilize or reduce their total energy consumption unless economic growth is less than this amount. But I'll show later that no contemporary economic policy maker in any country, and certainly no politician, is going to settle for such a middling growth rate. In other words, as long as we're addicted to strong economic growth, our total energy consumption will not go down, even if we steadily improve our energy efficiency.

To sum up, when it comes to energy, and particularly when it comes to conventional oil, we're constrained on every side: our appetite for energy is enormous and quickly growing; we're deeply dependent on oil to satisfy that appetite, yet oil's supply is tightening quickly; some alternatives to oil endanger national and international security, while others are technically or economically infeasible; and just about anything we do will have a major impact on the planet's natural environment.

Oil will become far scarcer and costlier. Conventional oil's peak will begin, Colin Campbell writes, "a period of recurring price surges, recessions, international tensions, and growing conflicts for access to critical oil supplies, as the indigenous energy supply situation in the United States and Europe deteriorates."[82] Put simply, oil's peak will have seismic consequences for the entire world. It will shift the ground underneath us.

Could this lead to a modern version of the Roman empire's fall—caused by escalating tensions as players on the world stage struggle to control oil supplies and as skyrocketing energy costs contort our economies? I'll show in the next chapter that social breakdown will become steadily more likely as our world's accumulating tectonic stresses—especially energy stress—combine to overload our societies.

EARTHQUAKE

OZENS OF KILOMETERS beneath the Colosseum, the mighty
forces of Earth's geology are at work. The African tectonic
plate, a thick slab of the planet's crust stretching from the
Mediterranean to Antarctica, is driving northward. Long ago, it
smashed into the even larger Eurasian plate. This monumental collision
crumpled and tore the edges of the two plates along a front line
extending the length of the Mediterranean basin. The collision also
buckled the crust upward to form mountain ranges like the Alps,
Apennines, and Carpathians. Under Italy, the crust has been pushed
downward, where it melts into magma and occasionally bursts back to
the surface in spectacular volcanoes.[1]

All this churning beneath the Italian peninsula has left its mark on
the civilizations that have come and gone on the land above. It has sometimes brought disaster, as in 79 CE, when Mount Vesuvius's eruption
buried Pompeii. But it has also created a land of fertility, geographic
variety, and great abundance. Volcanism forged the stone so characteristic of Rome's magnificent buildings: subterranean magma heated water
deep underground, which then dissolved limestone lying in layers of
rock above. As the superheated water bubbled to the surface, it cooled,
and its burden of dissolved calcium carbonate precipitated out to form
the vast beds of travertine that are still mined in places like Tivoli today.

So it's a bit ironic that the very seismic activity that generated much
of the rock that Romans used to build the Colosseum has also harmed

the building—repeatedly and sometimes grievously. In the past two thousand years, more than a dozen significant earthquakes have struck Rome. Experts believe that temblors seriously damaged the Colosseum in 443 CE, 508, 801, 847, 1349, 1703, and 1812. The 847 quake probably destroyed the columns around the upper floor of the amphitheater that supported its roof, and the 1349 quake toppled much of the southern wall that forms the building's external ring.[2] In addition to these disasters, the Colosseum was plundered for its stone for a thousand years, often with the pope's authorization and even encouragement. The once-grand amphitheater was reduced to a ruin. Restoration began only when Napoleon briefly controlled the city in the early nineteenth century.

If the builders of the Colosseum had known what we know now about earthquakes, they might not have constructed it on its current site. The southern part of the building sits on soft alluvial material that amplifies ground motion, which explains why, over the centuries, that part has sustained far more earthquake damage. Still, while we know much more about earthquakes than the Romans did, we're little better at predicting when they'll occur. Taken together, Earth's geophysical processes are a complex system, and the earthquakes that it produces are awesome examples of "threshold events."

Deep under Earth's surface, tectonic plates grind into each other a centimeter or two a year, and enormous pressure builds up incrementally and invisibly. At some point, the pressure crosses a critical threshold, something snaps, and an earthquake occurs. But how and when it occurs—whether, for instance, the pressure is released through a long series of micro-quakes or a Colosseum-shattering disaster— depends on many subtle and often unknowable aspects of local geology: the type and porosity of the rock, the extent and direction of the rock's fractures and fissures, the amount of water in these spaces, and even the types of compounds dissolved in this water.[3]

Earthquakes aren't just geophysical events of special interest to those of us who live on major fault lines—whether in Italy or many other places such as Japan, Indonesia, California, and British Columbia. They are also instances of a more general kind of system breakdown—a kind that's important to us today, as we try to understand the risk of social breakdown. The population, energy, environmental, climate, and economic stresses affecting our world are just like tectonic stresses: they're deep, invisible, yet immensely powerful; they're building slowly; and they can release their force suddenly without warning.

The analogy between earthquakes and social breakdown is surprisingly apt. For example, earthquakes occur when two adjacent tectonic plates run into each other. Sometimes such collisions aren't catastrophic, because one plate slips beneath or alongside the other without too much trouble. Other times, though, the plates lock together along the fault line between them. Then the pressure building between the plates mounts and is eventually released in one great burst. Similarly, at the human level, our institutions and political and economic systems can lock up or become rigid, which prevents the release of stress and keeps societies from adapting to new circumstances. For instance, racial segregation remained widespread in the United States even after the U.S. Supreme Court declared state segregation laws illegal in the 1950s. Eventually in such situations, frustrations can erupt in events like the U.S. civil rights movement in the 1960s, the recent riots by black and Muslim immigrant communities in French cities, or the popular revolutions in Eastern Europe in 1989—revolutions that were born of people's deep frustrations and that quickly brought the collapse of the Soviet east bloc.

Seismologists have recently learned that major earthquakes often relieve stress in one region while boosting it in others, so, over time, earthquakes often come in chains, as the first triggers a second, which triggers another and another. The final earthquake in the chain might occur a very long way from the first—in fact, the December 2004 earthquake off Indonesia that caused the devastating tsunami in the Indian Ocean set off smaller quakes as far away as Alaska.[4]

Seismologists call this phenomenon "stress triggering," and we can see something similar in social systems: a disturbance in one part of an economy or society can dramatically—and often unexpectedly— increase stress in other parts, leading to cascades of economic or social change. When Hungary's new reformist parliament tore a hole in the Iron Curtain by opening the country's borders with Austria in August 1989, tens of thousands of East Germans saw their chance to escape to the West if they could reach Hungary. This precipitated a remarkable chain of events: within months, all border points between Eastern and Western Europe had been opened, the Berlin Wall had been ripped down, and one east bloc regime after another—from Czechoslovakia to Romania—had crumbled.

But there's one similarity between earthquakes and social breakdown that's particularly key for us here. Just as seismologists can't forecast a geological earthquake's precise timing or character, social scientists can't forecast the precise timing or character of a major social earthquake. It might happen tomorrow, two decades from now, or never. Yet seismologists can still say useful things about the varying risk of earthquakes from one region to another or from one time to another.[5] In the same way, we can say something useful about the risk of sociopolitical earthquakes at different times and places. Understanding these risks gives us a chance to make sensible, even life-saving judgments about which futures are plausible and which are wholly unlikely. The ability to make such judgments is a vital attribute of what I call the prospective mind.

One feature of earthquakes is an especially intriguing tool for prediction—and a potentially powerful one too, because it could help us prepare in advance. About half of all earthquakes are preceded by one or more foreshocks—small tremors that happen up to three days before the quake. "Foreshocks don't occur before all earthquakes," notes a leading seismologist, "but they occur commonly enough that [they suggest] a larger event might be coming."[6] There's a problem, however. As a prediction tool they generate far too many "false positives"—too many false alarms that an earthquake is on its way.[7] Still, foreshocks are fascinating, because if seismologists could better understand them, they might be able to warn people to move to safer ground, evacuate vulnerable buildings, or store extra supplies against the devastation that could follow.

Our social systems often exhibit foreshocks too. The Eastern European revolutions of 1989, for instance, were preceded by the rise of the independent trade union in Poland known as Solidarity and by Soviet President Mikhail Gorbachev's policies of "glasnost" and "perestroika." In this light, in the early twenty-first century, diverse events like the Iraq war, the 9/11 attacks, the 2005 urban riots in France, and hurricane Katrina might be seen as foreshocks of a coming global breakdown.

The science of earthquakes, then, can help us understand sharp and sudden change in types of complex systems that aren't geological—including societies. Future energy scarcity, especially, could shake our world from top to bottom, by making it far harder for us to cope with the other stresses converging on our societies. Peak oil could aggravate the impact of the world's population imbalances by weakening economic growth and worsening unemployment in poor megacities and in immigrant ghettos in rich countries.

But the analogy between earthquakes and social breakdown has its limits. It focuses our attention on a single dominant cause of breakdown, for example, because an earthquake is caused mainly by one factor—a buildup of pressure along a fault line. In my travels to investigate the problems we face around the world, an experience in California helped me see that if we're really going to understand social breakdown, we need to grasp what happens when diverse stresses combine—when, in other words, there are multiple causes of breakdown.

Negative Synergy

In the sun of a late-November afternoon, I was standing on a ridge overlooking the city of San Bernardino, a hundred kilometers east of Los Angeles. It was Thanksgiving Day 2003, just six months after my visit to the Roman Forum.

An American flag flapped in the breeze. It was tied to a length of charred pipe that someone had jammed into a mound of twisted debris. Around this lonely symbol of pride and perseverance were the blackened remains of a refrigerator and a hot-water tank; a naked, ash-covered brick chimney; and the concrete foundations of the house that once stood there. Beyond this perimeter was a forest of dead pine trees, their burned, needleless branches reaching up toward the sky as if appealing for salvation.

Just a month before, wildfires had ripped across Southern California—from the suburbs of San Diego northward three hundred kilometers to Simi Valley. Fires had ravaged the San Bernardino Mountains, including the ridge where I was standing. Hundreds of houses had been destroyed, both in these mountains and within the city's residential zones nestled below.

Southern Californians live with the constant danger of fire on the outskirts of their cities. After summer's withering heat, forest and bush are tinder dry, and fierce Santa Ana winds can turn any spark into a conflagration. But firefighters had never seen anything in their lives like this inferno. Flames as high as thirty-story buildings—extending across fronts dozens of kilometers long—raced over hillsides, up valleys, and through suburban tracts, devouring everything in their path. By the time the fires had been put out, over three thousand homes had been destroyed, twenty-four people had died, and the total cost ran into billions of dollars.

Nothing conveyed the pain better than the naked chimneys—eerie sentinels left behind amid the wrecked houses on both sides of me. They must have been grand, these houses perched on the mountain's edge, boasting splendid views of the city from their decks and patios. At each site, I could see that families had come back to collect a few artifacts of everyday life, carefully plucking memorabilia from the ashes—a pot, a ceramic bowl, the melted remains of a prized chandelier, a vacant photo frame—and putting them to one side in little clusters, to be picked up later.

And then, looking across the city, its dead-straight streets converging to a vanishing point in the distance, I saw an immense brown cloud moving in from the west, as if someone were pulling a dirty curtain across a stage. The afternoon winds were whipping across the charred hillsides, lifting the newly exposed soil high into the air, and churning it into a dust storm.

As I observed this scene, I thought about another Californian conflagration some six hundred kilometers to the northwest and almost a century before—the great fire in San Francisco in 1906 that I described in this book's opening pages. The two episodes were quite similar in their staggering economic cost.[8] And they were similar in another way too, because both, I realized, were examples of *negative synergy*.

Synergy happens when people, things, or events combine to produce a larger impact than they would if each acted separately. We've all heard about the phenomenon; business consultants constantly babble about exciting synergies achieved when people work together on a new project. We tend to assume that synergy is always positive and beneficial, but it can just as easily be negative and harmful. For instance, a person might be able to knock back a few drinks and still drive quite competently under normal weather conditions; the same person, without the drinks, might be able to navigate a car through a sudden snowstorm. But if the person combines the drinks and snowstorm, the result might easily be fatal. Similarly, in the aftermath of the 1906 San Francisco earthquake, sparks and fuel combined to cause an explosion of fires across the city. Either the sparks or the fuel alone wouldn't have been enough—but together they produced calamity.

Policy makers, social scientists, and commentators almost always overlook the potentially ferocious power of negative synergy, perhaps because they don't fully understand its implications.[9] Yet its implications were clear on the ridge overlooking San Bernardino. A month earlier, when the wildfires were burning, the media had given them lurid coverage, and everyone reacted as if they'd burst from nowhere. But in reality, they were just the latest chapter in a story of converging stresses that had been ignored till then and that has received little attention since.

Three developments combined to cause the conflagrations, any two of which probably wouldn't have been enough by themselves. First, in previous decades suburban building had expanded into the bushlands and forested zones across California, including the San Bernardino Mountains above Los Angeles. Many people had built their homes in those attractive edge zones where cities meet nature. Second, in 2003, several years of harsh drought—including, in 2001–02, the driest year in more than a century—had desiccated Southern California, sucking the moisture out of the soil and weakening trees and other vegetation. And third, an infestation of bark beetles had exploited the weakened trees and laid waste to vast stretches of pine forest. Today, experts estimate that 90 percent of Southern California pine forests will eventually die, including the entire San Bernardino National Forest.[10] Bark beetle infestations are in fact chewing their way through pine forests from Alaska and British Columbia to Arizona. Some experts think these infestations are an early sign of human-induced climate change, because warmer temperatures have extended beetles' life spans and geographic range.[11]

I could see the infestation's grim results. Farther back from the ridge—beyond the point where the fire had finally died out—needles from dead and dying pines were still piled in thick layers across driveways, gardens, and sometimes right up to the foundations of unburned houses. In some places, this layer of explosively inflammable duff was a third of a meter deep. I noticed that the ridge, trees, and blanket of needles were all bone-dry. A single spark would produce another inferno. The people still residing in the houses on that hillside were living inside a giant tinderbox.[12]

The October 2003 fires were a dramatic breakdown of the human/nature system in Southern California. Two severe, simultaneous, and causally related stresses—drought and a beetle infestation—synergistically combined with ill-advised patterns of residential

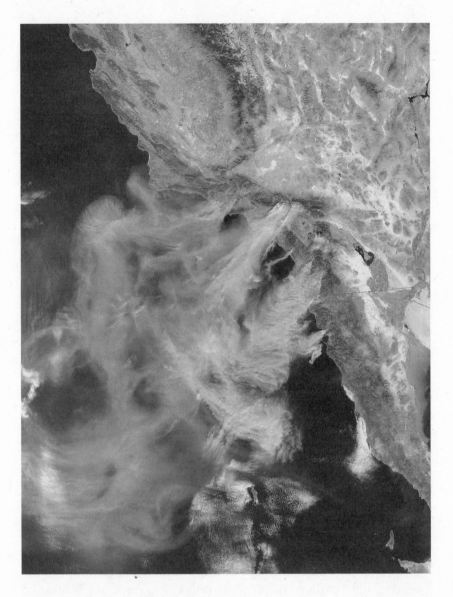

Smoke from the October 2003 wildfires in southern California, as seen from space

construction to create a nightmare of wrecked homes, devastated landscapes, and collapsing property values. One might think that such an outcome would encourage residents to question the wisdom of living in such a spot.

Yet everywhere I looked new telephone poles had been rigged with power lines and lots were being cleared of debris for reconstruction. What defiance! But then it occurred to me that unlike the Romans when they built the Colosseum, the residents of the ridge now knew for sure that they were constructing their buildings in a danger zone. When, I wondered, does defiance become denial?

Overload

When I talk about the breakdown of complex systems like the one along the San Bernardino ridge what, exactly, do I mean? We typically use the word "breakdown" when something's regular function is disrupted or its movement stops—when we talk about our car's breakdown, for instance. We also use it to refer to a sudden change in normal affairs. Here I'll use the word a bit differently—to refer to a rapid loss of complexity. The breakdown of a system—whether an ecosystem, economy, or an imperial government like Rome's—simplifies its internal organization and reduces its range of potential behaviors.[13]

Think, for example, of what happens when someone experiences a psychological breakdown. Their activities and goals become much simpler. Life becomes a matter of survival, of satisfying basic needs on a day-to-day basis. The same is true when ecological, technological, and social systems break down. The San Bernardino fires simplified the region's ecology—wiping out many of its plants, animals, and ecosystems—just as it simplified the human society that lived there (after all, most people left). And during the blackout across eastern North America in August 2003, we lost the communication and transport technologies that make possible our high-velocity interactions and our multiple professional, commercial, and social roles. Almost all business meetings, financial transactions, and shopping excursions abruptly stopped. Our range of options was radically reduced and our goals radically simplified—first and foremost, we wanted to get home. And if we were already home, we wanted to secure basic necessities—water, food, and light.[14]

People often use the words "breakdown" and "collapse" synonymously. But in my view, although both breakdown and collapse produce a radical simplification of a system, they differ in their long-term consequences. Breakdown may be serious, but it's not catastrophic. Something can be salvaged after breakdown occurs and perhaps rebuilt better than before. Collapse, on the other hand, is far more harmful: the damage endures— it may even be permanent—and there's far less knowledge, wealth, or information left behind to use in a process of renewal.

What factors make a society more likely to experience breakdown or even collapse? Historians and social scientists have endlessly debated this question, and they may never reach a conclusion.[15] All the same, a large body of research now points to the rough answer I suggested in chapter 1: a society is more likely to experience breakdown when it's hit by many severe stresses simultaneously, when these stresses combine in ways that magnify their synergistic impact (just as happened in the San Bernardino mountains), and when this impact propagates rapidly through a large number of links among people, groups, organizations, and technologies.

When a society has to confront a bunch of critical problems at the same time, it can't easily focus its resources on one and then move to the others. Unfortunately, the key stresses that I've pinpointed in this book, and that are affecting our world right now, are all getting worse at the same time. Some of them, like the demographic imbalance between rich and poor countries and environmental damage in poor countries, are already severe. Others, like energy scarcity and climate change, while perhaps not yet severe, will probably pass critical thresholds before too long. They may not be bad now, but in the future they likely will be.

At the heart of my argument is the idea of overload. A society over-loaded with stresses breaks down.[16] Whether a society is overloaded depends not only on the nature of the stresses it encounters but also on whether it can manage or adapt to them. Societies vary a lot in their ability to cope with stresses. At one end of the spectrum, Western capitalist democracies are probably the most adaptive societies in history. During the past century alone, they've spent unimaginable quantities of blood and treasure in two world wars, recovered from a deep economic depression, absorbed untold numbers of immigrants, and won a hideously expensive, half-century political competition and arms race with the Soviet Union. They emerged from the century not just intact but wealthier and more powerful than ever.[17] At the other end of the spectrum, we find societies— including many in sub-Saharan Africa and some in Asia and Latin America—that have much lower ability to manage or adapt, because of poverty, environmental damage, low education levels, chronic internal violence, and weak and corrupt governments. A few, like Somalia and Haiti, have completely succumbed to the stresses pounding away at them: these countries may still exist as a matter of international law, but they have, for all intents and purposes, ceased to exist as coherent societies.[18]

In general, the greater the number and severity of stresses affecting a society—and the more they combine synergistically—the greater the chance of social breakdown. But two other factors also affect the impact of stresses on a society. And it was an earlier experience in California that helped me see why these two factors are critically important.

Connectivity and Speed

I was driving a brand-new rental car in the fast lane of a divided highway heading south from San Francisco. For once on California's choked roads, the traffic was moving quickly, and I was speeding along at about 130 kilometers per hour. Less than ten meters behind me, an SUV hugged my bumper, followed close behind by a string of other cars. To my right, between the highway's shoulder and me, were two lanes of bumper-to-bumper vehicles.

I had pulled into the fast lane to pass a semitrailer, and I was now almost halfway down the rig's length to my right. But I suddenly realized my car's engine had gone dead.

I pumped the accelerator. The engine didn't show a hint of life, and the pedal seemed to have gone to mush. In an instant, I realized I was in a potentially fatal situation. As I shed speed, the SUV closed in behind me; I could see the vehicle's nose drop as the driver snubbed the brakes. All escape routes to my right appeared blocked. And, without engine power to drive the car's power steering, my wheel had gone leaden.

I don't really recall exactly what happened next. I remember realizing that I mustn't touch my brakes because my momentum was precious—I needed it to get off the road. In any case, there was a tailgater hard on the rig's heels, which meant I couldn't get behind the truck. So I use my remaining momentum to overtake the semitrailer and squeak into the small space in front of him. The truck's horn growled behind me, and all I could see in my rearview mirror was his grille. From there, even though I was losing speed quickly, I made the jump to the right lane and then to the narrow shoulder.

Coasting to a stop, I found that there was barely enough room to get my car off the road. Traffic roared by so close that I couldn't open the driver's-side door. I flipped on the emergency flashers, struggled out the passenger's door, and tumbled down the highway's steep embankment, my heart racing.

Four hours later, after I'd made many calls to the rental agency and California Highway Patrol, a flatbed truck arrived to take the wretched car away. It all made for a good story when I finally arrived—in a different car and many hours late—at my destination, a conference on Monterey Bay. I wouldn't have thought much more about the incident if it hadn't been for something that happened at the conference itself.

It was May 2001, and the conference was an annual event of the Northern California World Affairs Council. About one thousand delegates had converged on the magnificent Asilomar conference center to discuss the dramatic twists and turns of international affairs in the early years of the new millennium. I'd been invited to join a plenary panel on globalization, and I recall that the large hall was packed. The audience wasn't to be disappointed, because the panelists' viewpoints differed sharply, and fur was soon flying. Toward the end of the session, one of my fellow panelists, a gentleman who'd made a fortune in Silicon Valley during the dot-com boom, said something that especially caught my attention.

"The more connectivity in our world, the better," he declared.

In my experience, absolutist, unqualified statements like this one are almost always wrong. And when they're converted into government and corporate policy dogma—as has happened over the past couple of decades to the general idea that greater connectivity among people, technologies, economies, and societies is always better—we're in real trouble. There is, of course, no doubt that our world's soaring connectivity and the ever-greater speed with which material, energy, and information move along our world's connections can often lead to good things. But sometimes they don't; in fact, sometimes they multiply a society's stress. Whether or not this happens depends on the specific characteristics of a society's connectivity and on the specific characteristics of the stresses the society encounters. On these matters, the devil is in the details.

Together, connectivity and speed are, in fact, the first of the two "multipliers" I mentioned in chapter 1 (the other being the escalating power of small groups to destroy things and people). These multipliers combine with stresses to make breakdown more likely and, when it happens, more disruptive.[19]

To better understand how and why connectivity and speed work this way, we'll need to take a brief detour into recent research on the implications of connectivity in complex systems. To begin, it's useful to think of all complex systems—including the myriad economic, political, social, and technological systems in our own societies—as networks. They consist of sets of nodes and the links connecting those nodes. Nodes can be machines, people, groups, organizations, and even whole countries, while links can be anything that carries material, energy, or information between these nodes. The U.S. economy, for instance, is made up of nodes like corporations, factories, business associations, banks, labor unions, and urban agglomerations; it's also made up of links among these nodes in the form of fiber-optic cables, electrical grids, gas pipelines, highways, and rail lines.

As societies modernize and become richer, their networks become more complex, interconnected, and faster: they add more nodes, they increase the number of links among the nodes, and they boost the speed at which stuff moves from node to node along the links. The new connections run horizontally as they link similar entities together within a single level of social organization—for instance, people are linked with people, cities with cities, and countries with countries. They also run upward and downward across the levels—for example, a person becomes linked to a city, which is in turn linked to a country, which is finally linked to a vast array of international organizations and institutions.[20]

Researchers have learned a great deal about the benefits and costs of greater connectivity and speed.[21] Greater connectivity often boosts economic productivity by creating larger markets that allow companies to reap the benefits of economies of scale and that encourage people and companies to specialize. Also, when people are better connected, they can usefully combine their diverse ideas, skills, and resources. Greater speed means more goods and services can be produced, transported, sold, and bought in a given period of time. Together, greater connectivity and speed often make economies and societies more resilient to shock because they can respond faster and draw from their larger networks a wider range of skills, resources, capital, and goods and services.

But that's not the end of the story. At the Asilomar conference, the Silicon Valley entrepreneur's bold claim struck me as particularly

dubious, as I had just experienced the downside of hyper-connectivity
and speed on the California highway. We often hear of multi-vehicle
pileups caused partly by people following each other too closely and too
fast. My car and its surrounding cohort of vehicles—all roaring down
the highway together—weren't physically linked to each other, but we
were still connected through information flows (via our eyes and ears)
and our mutual vulnerability and interdependence. Our high speed and
dangerous nearness tightened this connectivity, so we had almost no
room for error, accident, or mechanical failure. In other words, we had
little slack or buffering capacity to compensate for surprises. Together
we'd become what experts call a "tightly coupled" system.[22] When something went wrong at this system's heart—when my car's engine failed,
for example—the consequences could easily have been catastrophic for
many people as well as me.

Tight coupling at high speed contributes to multi-car accidents.

So the first cost of greater connectivity is that damage or a shock in one part of the system—the failure of a machine (like my car engine), the release of a computer virus, or a local financial crisis—can cascade farther and faster to other parts of the system. This is especially true when the nodes in the network, or the elements in the system, are packed so closely together that the links among them are very short—that is, when they're tightly coupled. Then problems with one node or element can ramify outward before anyone can intervene. Such domino effects happen not only in multi-car pileups but also in telephone, air-traffic, and financial systems—and, as we saw in the 2003 blackout in North America's electrical grid.

Our world's tight connectivity also promotes the rapid spread of disease. In fact, we are now seeing a negative synergy between the massive size of the human population and its internal connectivity that helps new diseases—like HIV/AIDS, severe acute respiratory syndrome (SARS), and perhaps soon avian influenza—develop and propagate around the planet faster than ever before. Collectively, humankind now makes up one of the largest bodies of genetically identical biomasses on Earth: all of us, taken together, weigh nearly a third of a billion tons. Combined with our proximity in enormous cities, and our constant travel back and forth across the globe, we're now a rich environment— just like a huge Petri dish brimming with nutrients—for the emergence and spread of disease.[23]

Connectivity harbors other risks too. As we create more links among the nodes of our technological and social networks, these networks sometimes developed unexpected patterns of connections that make breakdown more likely. They can, for instance, develop harmful feedback loops—what people commonly call vicious circles—that reinforce instabilities and even lead to collapse. A stock market crash or financial panic is such a vicious circle, because selling drives down prices, begetting fear in the market and more selling, which lowers prices even more.[24]

Also, the new links we create can connect previously separate systems or system components, so failures or accidents that would before have been isolated from each other now combine in unexpected and harmful ways.[25] This is the best explanation for the terrible Three Mile Island and Chernobyl nuclear-reactor accidents and the Challenger and Columbia space-shuttle disasters. In each case, while the designers and managers may have understood the system's bits and pieces, they didn't fully understand what could happen when all the bits and pieces combined—in other words, they didn't really understand the system as a whole. The systems became so complex and interconnected that they exhibited emergent properties—the whole became more than the sum of its parts. So those in charge didn't anticipate all possible combinations of component failures or possible negative synergies of combined failures, and tragedy was the result.[26]

These were the roots, too, of the 2003 blackout: the deregulation of the North American power grid in the 1990s caused long-distance electricity sales to skyrocket, which stimulated a surge in connectivity between regional electricity production and distribution systems that had previously been isolated from each other. At the time of the blackout, the new integrated system included six thousand power plants run by three thousand utilities overseen by 142 regional control rooms. The intricate rules developed decades earlier to manage a grid in which most power was generated reasonably close to its consumers were suddenly obsolete.[27] Now technicians had to have, as one expert said at the time, the reflexes of "a good combat pilot managing an aircraft that has been badly damaged." On the day of the blackout, even those reflexes weren't good enough.[28]

Although researchers have mainly focused on technological systems, this kind of synergistic failure is just as likely in highly connected social systems. In fact, the dividing line between technological and social systems is never distinct, and almost all the systems we rely on—from electrical grids to banks to governments—are intricate combinations of machines, people, and organizations. In a world of ever-increasing connectivity and speed, unanticipated interactions among previously separate systems happen more often, as do unanticipated combinations of failures within systems. And the likelihood rises that some of these combinations will cause catastrophe.

A Clausewitz of Complexity

Greater connectivity harbors another peril too: it can increase our vulnerability to terrorism. To see why, we need to delve just a bit deeper into the new science of networks.

In recent years, scientists have discovered that there are two main types of network. Specialists call them "random networks" and "scale-free networks," for technical reasons we needn't go into here. For our purposes, the critical thing is that these two types have different patterns of links connecting their nodes.[29]

A random network looks like the U.S. interstate highway system, in which the nodes are cities and towns, and the links are highways. In such networks, most nodes have a moderate numbers of links to other nodes, while a few nodes have a small number of links, and a few others have a large number of links. No node has a very large number of links to other nodes. A scale-free network, on the other hand, looks a lot like the U.S. air-traffic network in the 1990s.[30] Most nodes have a very small number of links, while a very few, called *hubs*, have a huge number of links to other nodes.[31] Although researchers long assumed that most networks were like the interstate highway system, recent study shows that a surprising number of the world's networks—both natural and human-made—are more like the air-traffic system. These scale-free networks include most ecosystems, the World Wide Web, large electrical grids, petroleum distribution systems, and modern food-processing and supply networks.[32]

Such differences have huge implications for the resilience of the network. In a random network the loss of a small number of nodes can cause the overall network to become incoherent—that is, to break into disconnected subnetworks. In a scale-free network, such an event usually won't disrupt the overall network because most nodes don't have many links. But there's a big caveat to this general principle: if a scale-free network loses a hub, it can be disastrous, because many other nodes

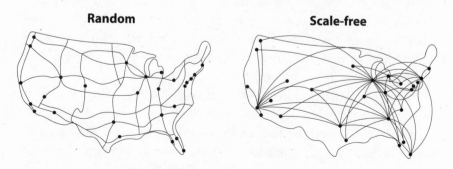

The world's networks can be divided into two general categories.

depend on that hub. An ecosystem, for example, will have a certain number of "keystone species"—species that provide vital services, like pollination, to a wide range of other species—and these keystone species are essentially hubs in the ecosystem's larger network of species. If enough of these hubs are lost, the ecosystem can collapse.[33]

Scale-free networks are particularly vulnerable to intentional attack: if someone wants to wreck the whole network, he simply needs to identify and destroy some of its hubs. And here we see how our world's increasing connectivity really matters. Scientists have found that as a scale-free network like the Internet or our food-distribution system grows—as it adds more nodes—the new nodes tend to hook up with already highly connected hubs. New e-mail users on the Internet tend to connect themselves to well-established server computers, and new farms tend to sell their food to large and already dominant whole-sale food distributors. Also, as already existing nodes create more links among themselves, they're more likely to attach themselves to the biggest and most connected hubs. We can see this most clearly on the Web: dominant Web pages—including massive hubs like amazon.com, google.com, or nytimes.com—tend to receive the large majority of all new links from other pages. So in a scale-free network, greater connectivity tends to make already dominant hubs even more dominant. In doing so, it can actually make the network more vulnerable to attacks directed against those hubs.

Once a hub is damaged, intentionally or otherwise, its problems can spread quickly far and wide. For example, cities are hubs in the global network of human population, and they can play a powerful role spreading new diseases. The virus that causes SARS first appeared in early 2003 in Guangdong in southern China—a relatively poor, densely populated region where some people live in close contact with animals that are sometimes diseased. From there, wealthy and mobile residents and visitors carried it to Hong Kong. Then other travelers carried it on planes to cities around the world, including Toronto.

The way SARS spread shows what could happen with tomorrow's new pathogens. A key danger is that a highly contagious and virulent new disease—something like SARS, except worse—will work its way into the densely packed squatter settlements of the megacities in poor countries. In the vast slums of São Paulo, Delhi, Dacca, Calcutta, Lagos, and Mexico City, health-care facilities are rudimentary. Respiratory and intestinal infections are already endemic, in part because of appalling urban air and water pollution, so selective quarantine is impossible. And members of local elites, who regularly travel around the world, often live nearby. Once a tenacious pathogen is established, poor urban zones could become what specialists call "epidemiological pumps"—that is, they would be hubs that are a permanent disease reservoir and continually pump the pathogen back into the planet's human population.[34]

Almost certainly, malicious individuals and small groups, including terrorists, are starting to understand how to exploit our interconnected and high-velocity networks to multiply their disruptive power.[35]

Sometimes we see the results of this multiplication effect in a very personal way—especially when we have to deal with viruses on our home or office computers: a single kid hacking away in his basement can cause disarray in computer systems around the world. Yet this kind of thing is a minor nuisance compared with the chaos that could follow a major attack on the scale-free networks that provide us with the goods and services we need to survive—especially our energy, information, and food.[36] Our energy systems, which encompass everything from networks of gas pipelines to the electricity grid, have countless hubs such as oil refineries, tank farms, and electrical substations, many of which are easily accessible. When energy demand is at peak, as it often is in air-conditioned North America in mid-summer, the electricity grid especially is extremely tightly coupled. An attack against key hubs—such as a number of substations—could have a devastating impact.[37] Our food-supply system, too, has countless hubs, including huge factory farms and food-processing plants, with many connections to other nodes and then to all of us who eat the food. Unprotected grain silos dot the countryside, and trains made up of railway cars filled with grain often sit for long periods on railway sidings. Attackers could easily break into these silos and grain cars and lace the grain with contaminants—which would then diffuse through the food system.

But we shouldn't exaggerate the risks. Terrorists have to be clever to exploit the vulnerabilities of our networks. They have to attack the right hubs in the right networks at the right times, or the damage will remain isolated and the overall network will be resilient. We can also protect ourselves by introducing ways of tracing attacks, as we do already by using batch numbers on drugs and food products; these numbers help authorities find items that have been contaminated and identify people who had access to them. Another protective factor is a network's redundancy—that is, its ability to offload the functions once served by damaged hubs onto undamaged hubs. But strategies like improving tracing and redundancy aren't foolproof, and the incentives for terrorists are large: if they succeed in attacking our complex networks, they can spark cascading failures causing immense hardship.[38] In fact, according to Langdon Winner, a theorist of politics and technology, the first rule of modern terrorism might be "Find the critical but nonredundant parts of the system and sabotage . . . them according to your purposes." Referring to the great Prussian military theorist Carl von Clausewitz, Winner concludes that the science of complexity awaits its own Clausewitz "to make the full range of possibilities clear."[39]

We also create extraordinarily attractive targets for attack when we concentrate high-value assets—people included—in geographically small locations. When we build larger factories, we generally lower cost per unit of output, and when we concentrate expensive equipment and highly skilled people in one place, we can access and use them more easily and efficiently. Bringing them all together also creates those valuable synergies among people, things, and ideas that are an important source of our wealth. This is one reason we build places like the World Trade Center.

But terrorists are learning they can cause a huge amount of damage in a single strike against such a target. On September 11, 2001, a complex of buildings that took seven years to build was destroyed in an hour and a half, killing almost three thousand people, obliterating fifteen million square feet of office space, and exacting upward of $20 billion in direct costs. A major telephone switching office was destroyed and another heavily damaged, while some key transit lines were buried under rubble. The attack also immobilized the Bank of New York, one of two banks that provide clearance services for Wall Street's fixed-income transactions, and an institution vital to the market for U.S. Treasury securities.[40]

Yet, despite the horrific damage to the area's infrastructure, the immediate operation of the financial system, and the New York economy, the attack did not cause catastrophic failures in U.S. financial, economic, communication or other networks. The World Trade Center may have been an attractive target, but it was not—as it turned out—a critical, nonredundant hub.

The attack nevertheless had a crucial effect on another kind of network: a tightly coupled and sometimes unstable psychological and emotional network. We're all nodes in this particular network, and the links among us consist of Internet connections, satellite signals, fiber-optic cables, talk radio, and twenty-four-hour seven-days-a-week news television and radio. In the minutes following the attack, information about it flashed across this network. We then sat in front of our televisions for hours on end; we plugged phone lines checking on friends and relatives; and we sent each other millions upon millions of e-mail messages—so many, in fact, that the Internet was noticeably slower for days afterward.

Along these links, from TV and radio stations to their audiences, and especially from person to person through the Internet, flowed raw emotion—grief, anger, horror, disbelief, fear, and hatred. It was as if we were all wired into one immense, convulsing, and reverberating neural network. Thanks to new communication methods such as the Internet and twenty-four-hour-news television, we've created a psychological network that acts like a huge megaphone, vastly amplifying the emotional impact of terrorism. As a result, the ultimate, indirect effects of a terrorist act can now be incomparably greater than its direct effects.

So, the biggest impact of the September 11 attacks wasn't the direct disruption of America's financial, economic, communication, or transportation networks—physical things, all. Rather, by working through the psychological network we've created among ourselves—a network that extends around the planet—the attacks' biggest impact was their shock to our subjective feelings of security and safety. Such shocks don't remain subjective: they soon have huge, real-world consequences. Among other things, when we're scared, insecure, and grief-stricken, we don't buy a lot of stuff. We aren't ebullient consumers. Instead we behave cautiously, and we save more. Consumer demand drops, corporate investment falls, and economic growth slows.

In the end, thanks to the multiplier effect provided by our highly connected information and emotional networks, the 9/11 terrorists may have achieved an economic impact far greater than they ever dreamed possible. The total cost of lost economic growth and decreased equity value around the world ultimately exceeded $1 trillion—and that total doesn't even include the increased spending on security measures and the later Afghanistan and Iraq wars.[41] Since the cost of the attack on the World

Trade Center to Al Qaeda was probably only a few hundred thousand dollars, the terrorists multiplied their impact well over a million-fold.

We could see the same amplification effect in the American public's response to the September and October 2001 anthrax attacks. Only five people died and fewer than two dozen were infected, probably as a result of no more than ten anthrax-laced letters mailed through the postal system. Yet the 9/11 attacks had so frightened the media and public that they reacted almost hysterically to each piece of news about anthrax. Again, we can be sure that terrorists around the world took note.

Boundary Jumping

"If there's another major attack, people will leave the city in droves."

Andrew, a colleague of mine in New York City, was sitting in his office in a building not far from Grand Central. It was October 2001, and I'd phoned him from Canada to discuss some business. But our conversation quickly turned to the city's fevered mood. After the attack on the World Trade Center and the string of anthrax letters, New York's normally thick-skinned inhabitants seemed near their breaking point.

Of course, another attack did not occur, so we'll never know just how close New York's citizens came to leaving the city en masse. But Andrew clearly had the idea that social breakdowns are like earthquakes: stress accumulates over time, then some kind of external trigger releases the stress to produce a sudden reorganization of the system. The psychological pressure on the people of New York, Andrew implied, had reached just such a threshold.

As stress accumulates in a system, there's a dangerous buildup of something akin to what physicists call "potential energy." The tectonic plates grinding into each other along the coast of California or under the Mediterranean basin store enormous potential energy in the ground. In the same way, prior to the 2003 California wildfires, drought and infestation combined to cause a buildup of highly inflammable material in residential areas—material, again, with huge potential energy. The accumulated energy may eventually be released in a single mammoth event—as happened in the earthquake that devastated San Francisco. Or it may be released in a series of smaller, rapid-fire shocks. Such staccato shocks—any one of which could be managed on its own—can overload a system and causes its breakdown, just the way a boxer sends his opponent reeling with a series of sharp blows.

This kind of thing is actually quite familiar to most of us, although probably not personally. Most of us know of someone who, through bad planning, carelessness, or simply sheer misfortune, has confronted a string of life crises in rapid succession. They might have lost a job, suffered a severe illness, and had their marriage fall apart—all around the same time. Perhaps these events were causally linked (the loss of the job might have precipitated the illness and the marriage problems), or perhaps they were entirely independent. In either case, the result has been catastrophic: although the person might have coped with two of these events, the combination of three pushed him or her over a threshold. Such unlucky people can end up on the street, without social support and without a home.

Some social scientists propose something similar to explain large social crises, including history's great revolutions in England in the mid-seventeenth century, in France in the late eighteenth century, and Russia in the early twentieth century. The scholars argue that revolutions happen when inflexible societies experience multiple shocks—or body blows—at many levels simultaneously or in quick succession. As Jack Goldstone, one of the world's leading theorists of revolution writes, "Massive state breakdown is likely to occur only when there are *simultaneously* high levels of distress and conflict at *several levels* of society—in the state, among elites, and in the populace."[42] In revolutions, he writes, "there is a crisis of national government, but there are also crises of local government. There are conflicts with the state, but also regional conflicts and even conflicts within families. There are elite rebellions, but also a variety of rural and urban popular movements."[43]

Accumulated stress can also, over time, make a system less supple or resilient. In other words, just like a stick that has dried out and become brittle, the system becomes more likely to break if exposed to too much outside pressure or a sudden sharp shock. Ecosystem collapse often occurs this way: when pollution, overharvesting, or some other long-term stress severely damages a fishery, forest, or grassland, the ecosystem can lose much of its resilience and become more vulnerable to breakdown.[44] The overfishing of the great cod stocks off eastern North America in the early 1990s, for instance, weakened the stocks so much that they may not have been able to cope with otherwise normal fluctuations in ocean salinity and temperature.[45]

In history, we can find many examples of civilizations that have been pushed over the edge to collapse by the combination of multiple stresses and weakened resilience. A good example is classic Mayan civilization, which flourished in and around Mexico's Yucatán peninsula from the third to the eighth centuries. In this case, a key long-term stress was an expanding population that became too large for the region's resource base, given the limited productivity of Mayan agriculture. Once peasant farmers had fully exploited the fertile valleys, they moved up the nearby hillsides, cutting down the forests to get wood for fuel and land for farming. But the hillsides' soil was thin, acidic, and soon depleted of nutrients. And because it was no longer stabilized by the root systems of forests and vegetation, it washed down into the valley bottoms, plugging drainage and irrigation systems and lowering crop yields.

The scarcity of good cropland and the loss of food supplies brought about constant warfare among the Mayan kingdoms. Rulers and political elites focused on short-term gains: enriching themselves, extracting wealth from peasants, fighting each other, and building monuments to their glory. Then another potent stress was added to this mix: between 750 and 800 CE, Central America was hit by the most severe and prolonged drought of the millennium. But by this time overpopulation, cropland and energy scarcity, and chronic warfare had so debilitated many of the kingdoms that they had little reserve capacity to cope. Some kingdoms imploded, and eventually the region's population declined by over 90 percent.[46]

Rome is another example of a civilization that succumbed to multiple stresses and weakened resilience. Some scholars argue, however, that such a conclusion about the fall of Rome is fundamentally flawed because the western empire never really collapsed—it merely transformed itself into something else.[47] Much Roman culture and some Roman institutions sustained themselves (via, among other things, the Holy Roman Empire) through the Middle Ages and even up to the modern era. Also, these scholars continue, we can't really fix the exact peak of Roman power, wealth, and accomplishment, so we can't establish the precise start of its decline. And if getting the timing right is hard, so is specifying the geographical location of Roman decline: during periods when parts of the empire were clearly in trouble, other parts were often expanding and prospering.[48]

Yet if collapse involves a severe loss of complexity, it seems that in some ways Rome did indeed collapse. Recent archaeological excavations show that, particularly in the west, social complexity ebbed, as did urban populations (as we saw in chapter 3), large-scale administration of people and territory, and interregional travel and trade.[49] People often reverted to simpler technologies: for instance, they stopped using wick lamps and returned to the open-saucer lamps common in Paleolithic times.[50]

By modern standards of social change, though, this transformation didn't happen fast. We can't point to an episode or moment when the western empire experienced a sudden, sharp failure in structure, organization, and function. If we measure the process by the decline in the empire's total territory—a reasonable gauge of its subsiding administrative capacity—the process unfolded over nearly a century.[51] As was true in the waning days of many ancient polities, the western Roman empire's fall was a slow-motion crisis (which is why many scholars prefer to call it a decline rather than collapse).[52]

But then, by modern standards, almost everything happened far more slowly in Roman times. The very slowness of Rome's crisis underscores a lesson for us today. If social breakdown occurs now—whether it encompasses one country, many countries, or the whole world, and whether its consequences are moderate or severe—we can be sure it won't unfold at the same leisurely pace as seventeen hundred years ago. The underlying mechanisms may be the same—a combination of accumulated stresses, weakened resilience, and multiple shocks. But today our global social, technological, and ecological systems are so tightly linked together, and they now operate at such velocity, that the duration of any future breakdown or collapse is likely to be dramatically compressed.

If we use the contraction of an empire's total territory as a crude indicator of the empire's decline, we find that ancient and early-modern empires usually (though not always) declined over centuries. Empires as diverse as the Islamic caliphate, the Ming dynasty, and the Mongol empire took about two centuries to vanish, and Byzantium declined over a thousand years. In striking contrast, almost all modern empires seemed to disappear virtually overnight: the Spanish, Ottoman, French, British, and Soviet empires all disintegrated within decades. Sometimes, as in the case of the Soviet empire, they vanished in a few years.[53]

We don't usually mourn the disappearance of empires.[54] Most people assume it's a good thing when colonizing societies vanish, for the freedom

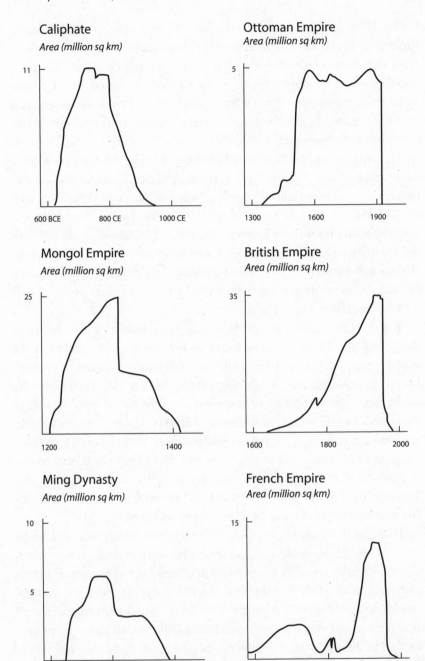

Caliphate
Area (million sq km)

11 —
600 BCE   800 CE   1000 CE

Ottoman Empire
Area (million sq km)

5 —
1300   1600   1900

Mongol Empire
Area (million sq km)

25 —
1200   1400

British Empire
Area (million sq km)

35 —
1600   1800   2000

Ming Dynasty
Area (million sq km)

10 —
5 —
1300   1500   1700

French Empire
Area (million sq km)

15 —
1600   1800   2000

Modern empires have collapsed more quickly than early-modern empires.

of the subjugated peoples if not for their prosperity. Still, the evidence of the quickening tempo of imperial breakdown hints at a critical conclusion, and one particularly germane to our own time: as human societies' connectivity and speed increase, social breakdown, when it does happen, generally happens faster.

Our societies' rising connectivity and speed have a final disturbing effect. In the past, cascading failures usually occurred within single systems—like electrical grids or banking systems—but now these failures are more likely to jump system boundaries. If, for example, terrorists use a new genetically engineered organism to contaminate a Western country's food supply, the disruption will spread in a flash beyond the food system to our larger economic and political systems. Because today's communication technologies vastly multiply our emotional reaction to shocking events—something we saw in full force in the wake of the 9/11 attacks and anthrax letters—this kind of terrorism could easily cause a financial panic and even civil disorder, despite the fact that the threat posed to any one person from the attack might be very small. The scale, connectivity, and speed of our modern food-supply system could also spread a new bio-terror organism far and wide before authorities have figured out what's going on, increasing the chances that it would jump from the food system to affect ecological systems in nature.

The 2003 SARS outbreak may have been a foreshock of this kind of boundary-crossing phenomenon. The virus swept around the world in weeks and sparked instantaneous social and economic emergencies in places as diverse as Hong Kong, Singapore, Vietnam, and Toronto—the almost immediate cost to Asian economies, largely because of a sudden drop off in air travel, was between $11 and $18 billion.[55] With SARS we were lucky because the virus, contrary to popular perception, wasn't very contagious or virulent; next time we may not be so lucky.

If our societies are already brittle because accumulating stresses have eroded their resilience over time, what starts as a local and seemingly manageable breakdown could jump boundaries and quickly spread around the globe, and might even trigger a collapse of global economic and political order. Such an outcome would be a tangible example of what, in chapter 1, I called "synchronous failure"—an event caused by multiple, simultaneous, and synergistic stresses that together generate multiple, simultaneous, and synergistic failures. We can't see the future, of course, so we can't possibly know whether such a thing might occur and, if so, what exactly it would look like. In fact, the whole notion might seem pretty outlandish. But then again, we've never before lived in a world like the one we've created today—in which we can disrupt the planet's most fundamental natural processes, carry a new disease to distant continents in days, and move terabytes of information across the planet in a second—and in which half a dozen of us, with the right materials, could destroy an entire city.

"It all looks beautifully obvious—in the rear mirror," wrote the novelist and social philosopher Arthur Koestler. "But there are situations where [one] needs great imaginative power, combined with disrespect for the traditional current of thought, to discover the obvious."[56]

FLESH OF THE LAND

I N A DISTANT PART of Canada, along the wild western coast of
Hudson Bay, scientists have noticed an alarming trend.

Each year in September, they tranquilize a random selection of
the region's polar bears and measure their weight and length. From
combining these two measures, they create an index of the bear's overall
health. For female bears, the index has declined 17 percent in the nearly
twenty years of the study; for male bears, the decline has been more than
20 percent.[1] This change is bad news for the bears. It also reveals something critical about our changing relationship with nature.

Polar bears have always had to prevail in a battle against a formidable thermodynamic reality—a natural environment that has limited
food and that's often so cold it sucks heat out of their bodies. Accumulating
enough high-quality energy to survive and reproduce is no easy task,
and the bears begin as soon as they emerge from their dens in early
March, after the long subarctic winter. Fortunately for the bears, about
the same time ringed seals give birth to their pups in small snow lairs in
the ice that covers much of Hudson Bay in winter and spring. When
these pups are weaned about six weeks later, they're 50 percent fat and
dangerously naive about predators. The ravenous polar bears roam
across the ice searching for the tasty morsels; and in the short time
before the ice breaks up in July and the pups are no longer accessible,
the bears manage to stuff into their bodies a large portion of their
annual caloric intake.

But the northern environment is changing fast—perhaps faster than the bears can adapt to it. Over the past fifty years, the climate in the region has warmed about 0.3 degrees Celsius each decade; and the ice has begun breaking up earlier—now, on average, about two to three weeks earlier than thirty years ago. After breakup, the seals, including the all-important pups, are no longer accessible to the bears. With the shorter hunting season, pregnant female polar bears, especially, store less fat for the long winter to come and are thin by the following spring. This means they have more trouble bearing and raising healthy cubs.[2]

Warmer weather is having another unfortunate impact on polar bears. Rains are starting earlier in the north, even at the beginning of March. They saturate the snow that was laid down over the long winter, making it soft and heavy. Under its weight the bears' dens sometimes collapse, which can crush to death the still-sleeping animals and their cubs.[3]

Few of us have met polar bears in the wild, but we have potent images of them in our minds—perhaps of single, peripatetic bears wandering over the North's endless ice fields, far, far away from our influence. But that mental image is misleading. It turns out that every one of us is doing things each day that intimately affect just about every bear in the Arctic region. They may be immensely strong and brilliantly adapted to their brutal environment as it was till recently, but sometimes they're not strong enough to withstand our collective assault.[4]

Whether or not we're aware of it, our influence insinuates itself into practically every niche and relationship in Earth's web of life—disrupting ancient patterns of plant and animal behavior and key ecological relationships from the deep seas to the high mountains and from the tropics to the poles. When we burn gasoline in our cars or coal in our electrical plants, when we produce concrete to construct buildings and highways, and when our forests burn, the carbon dioxide that is emitted affects the global climate. And as Earth's atmosphere warms—much faster, it turns out, toward the North Pole—myriad relationships between species and their environments are being altered.[5] An alpine shrub migrates up mountain slopes to where it's cooler, a species of bird lays its eggs earlier in the spring on an Antarctic island, and sardine catches plummet in an African lake.[6] This list is now endless, but even so ecologists still know about only a tiny fraction of these shifts. Almost all of them are unseen; and given the complexity of the environmental systems involved, only a few can be predicted with any accuracy.

Yet just like the proverbial canary in a coal mine, these strange developments—polar bears crushed in their dens, frogs born with extra legs in Minnesota and Australia, bees and other pollinators vanishing from our orchards—are telling us something is going haywire with our larger global environment.[7] And no matter how exceptional we think we are—no matter how much we think we can isolate ourselves from, or rise above, nature—we're still intimately entwined with nature's processes.[8]

We've seen how the tectonic stresses arising from population imbalances and energy scarcity increase the risk of social breakdown—in our communities and countries, as well as globally. Escalating damage to the natural environment also threatens us. Our environmental problems are now so considerable that I'm going to divide them into two categories: in this chapter I'll mainly discuss problems arising from damage to the lands, freshwater, forests, and fisheries, and in the next chapter I'll discuss problems arising from climate change. These two categories are of course connected in many ways, but they are distinct enough—and each is critical enough on its own—that I consider them separately as the third and fourth tectonic stresses.

Stages of Denial

Not everyone considers environmental stresses to be particularly serious or recognizes how dangerous they can be. Some people say that there's no environmental crisis and point to the real decline in water and air pollution in rich countries in recent years. These skeptics also assert that environmental activists exaggerate evidence of soil erosion, deforestation, and water shortages in poor countries. And finally they dispute the claim that we face serious environmental problems that encompass the planet, like climate change. The whole environmental issue, they say, has been overblown by a bunch of softheaded ecologists and left-wing anticapitalist pressure groups.

One of the most vocal representatives of this group is Bjørn Lomborg, a Danish political scientist and statistician. Lomborg's 2001 book, *The Skeptical Environmentalist: Measuring the Real State of the World*, has been music to the ears of many commentators—especially in the pages of conservative print media like *The Wall Street Journal* and *The Economist*—who find environmental concerns an irksome intrusion into business as usual.[9] In over five hundred pages of text, charts, and endnotes (the latter ostentatiously numbered from 1 to 2,930), Lomborg argues that environmentalists usually overstate their claims, abuse statistics, and use data selectively. The popular media pay attention to their claims, he says, because they're systematically biased in favor of bad news. Contrary to the story environmentalists and the media tell, Earth's environment is in an acceptable state, resources are abundant, and humankind isn't a danger to the planet. Most important of all, our quality of life has improved immeasurably in the past century and will continue to improve indefinitely into the future.[10]

Lomborg does make some telling points: some environmentalists do wildly exaggerate their case and handle evidence sloppily, and popular media generally highlight bad news over good. But this is an old story, and Lomborg's book is just the latest installment in a long history of environmental skepticism that goes back to the birth of the environmental movement. Also, over and over again he makes exactly the mistakes that he claims environmentalists make.[11] Critiques of his book by some of the world's leading demographers, biologists, climate scientists, and energy experts have generally been devastating.[12] So why, then, should we pay attention to him? Surely we can just dismiss him as an often mistaken, sometimes incompetent, and occasionally deceitful advocate of an anti-environmentalist agenda.[13]

Unfortunately we can't. Lomborg's arguments and those of other skeptics are immensely seductive because they appeal to our natural complacency and resistance to change. These arguments tell us that we don't have to worry about environmental damage because it doesn't threaten our prosperity, national security, or our own lives, let alone our children's future. So they give us an excuse to avoid the expense and work needed to protect our environment. In other words, the skeptics' arguments make denial easy.

Those of us who succumb to denial will often, I believe, go through three psychological stages.[14] First we might try *existential denial:* in this case, we'll say the environmental problem in question—for instance, climate change—simply doesn't exist. But if the weight of evidence becomes impossible to ignore, we can turn to *consequential denial.* Here, we'll admit the problem exists but say it doesn't really matter, either because it doesn't affect us significantly—so what if climate change hurts distant polar bears—or because we can easily adjust to its effects. Finally, if we can't credibly deny both the problem's existence and its consequences, we might say we can't do anything about it. This is *fatalistic denial*. We'll just throw up our hands and declare we're not going to think about the problem; even if it wrecks the world, so be it. For the die-hard environmental skeptic, fatalistic denial is a last and all-but-impenetrable line of psychological defense.[15]

Of course, sometimes an environmental problem genuinely isn't serious, and skepticism is entirely justified. Other times the weight and immediacy of evidence encourages us not to deny the problem's consequences but instead to do something about it. In rich countries in the twentieth century, polluted harbors and urban smog became so obvious and clearly harmful that even people unreceptive to environmental concerns eventually admitted that remedies had to be found. But for the most difficult environmental problems—those that, like climate change, are shrouded in scientific complexity, impose many of their costs far from home, and challenge strong vested interests inside our societies—denial is often irresistible.

Beyond the Horizon

If we're going to move beyond skepticism and denial to get a better grasp of our world's environmental situation, we can start by making two critical distinctions: first, between environmental problems in rich countries and those in poor countries; and, second, between damage to the environment in a particular place—to a region's cropland, forests, or rivers, for instance—and damage to global environmental systems like Earth's climate. Skeptics such as Lomborg often muddle the two distinctions in ways that make their arguments seem far stronger than they really are.

For instance, skeptics often point out that rich countries like the United States, Canada, Britain, Japan, and Germany have solved many of their local and regional environmental problems.[16] In the U.S., a few decades ago, the use of gasoline and paint containing lead meant that many children had lead poisoning, and huge flows of urban and industrial waste so polluted waterways that harbors like Boston's were biologically dead. Today lead levels are a tiny fraction of what they were, and lead poisoning is rare, while Boston's harbor is alive with fish, lobsters, and harbor seals.[17] And, in the 1960s and '70s, thick smog regularly smothered Los Angeles, but now the air, though still polluted, is clear enough that the city's residents can regularly view the San Gabriel Mountains to the east.[18] Examples like these have led some analysts to conclude that a country damages its environment less once it's wealthy. "[In] the longer run," says the economist Wilfred Beckerman, "the surest way to improve your environment is to become rich."[19]

Why? First of all, the world's rich countries have passed through the dirtiest stage of industrialization—a stage during which they generated much of their wealth in huge plants such as metal smelters, factories, and pulp mills that belched smoke and effluent into the air and water. Now, more of their wealth is produced in service and knowledge-based industries that are far cleaner. Also, rich countries have free markets, so as environmental damage makes some natural resources scarcer, prices rise, which can motivate entrepreneurs, scientists, and consumers to clean up pollution and conserve essential resources. For example, when water prices in a drought-prone region reflect water's real scarcity, farmers, homeowners, and corporations learn to use less. And precisely because rich countries have lots of money, they can better afford to clean up the natural environment or improve resource-use efficiency, if they decide to do so.

Politics in rich countries also plays a key role. Once people are wealthy, they don't have to worry as much about securing the basics of life, like food and shelter, so they become more interested in solving problems that affect their quality of life. They become especially concerned that the surrounding air and water might make them sick, because it's contaminated, for instance, with toxic metals such as lead. Because they live in democracies, they can lobby their politicians to clean things up. This pressure works well when the problems are highly visible—when, for instance, people can see the smog in the air or brown sludge in the rivers. Solutions are usually technical fixes implemented at the "end of the pipe"—that is, they are devices that treat the pollution *output* of factories, cars, and the like. Cities build sewage treatment plants, utilities install scrubbers to clean ash and sulfur from power-plant emissions, and auto manufacturers put catalytic converters on car tailpipes. These solutions don't reduce the economy's resource inputs.[20] And they don't require changes in social relations—for example, in the distribution of wealth and political power between groups—or in the basic operation of capitalism.

Rich countries have indeed solved some of their visible, local environmental problems. But in the meantime their impact on the global environment has in many ways grown—and grown a lot. The driving force is affluence: as people become wealthier, they generally own and consume more stuff—from bigger houses to larger plates of food at each meal—and they generally travel more, both locally and across long distances. Between 1970 and 2002, the floor area of the average American house grew nearly 50 percent (from 140 to 207 square meters), even though the number of occupants per house fell; and between 1977 and 1996, the weight of the average American cheeseburger grew over 25 percent, and the volume of the average soft drink grew more than 50 percent.[21] In 1970, Americans owned less than half a car or truck a person; now they own almost 0.8 vehicles a person, and on average they drive *each* vehicle more—about 20 percent more miles annually.[22]

All this consumption and movement needs a lot of energy. The energy is produced mainly by burning oil, coal, and natural gas, which generate carbon dioxide. In most rich countries, wealth has grown faster than energy efficiency, which means total energy use has risen too, as have total emissions of carbon dioxide.

It's as if those of us living in rich countries have intentionally pushed our environmental impact beyond the horizon, so we can't see it anymore. Sometimes we do this quite literally: our environmental regulations encourage dirty industries to move to other parts of the planet, often to poor countries with weaker regulations and governments. We like to consume the products of these industries, but we don't want to live with their mess. We also extend our reach for natural resources far beyond our borders, sometimes because we've already depleted our local supplies, and other times because we want to protect the local supplies. We've wrecked many of our fisheries along the coasts of North America and Europe, so we bring exotic species (like orange roughy) from distant oceans to our dining tables, which helps deplete faraway fisheries too. The Japanese prize high-quality wood, but they also strictly safeguard their own forests, so tracts of virgin tropical forest outside of Japan—in Indonesia, Malaysia, the Philippines, and elsewhere—are logged to feed the voracious Japanese appetite for wood. In a globalized, interconnected world, such are the advantages of wealth.

Sometimes, though, we push our environmental impact beyond the horizon in less obvious ways, especially by converting that impact from something we can easily see to something we can't see. For instance, we clean up the smog and brown sludge around us by using sophisticated technologies that themselves consume lots of energy. Catalytic converters reduce the smog-producing chemicals coming out of our cars' tailpipes, but they also make our cars less fuel-efficient. This means that, all things being equal, our cars burn more gas per kilometer and emit more carbon dioxide. In the process of dealing with a local problem—urban smog—we've increased our impact on Earth's atmosphere and climate.

In such ways we shift the risks and dangers associated with our environmental impact away from us in space and time. We solve our local environmental problems in the present by doing things that increase the risk for everyone around the world and for generations in the future too. And because, now, we don't readily see the global and long-term impact—because we can't readily connect, for instance, our emissions of carbon dioxide today with climate change that might affect people or ecosystems far away and many years in the future—we're much less inclined to change our behavior. As long as it's out of sight, it's out of mind.

Squeezed in a Vise

Conditions are very different in poor countries in Asia, Africa, and Latin America. It's true that some of these countries—especially those that are large and rapidly industrializing, such as China, India, and Brazil—are now major producers of carbon dioxide, just like rich countries. But almost all poor countries also have serious, and often rapidly worsening, local and regional environmental problems, including badly polluted rivers and lakes, denuded forest land, and degraded agricultural soil. This environmental damage can cripple local communities and economies and, in the long run, erode a society's stability.

It's easy for those of us in rich countries to forget that nearly half the people on Earth take their water from local wells, ditches, and streams; use their own or their landlord's farmland to grow their food; and gather wood, charcoal, straw, or cow dung to heat their homes and cook their food. Almost all these people, who number some 3 billion, are extremely poor and live in the rural zones of countries that are themselves poor. They often can't access the technologies, know-how, and infrastructure that would help them protect their natural resources. Fuel-efficient cookstoves and drip-irrigation systems may be too costly, not available in their regions, or not compatible with local customs. Infrastructure—such as the picturesque terracing that we see in photographs of places like Bali and that can slow soil erosion on steep hillsides—is often too expensive to build and maintain.

Making matters worse, in poor countries a small and powerful minority almost always controls a large portion of the land, water, and forests, while most people have access to very little. The wealthy elites bend the laws governing natural resources so that they can extract exorbitant profits. In Brazil, Indonesia, Thailand, Myanmar (Burma), and Ghana, for instance, compliant governments grant their cronies the rights to immense tracts of forest land. Logging companies then clear-cut the forests and export the logs. They have little interest in replanting or protecting the land, so they leave behind ecological devastation and the shattered remnants of communities that have depended on the forests for generations.

Meanwhile, large numbers of rural poor people who can't get the cropland, water, and fuel wood they need—because these resources are degraded or because elites control them—move to the slums now engulfing most cities in poor countries, as we saw in chapter 3. Many others stay in the countryside but move to economically marginal zones like mountain slopes, tropical rain forests, and semidesert lands. Sometimes they'll follow in the footsteps of loggers who've already ransacked the forests and then settle the damaged and vulnerable lands that the loggers have left behind. Eventually the local natural resources in these marginal zones—fragile and limited to begin with—can't support the rapidly growing population. The result is heartbreaking environmental damage. To clear cropland and gather firewood, people cut down any remaining trees and eventually all shrubs and bushes; their goats browse the grass and surviving seedlings down to the ground; and finally seasonal rains wash the loosened soil away.[23]

In my travels in Southern Africa, India, the Himalayas, the Philippines, China, and Mexico, I've seen tens of thousands of square kilometers of land stripped of virtually all vegetative cover and cleaved by washouts and erosion gullies meters deep. As the soil disappears, only the rocks stay behind: fields are littered with boulders, and the underlying bedrock pushes its way to the surface. It's as if the flesh of the land had been peeled away, with only the bones left to bleach in the hot sun. And the sun is very hot because once the trees, shrubs, and topsoil are

As the soil washes away, only the rocks stay behind.

gone, the regional climate changes. Without plants to cycle water between the ground and the atmosphere, rains diminish and the land dries out—often turning to desert.

Also, in many poor countries, a negative synergy of high population density, rapid urban growth, and dirty industrialization has taken a staggering environmental toll. These countries are, not surprisingly, usually most crowded in those areas that had the best soils, water, and forests originally. In ancient times, people settled these areas first, and they used the rich local resources to build their towns and cities, so population density is highest in Egypt along the Nile and its delta (just as it was in Roman times), and in South Asia along the fertile basins and deltas of the great northern rivers, the Ganges and the Brahmaputra. Today, the natural environment in such densely packed zones is hideously degraded: forests were converted to farmland long ago; the farmland, in turn, is being devoured by metastasizing cities, or its soils are being depleted of nutrients and salinized, acidified, and water-logged through overuse and overirrigation. Aquifers are overdrawn, and lakes and rivers are laden with urban sewage and waste from poorly regulated industries like metal-plating shops, pharmaceutical factories, and fertilizer plants.

Poor countries don't have the advantages that allow rich countries to deal with such problems. Most are still in the dirtiest phases of industrialization; their cheap labor attracts high-polluting industries from rich countries. Their markets are often corrupt and so don't give entrepreneurs strong incentives to tackle environmental damage. They don't have the wealth to train scientists or to invest in advanced equipment to clean up pollution. Their citizens are generally still more concerned about the basics of life than quality of life. Also, because many poor countries (like China) aren't really democratic, their citizens can't exert effective political pressure for a cleaner environment, even if they're concerned about it. And finally, because governments often are either hell-bent themselves on rapid industrialization regardless of cost, or beholden to elites that have made their fortunes exploiting natural resources, environmental laws are rarely passed and even more rarely enforced.

When markets don't work properly, when capital and expertise are scarce, governments weak, and the democratic voice limited, societies don't deal well with serious and complex challenges like a declining environment. So poor countries often find that they're squeezed in a vise. On one side they have large and still-growing populations with rapidly rising economic expectations that depend heavily on an ever more degraded environment. On the other side they have persistent economic and political weaknesses that keep them from coping effectively with environmental problems. Eventually, as we'll see shortly, some of these countries can crack under the pressure.

The Anthropocene

So, when we distinguish between environmental problems in rich countries and those in poor countries—the first of the two distinctions I made earlier—we find a mixture of success and failure. In rich countries, some problems like dirty air and water have been addressed quite effectively, while in poor countries many environmental problems are getting worse fast. That's not the end of the matter, though, because to get a full picture of what's happening to our environment, we also need to distinguish between damage to the environment in a particular place and damage to global environmental systems. In other words, we need to look beyond local and regional environmental problems to see changes in the more general relationship between human beings and nature. And when we do, we discover a multidimensional environmental crisis that encompasses the whole planet.

The origin of this crisis lies within our species. We're now so large in our numbers and so powerful with our technologies that we've become a planetary force.[24] We have the raw power to alter the basic characteristics of Earth's biosphere—a life-supporting layer no thicker, proportionately, than the skin of an apple. Some specialists have coined a term for this new era—the *Anthropocene*. The interval of geologic time from the end of the last ice age, roughly eleven thousand years ago, to the present is known as the *Holocene*. The Anthropocene is the most recent part of the Holocene and is marked by large-scale human perturbations of Earth's ecology and its material and energy flows. Scientists differ as to when they think this era started, but most date it to the late eighteenth century, when human activities caused the atmosphere's levels of carbon dioxide and methane to rise sharply.[25]

In what ways are we now a planetary force? Consider, for example, one of the biosphere's deepest and most essential processes—the oxidation of organic matter, which happens when oxygen in our atmosphere breaks down the molecules in plants and animals to release energy, water, and carbon dioxide. It happens in our bodies when our cells metabolize our food to produce the energy we need to live, and on a forest floor when fallen logs and other debris rot. Oxidation is an essential step in the mighty cycle of carbon through our air, land, and oceans. But in the last few centuries human activities have accelerated natural oxidation one hundred-fold, releasing enormous amounts of carbon dioxide into Earth's atmosphere—carbon dioxide that contributes to global warming.[26]

When we burn billions of tons of fossil fuels like coal, natural gas, and oil to produce energy, we're rapidly oxidizing these organic substances. Logging and clearing of forests also accelerates oxidation by causing fires and leaving behind rotting debris. For instance, in the last months of 1997, wildfires swept through Indonesian forest land desiccated by a dangerous combination of drought and logging. They burned deeply into Indonesia's vast peat deposits—in some places twenty meters thick—smothering Southeast Asia in suffocating smog that caused health emergencies across the region and emitting as much carbon as all Earth's terrestrial plants and animals absorb in a whole year.[27] Other human-caused oxidation isn't as visible. Around the world, thousands of fires burn nonstop in underground coal seams, often triggered by poor mining practices. China alone has hundreds of these deep, human-created fires, and by some estimates they consume more than two hundred million metric tons of coal each year and produce nearly as much carbon dioxide as all the cars and small trucks in United States.[28]

We've also changed the planet's cycles of sulfur and nitrogen—elements both essential to life's functions (the proteins in our bodies contain both nitrogen and sulfur). Although pollution controls have cut emissions of sulfur sharply in rich countries in the past three decades, worldwide the burning of fossil fuels high in sulfur (like some coal) has quadrupled global emissions of sulfur dioxide into the air. The surge of sulfur has acidified lakes and damaged forests across swaths of North America, Europe, and China, leaching out nutrients, depleting calcium, and releasing toxic aluminum from the soil.[29] By making synthetic fertilizers and burning fossil fuels we have roughly doubled the amount of reactive nitrogen in our natural environment.[30] This nitrogen seeps into every corner of the biosphere and is incorporated into the molecules of virtually all life. In fact, half the nitrogen in Earth's green plant material, and about 40 percent of the nitrogen in your body, now comes from artificial fertilizer.[31]

And why should we worry about all this excess nitrogen? For a number of reasons. It upsets the chemical balance of soil on land, contributes to the warming of the atmosphere (nitrous oxide, emitted by our fertilized fields, is a potent greenhouse gas), and appears to be overfertilizing forests around the world, increasing their vulnerability to climate extremes and to parasites.[32] It also causes overabundant growth of algae and other aquatic plants in lakes and coastal waters. After these organisms die, they sink to the bottom and decay, sucking oxygen from the water, driving away fish and invertebrates, and killing many bottom-dwelling creatures. Oxygen-starved "dead" zones can now be found from New York's Long Island Sound to the Gulf of Mexico, the Florida coast, the Baltic Sea, and the lagoon of Australia's Great Barrier Reef. Since 1990 the number of such zones around the world has doubled.[33]

The most striking evidence that we live in the Anthropocene is the planet's changed landscape. When we build houses, office towers, and highways, and when we till our fields, excavate gravel, and erect dams, the 6.5 billion of us move countless millions of tons of dirt and rock—an amount that is now about ten times larger than the total moved through the natural action of wind and water around Earth.[34] Partly as a result, we've transformed and often degraded about *half* of the world's ice-free land.[35] Row-crop agriculture, cities with their suburban sprawl, and industrial zones alone cover about 15 percent of Earth's ice-free land—an area larger than the United States, Canada, and Mexico combined.

When we change Earth's landscape, we also wipe out our fellow species. Logging of tropical forests, especially, has destroyed habitat and boosted the rate of species extinction, which is now a hundred to a thousand times greater than it was before the Anthropocene.[36] We've extinguished a quarter of the planet's bird species and endangered a quarter of all mammals and reptiles.[37] And by inadvertently but carelessly transporting aggressive species from their normal habitats to new regions—species like the zebra mussel, the Asian longhorned beetle, and the fungus that causes Dutch elm disease—we've disrupted ecosystems and endangered native species far and wide.

Perhaps nothing better shows how we dominate the biosphere than our use of Earth's plant energy. In a famous paper published in 1986, the ecologist Peter Vitousek of Stanford University and his colleagues estimated that each year humans and their livestock consume directly (in the form of food, fuel, fiber, or timber) about 4 percent of the energy that land plants trap and store through photosynthesis—the process, essential to all life on the planet, by which plants use sunlight to create sugars out of water and carbon dioxide.[38] Four percent may not seem like much, but then they calculated that in the ecosystems we dominate—like cropland, pastureland, and cleared forest land—we manage or destroy an additional 27 percent. If we add the energy that land plants *would have* stored if their ecosystems had been left in their natural state, then, Vitousek and his colleagues estimate, we use, waste, or disrupt nearly 40 percent of all the energy that Earth's land plants could potentially store. The scientists conclude, "An equivalent concentration of resources into one species and its satellites has probably not occurred since land plants first diversified [some four hundred million years ago]."[39]

What does this abstraction mean on the ground? We can start with forests, because they are particularly vital to our survival. Trees keep the rains coming by sustaining the "hydrologic cycle" of water between the land and the atmosphere. They also absorb and trap carbon dioxide, playing a central role in stabilizing global climate. And wood and its products (like charcoal) are the main source of energy for over

2 billion people; in some regions of Africa, they provide up to 80 percent of the energy people consume.[40]

Today, the total area of forest in Europe and the United States is expanding, partly because paper companies are planting tracts of fast-growing trees for pulp production, and partly because people are moving from farms to cities, and the land they leave behind often naturally reverts to forest. But, on a global scale, these gains are swamped by losses in poor tropical countries, where logging, farming, and cattle ranching are gobbling up around fourteen million hectares of virgin forest a year, or about two hectares every five seconds.[41] Although this annual loss is only 0.7 percent of the world's total area of tropical forest, the increments add up fast: about half of the planet's virgin tropical forest and a quarter of its mangroves, which are critical to coastal fisheries in the tropics, are now gone.[42]

Brazil lost about twenty-five thousand square kilometers of jungle in 2002 alone—40 percent more than the year before and an area larger than New Jersey—as developers cut roads deep into the Amazon, and loggers, cattlemen, and soybean farmers followed in their wake.[43] An area of timber the size of Connecticut disappears each year in Indonesia, where the logging industry is wildly corrupt, and 80 percent of the timber trade is illegal and unlicensed. Indonesian wood exports are booming in response to demands for flooring in China, office stationery in Europe, and furniture in the United States. Experts estimate that the vast lowland natural forests of the Indonesian island of Kalimantan will be stripped bare by 2012; those in Sumatra will be gone by 2007.[44]

What about freshwater? Humankind now consumes as much water as would keep forty waterfalls the size of Niagara Falls running non-stop—every day, all day long, year after year. Each year we use between a third and a half of the planet's accessible freshwater runoff from rainfall and glaciers. Since this water is distributed unequally around the planet, many densely populated and water-scarce regions use virtually 100 per-cent of their available water. We have constructed more than forty-two thousand large dams, which, combined with smaller impoundments and other water technologies, now regulate nearly two-thirds of all the planet's rivers. An area the size of France is submerged under artificial reservoirs.[45] Some rivers, such as the Colorado, Nile, and Yellow, are so heavily exploited that little of their water reaches the sea, ruining the rich ecosystems and fisheries that their estuaries once supported. As

Each year, Indonesia loses an area of forest the size of Connecticut. Above, an Indonesian landscape after logging.

a result of both pollution and overuse of our rivers and lakes, about 40 percent of the world's population now lacks sufficient water for basic sanitation and hygiene, and nearly one out of every five people has not enough to drink.[46] If we extrapolate current trends, by 2025 about 3 billion people—or more than a third of the world's population—will live in countries with water stress or chronic water scarcity, a seven-fold increase since 1997.[47]

Our species is having a dire effect on Earth's oceans too. Nowhere else has our plunder of the planet's natural resources been so devastating, partly because it's hard to establish property rights for fish stocks, so fishermen don't have much incentive to conserve, and partly because we can't on a daily basis see overfishing's results concealed under the sea's surface.

Nearly half the world's major fish stocks are now fished to their maximum limit, about 30 percent are overfished, and many have collapsed. This depletion directly harms people because wild fish are the main source of protein for about 1.5 billion of us. As our human population has grown and catches of wild fish have stagnated, world per capita fish consumption, apart from that of China, declined 10 percent between 1987 and 2000 (despite a huge jump in the production of fish from aquaculture).[48] Our industrial fishing fleets now roam the oceans using the latest technologies to locate fish and scoop them from the water—laying waste to other sea life as they go.

Fishing boats first take the high-value and slow-growing predators like cod, swordfish, and tuna. So in many of the world's fishing grounds, we've now wiped out the uppermost layers of the ocean food web.[49] Since 1950, industrialized fishing has reduced the total mass of large predatory fish in the world's oceans by 90 percent. In nearly every case, after industrial exploitation of a particular fishery has begun, the fishery's population of large predators has collapsed within ten to fifteen years.[50] Decades ago, for example, the North Atlantic fishery was

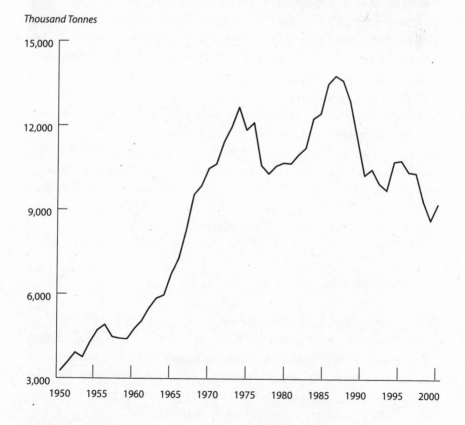

Thousand Tonnes

The annual global catch of cod, hake, and haddock has declined one-third from its peak.

world renowned for its immense stocks of cod, and fishermen would sometimes catch old, hundred-kilogram codfish as big as a man. Now the few fish left are small and fished out of the ocean before they're mature enough to reproduce. By the early 1990s, stocks of northern cod had declined 99.9 percent compared with the early 1960s—a rate of decline, one specialist notes, "almost unmatched among living terrestrial and aquatic species."[51] Because predators such as cod play a vital ecological role, their loss can cause a long chain of negative effects that disrupts the whole ecosystem.[52]

Once these top predators are depleted, fishing fleets move to smaller, bonier, and oilier species, like mackerel, herring, and anchovy.[53] In the tropical waters of Southeast Asia, fishing with dynamite and cyanide has killed thousands of square kilometers of coral reef that sustains coastal fisheries. As stocks along the world's coastlines decline, boats go farther and to greater depths to find their quarry—especially to the mountain ridges on the ocean floor where deep-water fish feed and spawn. There, using sophisticated sonar and global-positioning technologies, they strip-mine the seabed.[54]

And we shouldn't think that fish farming lessens our impact. Much of the diet of such "aquaculture" fish, including the Atlantic salmon we find in our supermarkets, is actually wild fish. So each year, factory fleets scour the oceans for 11 million metric tons of herring, sardines, and mackerel that are ground up and processed into fish food. Since it takes two to five kilograms of this food to produce one kilogram of farmed salmon, aquaculture—at least as it's currently practiced—actually encourages further plunder of wild stocks.[55]

Hollow Societies

When confronted with this kind of evidence of large-scale environmental damage, the skeptic might shift the focus to how people's lives are generally getting better: the statistics show very real improvement in average life expectancy, infant mortality, and food availability around the world in recent decades. Surely if people are healthier and wealthier, our environmental problems can't be all that bad.

But this argument is deeply flawed. That human well-being, narrowly defined, is improving does not mean that environmental damage isn't getting worse; nor does it mean it doesn't matter if it is worsening. At least for a while, we can generate great wealth for ourselves by drawing down nature's capital—by overusing our soils and forests, overfishing the oceans, and pouring immense quantities of carbon dioxide into the atmosphere.[56] Eventually, though, when nature's capital nears exhaustion, this reckless behavior will catch up with us, because overstressed ecosystems can lose their resilience and suddenly collapse. In some cases they already have, as we've witnessed with the North Atlantic cod fishery; in others we simply haven't yet pushed our natural environment beyond the critical threshold.

Also, the skeptics' statistics are usually averages—they summarize in numbers the condition of large populations and even, sometimes, all of humankind lumped together. Although these statistics can tell us important things, they don't give us a good sense of what is happening to individual people or even groups in their individual communities—in their cities, towns, and villages and on their farms.

A more fine-grained picture is vital if we want to see how local, regional, and global environmental damage can combine with ferocious force to relentlessly tighten that vise I talked about earlier—in which untold numbers of poor people around the world find themselves squeezed between, on one side, a degraded environment that doesn't provide an adequate living and, on the other, failed economies that don't supply other livelihoods. And it's the immense social dislocation and bitter frustration, anger, and resentment that comes when billions of people find themselves in such an impossible situation that's the essence of this third tectonic stress—from damage to our land, water, forests, and fisheries—and that's a very real threat to the social and political stability of nations and, ultimately, to world order.

There are, indeed, countless personal stories, almost all unheard and unseen. When we look closely we discover, for instance, that fish—almost all the fish—have vanished in Mexico's Gulf of California. During the decades before and after World War II, American and Japanese fleets hauled in all the tuna and sardines they could find. Then, when catches collapsed around 1990, unlicensed Mexican fishermen took over, vacuuming from the Gulf—for export to the United States and overseas—practically everything left behind. Thousands of sharks were killed just for their fins, which can sell for $700 a kilogram in Asia. Now the local fishermen in Bahía de Lobos, a destitute town on the coast, are practically starving. "There just aren't any fish anymore," says Teresa López, a villager. "Less and less every year for many years. Now we haven't enough to eat."[57]

We see, too, that the forests have disappeared from Haiti, producing an environmental nightmare of astonishing scope and severity, in what has become one of the poorest and most tragically ravaged countries in the world. When the Spanish and French first colonized Haiti in the late fifteenth and seventeenth centuries, it was prized for its abundant forests. Since then, the country has been almost entirely deforested, first by slave masters for their plantations, then by foreign loggers who stripped the land of mahogany trees in the nineteenth century, and more recently by desperately poor Haitian farmers clearing new land for their subsistence plots. Today, the country's few remaining trees, mostly high in the hills, are being chopped down to produce charcoal. And as forests have disappeared, erosion has followed, made worse by the land's steepness and the Caribbean's harsh storms. So much soil now washes off the slopes that the streets of Port-au-Prince, the slum-ridden capital, have to be cleared with bulldozers in the rainy season.[58]

Today, wretched poverty, sickness, and despair haunt Haiti. Nicole Bienvil lives in the town of Plaine Danger in remote Grand'Anse province, where she supports her five children and seven other relatives as a seamstress. She uses an old foot-pedal Singer sewing machine. Her youngest children have scabies, and she hasn't any food in her shack, partly because storms have washed away her family's garden. "It's erosion," she says. "There haven't been trees up there in the hills since I was a kid."[59]

––––––––––

Around the world, environmental damage—operating across scales from the local to the global—is hurting billions, making it harder for people to grow food, stay healthy, and be prosperous. And the costs are borne mostly by those who can least afford them, especially by those who live on eroded hillsides, in desertified wastelands left behind by logging, along coastlines where coral reefs and fisheries have been ruined, or in the measureless slums that ring every poor megacity.

The costs don't stop there. They ripple outward, undermining poor societies' general ability to produce wealth and taking a toll even in the booming economies of the developing world. The Asian Development Bank, an institution with solidly mainstream credentials, states unequivocally that "pervasive, accelerating, and unabated" environmental damage is hindering efforts to beat poverty in the Asia and Pacific region. "At risk are people's health and livelihoods, the survival of species, and ecosystems services that are the basis for long-term economic development."[60]

So environmental damage weakens poor societies from the bottom up and top down. Combined with other stresses—like friction between ethnic groups, external economic shocks, or unmanageable megacities—it can fray a society's social fabric; erode its community, urban, and government institutions; and foster civil violence, including riots, insurgency, guerrilla war, and even ethnic cleansing. We see these results everywhere in poor countries.[61] Consider these news stories from the past few years:

Fourteen Killed in Tana River Clashes

NAIROBI, Kenya (November 20, 2001)–Fourteen people were killed and thirteen seriously injured in Tana River District, Eastern Kenya, on Sunday when tensions between Orma and Pokomo communities over the use of land and water resources erupted into violence. . . . Some seventy people have now been killed over the last year as a result of repeated clashes between the communities, and Pokomo elders have claimed that the Orma have been accumulating firearms in preparation for more attacks.[62]

87 Orphans Will Be Told of the Killers Next Door

SANTIAGO XOCHILTEPEC, Mexico (June 4, 2002)–Eighty-seven children in this little village will grow up hearing how the people from the next village killed their fathers. The children will remember and retell the story of how a fight over a patch of land festered until it turned to killing. . . . All over the state of Oaxaca, where a quarter of the breadwinners earn less than $5 a day, people fight over who owns the land, the water, the trees. The killings of 26 men from Santiago Xochiltepec, population 650, took place Friday night on a darkening dirt road outside Las Huertas, population 700. The attackers lay in ambush with automatic weapons.[63]

Riots in Shanghai Suburb as Pollution Protest Heats Up
XINCHANG, China (July 19, 2005)—As many as 15,000 people massed here Sunday night and waged a pitched battle with the authorities, overturning police cars and throwing stones for hours, undeterred by thick clouds of tear gas. . . . [As] with many of the recent protests, the initial spark involved claims of serious environmental degradation. An explosion at the Jinxing Pharmaceutical Company this month in a vessel containing deadly chemicals reportedly killed one worker and contaminated the water supply for miles downstream.[64]

In Haiti, forest and soil loss contributes to a relentless economic crisis that erodes all public institutions, encourages pervasive corruption, and helps sustain vicious fighting between political factions; as criminal violence and kidnappings for ransom have soared, people try to escape the country any way they can—sometimes on boats as illegal refugees to the United States.[65] In the Philippines, cropland and forest degradation in the country's mountainous interior zones causes chronic poverty that's exploited by a persistent Communist insurgency.[66] In Rwanda, land shortages resulting from population growth and soil degradation were a major underlying reason for the bitter hatreds and violence that led to the horror of the 1994 genocide.[67] In the Darfur region of the Sudan, population pressure, land scarcity, and drought have encouraged attacks by Arab nomads and herdsman on black farming communities, producing hundreds of thousands of deaths and over a million refugees in circumstances that may yet rival the horror of Rwanda.[68]

This kind of persistent violence isn't always as conspicuous or dramatic as major old-fashioned wars between countries, where armies lined up on battlefields and blasted away at each other. But it can still have huge repercussions, with long-term implications for everyone in the world. It depresses business and trade, causes great flows of refugees within countries and from one country to another, and produces humanitarian disasters that—often too late—call upon rich countries' military and financial resources. It spurs many regimes to become more authoritarian and more aggressive with neighboring countries as they try to divert their public's attention from domestic problems.

What's more, environmental stress and the violence it sometimes causes can hollow societies out from inside, making them less resilient and, in turn, more vulnerable to catastrophic collapse. As central governments weaken or even disintegrate, they're more likely to lose control of peripheral regions to warlords, gangs, and secessionist groups. We've already seen what bad news this can be. When regimes in countries as different as Colombia, the Congo, Somalia, and Afghanistan have lost control of their territories, opportunistic criminals, drug lords, gunrunners, and sometimes terrorists have found easy havens for their operations. And when the health systems of poor countries fall apart, new infectious diseases that put the world at risk can take root and spread, as we've seen with HIV/AIDS in sub-Saharan Africa.

Those of us who live in rich countries may not be aware of the escalating environmental damage in poor countries. But it matters, far more than we realize, to our children and to ourselves.

CLOSING THE WINDOWS

I N 1983, in the middle of a backpacking trip across East Africa, a friend and I visited the Amboseli game park in southern Kenya.

We'd left from Nairobi in our tiny rented Suzuki late in the day, so we arrived at the park gates at dusk. After checking in with the lone warden at the gate, we drove on to find the campsite that was helpfully circled on our map. Night fell as we bounced along the gravel roads, using a flashlight to pick out the spot.

We met no one else, but we weren't alone. After driving half an hour, we stopped at a four-way crossing. As we turned left, our headlights illuminated an elephant blocking the road. We backed up, turned around, and started down another road. But elephants blocked that one too. No matter which way we went, our headlights revealed elephants—massive and formidable against the inky night.

We had managed to drive right into the middle of a large herd, and the animals seemed increasingly annoyed by our pipsqueak vehicle buzzing back and forth. Finally, we scooted our little car right under a bull's tusks. As we shot past, he reared on his hind legs and trumpeted—a mighty bellow right beside my head that left me babbling in alarm to my friend.

The herd safely behind us—or so we thought—we eventually found the camping area. By that time it was absolutely dark, so we quickly pitched the tent in our headlight beams, crawled into our sleeping bags, and fell asleep. But we didn't sleep for long. Around midnight, the herd

Kilimanjaro at dawn in April 1983

arrived at our campsite, and the pachyderms stayed with us until morning, foraging in the surrounding bushes and sometimes ripping branches from trees only a few meters from our tent.

A sleepless night was a small price to pay for what we saw early the next day, though. As we warily emerged from our tent, we found the elephants arrayed across the landscape, including two babies beside an acacia tree not fifty meters away from us. And to the south, just across the border with Tanzania, Mount Kilimanjaro soared six kilometers out of the African plain. The ancient volcano's ice-capped summit glowed in the dawn's pink light.

The sight was spellbinding. As we stood beside our tent in the cool morning air, surrounded by elephants and facing Kilimanjaro—so clearly defined by the emerging sunlight that it seemed we could reach out to touch it—we knew this was one of life's rarest moments. But at the time, we couldn't know just how truly precious it was. Someday we'd be able to tell our children and our grandchildren that we'd seen the glaciers of Kilimanjaro.

Kilimanjaro's Retreat

Since that morning, Kilimanjaro's famous ice fields have shrunk by half. Now they consist of a couple of square kilometers of fragmented glacier at the mountain's very apex, a residue that will be entirely gone by 2020.

As usual, scientists vigorously debate the causes. In 1900, the ice fields spread across twelve square kilometers. Their steady decline since then is at least partly due to a long-term drying trend in East Africa, because ice evaporates more readily when the air is dry. But scientists have also found strong evidence that over the past few decades, global warming has begun to melt the ice.[1]

This claim seems reasonable in light of what's happening to other ice fields around the planet. Practically everywhere, ice is melting faster than it's accumulating, and this melting is one of the strongest signals that significant global warming is now upon us. From the Rockies in North America and the Austrian Alps in Europe to the Tien Shan Mountains in Central Asia, the world's great glaciers and ice fields are retreating.[2] This harms more than pretty mountain vistas: almost 1 billion people—a sixth of the world's population—depend on mountain glaciers and winter snowpacks for water.[3] In South America, for instance, Bolivian glaciers and snowcaps have retreated 60 percent since 1978, threatening the water supply of the capital La Paz.[4] Ice fields across the Himalayas and Tibet are also shrinking.[5] When Edmund Hillary and Tenzing Norgay first climbed Mount Everest in 1953, they established their base camp at the foot of a glacier. That glacier has since retreated five kilometers. Rivers that originate in the Himalayas and are partly fed by glacial waters are essential to the lives of hundreds of millions of people in South Asia and China.

In the Antarctic, glaciers and ice shelves—thick sheets of ice extending from land and floating on the ocean—are thinning in many places. Temperatures along the Antarctic Peninsula that reaches north toward Argentina and Chile are rising faster than just about anywhere on Earth, and in the past dozen years several large ice shelves along the peninsula have shattered.[6] At the other end of the planet, in Alaska, glaciers have been shrinking since the 1950s, and the speed of this change has tripled in the past decade.[7] North of Fairbanks, melting of permafrost has buckled paved roads, and telephone poles along the roads have tilted every which way like tipsy toothpicks.[8] Along the state's

In Alaska, melting permafrost has caused telephone poles to tip.

southern coast, warm summers have helped unleash a beetle infestation on the Kenai Peninsula—similar to the one seen in Southern California's pine forests—that has killed nearly forty million spruce trees, the single largest loss of North American forest to insects ever recorded.[9]

Some of the most troubling news comes from Greenland. The island is covered by an enormous ice sheet that's about the size of Mexico and in some places about three kilometers thick—after Antarctica, it's the world's second largest body of ice. Scientists have recently found that the sheet's rate of ice loss has more than doubled in the past ten years, from 90 to 220 cubic kilometers annually. This year the ice sheet will dump into the ocean about 225 times the amount of fresh water that Los Angeles consumes.[10] Some of this ice loss comes directly from increased melting of the ice sheet because parts of Greenland have warmed almost 3 degrees Celsius in the past twenty years. But more ice loss comes from the movement of a dozen glaciers that flow from Greenland's highlands to the sea. At their sea edge, the glaciers break off into icebergs that float away to later melt in the open North Atlantic. In some cases, the glaciers' rate of movement has doubled just since the year 2000—to more than twelve kilometers a year—and their huge tongues extending into the sea have disintegrated into tens of thousands of icebergs. Scientists aren't exactly sure why the glaciers are moving so much faster, but they have an idea: when warm summer temperatures melt the glaciers' surface snow, the meltwater flows down crevasses to the glaciers' base, lubricating their slide into the sea.

In the Arctic Ocean, too, the warming seems to be accelerating: surface ice temperatures in the Arctic summer are rising an astonishing 1.2 degrees Celsius a decade, and the warming over the past twenty years has been eight times faster than it was over the past hundred.[11] The extent of sea ice in September 2005 was the lowest on record, and the regrowth of the icepack over the cold winter months up to March 2006 was also the smallest ever seen.[12] Many scientists believe that sometime this century Arctic sea ice will largely disappear in the summertime— something that hasn't happened in probably over a million years.[13]

All this melting ice around the world is one sign that Earth is warming quickly. In the next decades, further warming will likely hurt our societies in many ways, and for this reason I consider global warming—and the climate change it causes—to be the fourth tectonic stress that's raising the risk of social breakdown. Warming will increase the severity of extreme weather events like heat waves, droughts, floods, and large storms, including hurricanes; kill many forests by changing rainfall patterns and making insect infestations and wildfires worse; and raise sea levels, putting our coastal cities, farms, and infrastructure at risk of inundation. Warming will put enormous strain on our energy systems by boosting summer electricity demand for air conditioning while lowering water levels in many hydroelectric reservoirs. It will be harder for many regions to grow the food they need because most crops are susceptible to weather extremes. And it will also make us sicker by worsening urban air pollution and widening the geographical range of tropical diseases. All these changes will harm poor societies most—especially in combination with the other kinds of environmental damage I reviewed in the last chapter—because they have less ability to cope with tough challenges. But they'll harm our rich societies badly too: Hurricane Katrina's devastation of the U.S. Gulf Coast in 2005 was probably just a foreshock of what's to come.

The year 2005 may, in fact, turn out to be a psychological tipping point in the popular understanding of the perils of climate change. Katrina was, for one thing, an eye-opener for many people because the storm revealed the frightening implications of global warming. Although it's impossible to attribute any single weather event specifically to our planet's warming, Katrina was exactly the kind of catastrophe we're likely to see more frequently in a warmer world. But 2005 could turn out to be a tipping point for an even bigger reason: during the year and also in the first months of 2006, scientists around the world released an astonishing array of scientific studies that confirm our climate situation is rapidly becoming critical.[14]

Consensus

In its basics, global warming is quite simple, really: as the sun's radiation, including visible light, strikes Earth's surface, it's converted to long-wave infrared radiation, or heat, with wavelengths from about one to one hundred microns (a micron is a millionth of a meter). The laws of physics say that the amount of radiation coming in from the sun must be offset, over time, by an equal amount emitted as heat back into space. But trace gases in our atmosphere—especially water vapor—readily absorb heat at certain wavelengths, so the atmosphere stops some of the heat from escaping easily back to space, which keeps Earth's surface about 33 degrees Celsius warmer than it would be otherwise. This is a good thing, since the extra warmth allows life to thrive.

But what keeps the planet from becoming too hot? Ordinarily the atmosphere doesn't absorb as much heat at other wavelengths, so that heat escapes more easily to space, which maintains a balance between Earth's incoming and outgoing energy. It's as if the atmosphere has windows onto space that allow some heat to pass. Now, though, our enormous emissions of carbon dioxide, nitrous oxide, and methane are closing some of these windows because these gases absorb heat at the critical wavelengths that were previously largely unobstructed. Such gases—greenhouse gases—have tipped the planet's energy balance: at the surface of Earth, the incoming flow of radiation now exceeds the outgoing flow by nearly 1.0 watt per square meter.[15] This extra energy is raising the average surface temperature.[16]

At least that's the consensus today among climate scientists. Not surprisingly, though, some skeptics emerged early to question many aspects of the consensus.[17] For instance, United States Senator James Inhofe, the pro-business Republican who is chairman of the U.S. Senate's Environment and Public Works Committee and therefore a key player on the global warming issue, declared in 2003 that he was "becoming more and more convinced . . . that global warming is the greatest hoax ever perpetrated on the American people and the world."[18]

Because greenhouse-gas emissions come mainly from burning fossil fuels, and because fossil fuels are the industrial world's main source of high-quality energy, policies that really curtail emissions will affect practically every activity in the world's rich countries. They'll not only change each and every person's life in these countries, but they'll also challenge the political and economic power of many entrenched interest groups, from oil companies and car-manufacturing unions to lobbyists for the coal-mining industry. Even market-based strategies for dealing with carbon dioxide emissions (like systems for trading carbon-emission quotas, as provided for in the Kyoto climate treaty) will require—if they're properly implemented—reams of government regulation and elaborate national and international institutions that will intrude into every niche of business. Global warming, in other words, raises huge difficulties for laissez-faire versions of capitalism.

As a result, in rich countries few recent public-policy issues have been so controversial, and none has so galvanized conservative anti-environmentalism. Business groups and corporations that benefit from the fossil-fuel status quo—and the politicians who represent them, like Inhofe—are eager to hear and promote skeptical views. Some of these groups have used their substantial resources to finance an effort to discredit mainstream climate scientists and even the entire scientific community.[19] And, in the United States at least, the popular media have abetted the effort by insisting on presenting "both sides" of the issue. The media say that they're trying to give a "balanced" account, but by overrepresenting the views of a small minority of skeptical climate scientists, their reporting has paradoxically been biased toward the skeptical view.[20] That has helped create the impression that there's little scientific consensus on climate change, when in reality there's a lot.

Few climate-change skeptics are actually climate scientists, and, of those who are, only a handful work in leading research universities or institutes or publish in first-tier scientific journals. Most hail from other disciplines like economics and geography (one of the most prominent skeptics, the science-fiction novelist Michael Crichton, has degrees in anthropology and medicine), and a remarkable number haven't any scientific training at all.[21] Still, whether or not they have formal training in climate science, these skeptics have exploited the genuine scientific uncertainty surrounding global warming. The climate is one of the most complex systems in nature, and its behavior is fundamentally nonlinear and chaotic. Skeptics often misrepresent this complexity—and the scientific uncertainty that naturally comes of it—as evidence that all the science relating to global warming is "weak" or "shaky" and that we can't trust any scientific claims about global warming or climate change. It's only prudent, they say, to do more research before we adopt new policies or change our behaviors to deal with the problem.[22]

When climate-change skeptics make their arguments, they take advantage of a widespread misunderstanding of science. Journalists and laypeople commonly speak of the need for scientific "proof," but in reality science never proves anything decisively. To some degree, all scientific statements are couched in uncertainty, especially those about the behavior of complex biological, ecological, and climate systems.

Yet scientific uncertainty doesn't let us off the hook. In some circumstances, waiting to eliminate uncertainty before changing our policies or behavior might not be prudent at all. It might even be sheer idiocy if—as is true with global warming—the cost of delay could be extreme. In reality, the argument that we should minimize uncertainty by doing more research is often little more than a delaying tactic—an effort to maintain the status quo. And the status quo conveniently protects the power of dominant economic and political groups.

Confidence Game

All the same, as we try to pick our way through the minefield of the global warming debate, it helps to distinguish between what is known with high confidence, what is reasonably certain but still open to some doubt, and what is open to considerable doubt.

What can we say with high confidence?

Scientists are certain that humankind has changed the concentrations of key trace gases in the atmosphere. Since the beginning of the industrial era about 150 years ago, we've increased carbon dioxide levels by more than a third and methane concentrations by about 100 percent.

According to an important study released in late 2005, the atmosphere's level of carbon dioxide hasn't been so high in 650,000 years.[23] Today, this level is rising about 1.6 parts per million (ppm) a year, which is faster than at any time in the past twenty thousand years.[24] (The increases from 2002 through 2005, averaging around 2.18 ppm each year—2.53 in 2005 alone—were significantly above the long-term average of 1.6 ppm, raising concerns that Earth's natural mechanisms for absorbing carbon through its oceans and forests are beginning to break down.)[25]

There's a somewhat less complete but still solid consensus within the scientific community that the average temperature at Earth's surface has gone up by about 0.8 degrees Celsius since the last half of the nineteenth century. Recordings by thermometers at ground level show that Earth steadily warmed till around 1940, cooled a little till around 1970, and then warmed rapidly until the present (with 0.6 degrees of warming in the past thirty-five years alone). Some skeptics—including, in this case, a few climate scientists—say these records aren't accurate because of biases in the way temperature is measured at hundreds of sites around the planet and in the way these measurements are then synthesized into one average temperature. Some have also said that satellite data show little or no warming of the lower atmosphere and so contradict the ground-level data, but recent research shows this conclusion was wrong.[26]

Change in Earth's average surface temperature is only one indicator of global warming. There are many other indicators that skeptics rarely acknowledge. These include differences in temperature between winter and summer and between day and night; differences in temperature between Earth's surface and its upper atmosphere and between land and ocean; changes in the extent and thickness of Arctic ice and in the melting of mountain glaciers; changes in the frequency of extreme weather events, like droughts, floods, and major storms; shifts in the average dates of the first frost in winter; and shifts in the average dates of river-ice breakup, the first bloom of plants, the end of animal hibernation, and the onset of bird migrations in spring.[27]

When scientists use multiple indicators involving measurements of phenomena from thousands of locations all over the planet (and in space), they get a more detailed picture of what's happening to Earth's climate. That picture shows that the planet is warming fast. The consensus estimate is that the rate of warming since 1976 has been triple that of the past century as a whole and that the planet's average surface temperature is now increasing by around 0.17 degrees Celsius each decade. Nine of the ten hottest years since the 1860s (when people began accurately recording temperature) have occurred since 1995, and 2005 was statistically tied with 1998 as the warmest year on record.[28] Analysis of tree rings and other long-term data indicate that these were the warmest years in the past one thousand years.[29]

Now, warming of 0.17 degrees a decade may not seem very rapid— by itself it's an imperceptibly small change. But small changes add up over time, and only a few degrees of warming can produce huge shifts in global climate. Just five degrees separate the end of the last ice age from today's climate.[30] So the 0.6-degree additional warming at Earth's surface that we've seen since the 1970s could already be producing some noticeable effects. And, in fact, in 2006 the World Meteorological Organization (WMO) identified a rash of climate anomalies that occurred during the previous year. These included record temperatures in Australia; deadly heat waves in South Asia, southern Europe, and North Africa; severe drought in east and southern Africa, western Europe, and Brazil (especially in the Amazon basin); devastating flooding in southern China and eastern Europe, and a record-breaking monsoon in western India that affected 20 million people and killed eight hundred. The 2005 Atlantic hurricane season also saw the most tropical storms ever recorded in one year—twenty-seven, of which fourteen were hurricanes, including Katrina.[31]

It's not certain that global warming is a major or even a minor cause of these anomalies.[32] "You cannot attribute this to any single cause," says WMO Secretary General Michel Jarraud. "It's about the very complex interaction between all the elements that make up the very complex machine that is Earth. . . . But global warming is likely to lead to more frequent extraordinary events and greater intensity of these events."[33]

Back Casting

Concentrations of greenhouse gases in our atmosphere are increasing fast, and the Earth is getting hotter. There should be little doubt about these facts. But we must still address the biggest question of all: are the rising levels of heat-absorbing gases in Earth's atmosphere *causing* the observed warming at the surface? On this question, scientific

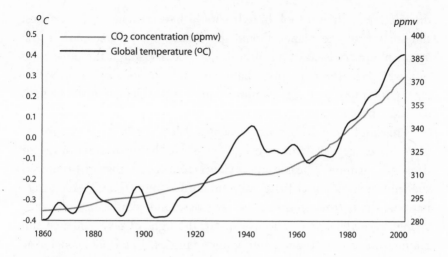

Since 1970, global temperature and the atmosphere's level of carbon dioxide have increased together.

uncertainty is significant, so skeptics are able to make some of their strongest arguments. Levels of carbon dioxide in the atmosphere increased steadily over the past century, they point out, but surface temperature didn't. Only after 1970 did both move upward together. This means, they conclude, that higher levels of greenhouse gases aren't the best explanation of the century's warming. More likely it was due to natural factors, like changes in the intensity of the sun's radiation.[34]

Most climate scientists think this view is wrong. They use sophisticated computer models of the global atmosphere and oceans to study how various natural and human-caused factors—whether higher concentrations of carbon dioxide, more solar radiation, volcanic dust, or something else—interact to affect the climate. Skeptics ferociously denounce these models (often suggesting that they're little more than numerical black magic), but today's models actually do an astonishingly good job of "back casting" Earth's climate over the past century, including the warming we've observed. They also accurately reproduce the dominant cycles and patterns of temperature change around the planet. Though it's true that the models can't yet precisely account for changes in regional and local climate, their resolution is steadily improving as computer power increases. And here's the really important point: the models tell scientists what warming due to higher carbon dioxide levels should look like. It should have telltale characteristics: the surface of the planet should get hotter, for instance, while the stratosphere cools, and warming should occur faster at the poles than in the tropics. And that's just what seems to be happening today, which is why it's important to track more than just the change in the planet's average surface temperature.

This kind of analysis strongly suggests that our emissions of greenhouse gases, especially of carbon dioxide, are the main cause of recent decades' warming. The consensus view is reflected in the 2001 report of the Intergovernmental Panel on Climate Change (IPCC)—the international body that coordinates, assesses, and synthesizes climate-change research around the world. "In the light of new evidence and taking into account the remaining uncertainties," the IPCC concludes, "most of the observed warming over the last fifty years is likely to have been due to the increase in greenhouse gas concentrations."[35] This consensus is echoed by other top scientific organizations, including the American Geophysical Union, whose Council stated at the end of 2003 that "it is virtually certain that increasing atmosphere concentrations of carbon dioxide and other greenhouse gases will cause global surface climate to be warmer."[36]

But does it matter? Skeptics often say that climate change could benefit us in some ways. For example, global warming makes the hydrologic cycle more vigorous—causing larger amounts of water to move more rapidly in the cycle between the surface of Earth and the atmosphere—so it could increase rainfall in some water-scarce areas.[37] Warmer temperatures and more carbon dioxide in the air (which stimulates growth of some plants) could combine to boost crop output in temperate zones and increase global wood supply.

No doubt there will be benefits. Yet harmful outcomes will probably be far more common and serious, especially for poor countries that can't easily adapt. Higher rates of evaporation will make many already arid regions prone to chronic drought. Warmer temperatures will cause mountain snowpacks to melt earlier in the spring, so less water will be available during the critical summer and autumn months when downstream farmers, cities, and towns need it most.[38] Severe floods, too, will be more common, especially in coastal areas, as higher sea levels combine with heavier rainfall and more frequent storms to drive people inland.[39] In tropical and subtropical zones, coastal communities will be vulnerable to increasingly destructive hurricanes, typhoons, and cyclones.[40] Almost any warming will cause agricultural output in the tropics to fall, while warming of more than a few degrees Celsius will hurt agriculture in midlatitudes too.[41] Higher sea temperatures will bleach and kill coral reefs, and wildfires will ravage forests, as we're already seeing around the planet.[42] More people will die of heat stress, of diseases carried by insects like malaria and dengue fever, and of waterborne diseases like cholera.[43]

Some companies, especially big insurance firms, are starting to plan for these risks in their business strategies. But no one can really be sure what harm global warming will cause. In 2001, the IPCC estimated that the planet's temperature will rise between 1.4 and 5.8 degrees Celsius by 2100 (a new IPCC report will be issued in early 2007, and it's likely to raise the upper limit of this temperature range). Yet the resolution of scientists' climate models is still too low to show exactly what will happen at regional or local levels; more fundamentally, scientists don't yet understand central aspects of our changing climate. For instance, will warming increase cloud cover and, if so, will the extra clouds counteract the warming, by reflecting sunlight back into space, or make it worse, by trapping more heat at Earth's surface?

We can be pretty confident, though, that we're forcing the global climate into a completely new realm. And there are three reasons—the momentum behind our rising greenhouse gas emissions, destabilizing feedbacks, and the potential for sudden and devastating climate shifts—why this realm could be a very scary place.

Momentum and Feedback

First, the relentless increase in our emissions has astonishing momentum. Slowing down or reversing this increase will be very hard, if not impossible. Even if world output of cheap conventional oil starts to decline, humankind will remain heavily dependent on fossil fuels, especially coal, for much of the coming century. If current trends of economic growth and technological change continue, by 2100—easily within the lives of our grandchildren—the world's total economic output will have risen almost tenfold (to more than $600 trillion in today's U.S. dollars), and global energy consumption will have quadrupled.[44]

Even assuming that new technologies will steadily reduce carbon dioxide output per dollar of GDP, by 2100 emissions of carbon dioxide from fossil fuels are likely to have tripled from today's level to over seventy-five billion metric tons a year.[45]

By far the majority of this increase will come from China, India, and other quickly industrializing poor countries. China is particularly important: it gets about three-quarters of its energy from coal, and burning coal produces more carbon dioxide for each unit of energy than burning oil or natural gas. For a while in the 1990s, experts were optimistic that China could limit its coal consumption and carbon dioxide output: the government had ordered the closure of tens of thousands of small and inefficient coal mines and power plants, and Chinese statistics showed the country's coal consumption fell as a result. New mines and plants were larger, more efficient, and cleaner.

But now it seems that many small mines and plants weren't closed after all, because local officials feared a sharp rise in unemployment. Official statistics show coal output rose a staggering 69 percent between 2000 and 2005, a growth rate of about 10 percent annually. According to one estimate, the country will almost triple its output of electricity from coal-fired plants by 2020. These coal-consumption trends will have an enormous effect on worldwide emissions of greenhouse gases: the International Energy Agency predicts that the increase in China's emissions from 2000 to 2030 will nearly match the increase from all rich countries.[46]

The long-term implications of this emissions momentum are alarming. Discussions of climate change usually assume Earth will experience, at most, a doubling of atmospheric carbon dioxide from preindustrial levels (that is, an increase from about 280 to 560 parts per million). We're already at 381 ppm, and experts expect we'll reach 560 somewhat after midcentury. Given what's happening in China and elsewhere, though, we're almost certain to shoot far beyond that doubling mark. By 2100, if we continue with current economic, energy, and technological trends, the carbon dioxide level in the atmosphere will be at least 700 ppm and rising. Many experts now think that a tripling of preindustrial levels is probable and that even a quadrupling—to 1,100 ppm—could occur sometime during the first half of the twenty-second century. That point may seem like the distant future, but really it isn't: it's possibly within the lives of our grandchildren and certainly within the lives of our great-grandchildren.

Today's computer models of climate are "tuned" to the dynamics of the atmosphere with its current mixture of gases, so we must be cautious when we interpret their estimates of what will happen if CO_2 quadruples. Scientists at Princeton University have run a climate model assuming such conditions. They've found that a quadrupling would raise the temperature across much of North America, Europe, and Asia by at least 8 degrees Celsius and across the Arctic by a stunning 12 to 14 degrees.[47] Because of natural time lags in Earth's climate, the full effect of a quadrupling of CO_2 wouldn't appear for several centuries. But eventually New York City's average temperature in July becomes roughly the same as Phoenix's average in July now, and the planet's climate as a whole becomes similar to that of the Late Cretaceous period—the last great age of the dinosaurs—between sixty-five and ninety million years ago.

We can debate whether humankind can cope with a doubling of carbon dioxide because there's room for doubt about its consequences. But there's no ambiguity about the ultimate implications of a quadrupling. Harvard University's John Holdren, one of the world's leading authorities on energy, carbon emissions, and climate change, puts it bluntly. "A quadrupled-CO_2 world would be a roasted world, with weather patterns and extremes of heat unlike anything yet experienced during the tenure of human beings on the planet. It would be a catastrophe for the human condition."[48]

The second reason why our climate situation could become particularly grim is that feedback loops may accelerate warming. A feedback arises when a change in one component of a system produces a string of effects that eventually loops back to affect again the original component. Some feedback takes the form of the kind of vicious circle we see in a financial panic or stock market crash, as I noted in chapter 5. Like any complex system, Earth's climate also has many feedbacks. Some of these, which scientists call "positive feedbacks," amplify and reinforce an underlying trend and so destabilize the climate, moving it in new directions. Others, called "negative feedbacks," are fundamentally stabilizing—they counteract the initial change and help return the climate to equilibrium. Not all positive feedbacks in complex systems are bad: while a vicious circle (such as a stock market crash) is a positive feedback that takes us in a direction we don't like, a virtuous circle (such as a stock market boom) is a positive feedback that takes us in a direction we do like.

Scientists don't understand many of the principal feedbacks in the Earth's climate, but they're concerned that positive feedbacks could inexorably drive the climate toward higher temperatures—and even produce a "runaway" greenhouse effect. Events in the Arctic, again, show how this could happen. Open water left behind by melting Arctic ice is much darker and absorbs about 80 percent more solar energy than the ice itself.[49] This extra energy makes Arctic waters even warmer, in turn melting more ice and increasing further the amount of open water that can absorb the sun's heat—a vicious circle that could change the energy balance of the entire Northern Hemisphere. Also, northern boreal forests store immense quantities of carbon in their trees. As the Arctic warms and the forests die from drought, fire, and pests, much of their carbon will be released into the atmosphere. And perhaps scientists' greatest source of concern lies under the frozen tundra in Siberia, northern Alaska, and northern Canada, where twenty-five hundred cubic kilometers of frozen peat reside in the permafrost.[50] As the permafrost warms and melts, the peat's organic material may decay and release huge amounts of methane, a powerful greenhouse gas. In 2005, Russian scientists warned that an area of peat bog in western Siberia the size of France and Germany combined—and potentially containing seventy billion tons of methane—had already begun to thaw.[51]

Climate scientists can identify some potentially counterbalancing negative feedbacks. For instance, the world's oceans absorb a lot of carbon dioxide, so as Arctic ice melts, the newly exposed Arctic water may actually reduce warming by sucking more carbon out of the atmosphere. In fact, the relative influence of positive and negative feedbacks is one of the most fiercely debated subjects among climate scientists. But in recent years, evidence has steadily mounted that dangerous positive feedbacks are outweighing beneficial negative ones, especially in the Arctic. In 2005, some twenty scientists published a paper in a top journal that made the case starkly. "There seem to be few, if any, processes or feedbacks within the Arctic system," they wrote, "capable of altering this trajectory toward [an ice-free] state."[52]

Walking toward a Cliff

The third reason we should be very concerned about our climate future is likely the most important, at least from the point of view of possible social breakdown in coming decades: the prospect of abrupt climate shifts or "nonlinearities." In a world where billions of people are tightly coupled to a steady stream of services from a stable climate—depending closely on regular rainfall to grow their food, for example—a sudden flip to a new climate regime would be a prescription for chaos.[53]

People commonly assume that we'll experience steady, incremental increases in global temperatures as levels of greenhouse gases climb. And if warming and climate change happen incrementally, the logic continues, we should be able to adjust reasonably well. But life isn't so simple, we've seen, at least not when it comes to complex systems.

These systems tend to have a number of stable states, or what experts call "multiple equilibriums." Think of a beach ball that has rolled into a hollow between some sand dunes. A moderate kick doesn't get the ball out of the hollow; instead it just rolls back down to the bottom. But a powerful kick will knock the ball over the nearby dunes and perhaps into another depression next door. The ball is analogous to a complex system, and each hollow between the sand dunes is analogous to a stable state or equilibrium for that system. Although the ball can't shift from one hollow to another easily, sometimes just the right combination of factors—or powerful kicks—can make it happen. When such a shift does occur, it usually happens very quickly and is a complete surprise; to use chapter 1's terminology, the ball exhibits a threshold effect as it crosses the lip of dunes from one hollow to another. And just like the beach ball in a sandy hollow, a complex system once it arrives at a new equilibrium can be very resistant to further change.[54]

Environmental systems sometimes flip from one equilibrium to another, and these flips can devastate people and societies.[55] For example, the loss of the cod fishery threw the economy of the Canadian Maritimes, especially that of Newfoundland, into a long and deep slump and caused tens of thousands of people to leave the region. Sometimes such flips happen because a positive feedback drives the system past a key threshold: for instance, once a fishery starts to decline, the remaining fish can become steadily more valuable, increasing everyone's incentive to strip the ocean of the last fish—and at some point, when the stock is pushed below the minimum needed for reproduction, the whole fishery collapses. It's virtually impossible to predict precisely when such threshold events will occur, so it's very difficult to plan for them. We find it hard to adapt our institutions and technologies to these events because they happen so quickly and can upend our world so completely. We're far better at adapting to slower, incremental change.

How does this relate to the global climate? Many scientists now believe that the planet's ocean-atmospheric system has multiple equilibriums, which means that climate change caused by global warming won't always occur steadily and incrementally. Instead, sometimes it will happen in sharp jumps as Earth's intimately interlinked currents of air and ocean water flip—all together—to new patterns of behavior.[56] Ice-core records from Greenland and elsewhere going back more than one hundred thousand years show that the planet's climate has reorganized itself dozens of times. In the process, large regions have warmed or cooled 10 degrees Celsius or more in as little as a decade.[57] And once ocean currents, trade winds, and storm tracks have reorganized themselves, the new pattern can persist for centuries, even millennia, just the way the beach ball stays resolutely at the bottom of its hollow.

Some people dismiss discussions of abrupt climate change as the stuff of science fiction or silly Hollywood movies, but the danger is all too real. Indeed, it could pose the greatest environmental challenge our species has faced in the past ten thousand years. And we may be far closer to such a threshold than we realize: scientists are particularly concerned about the possibility of a shift in the Atlantic Ocean's currents.

The story goes like this. In the Atlantic's tropical regions, heat from the sun rapidly evaporates the ocean's water. Because the evaporated water is fresh (the salt in seawater doesn't evaporate), the seawater left behind becomes saltier. Some of this relatively salty water then flows northward in the Gulf Stream. This mighty current, which is about seventy-five times larger than the Amazon River, contains huge amounts of heat; and when it reaches the North Atlantic, it releases some of its heat into the atmosphere—warming Northern Europe and to some extent eastern North America. The cooled tropical water, being salty and so relatively dense, then sinks into the depths of the ocean and flows back toward the tropics, eventually winding its way to the South Atlantic.

The North Atlantic is one of only two places on Earth where large quantities of water descend from the ocean's surface to its depths (the other is near Antarctica). The process is a key element of a planetwide flow of water—often called the Great Ocean Conveyor—that snakes from the Atlantic through the Indian Ocean and into the Pacific Ocean, and plays a vital role in moving the sun's heat energy around Earth, especially from the tropics to higher latitudes.

But the whole phenomenon depends on vast quantities of water sinking in the North Atlantic, and the extent to which this happens depends on the water's density, which is affected by its salinity.[58] Global warming seems to be reducing this water's salinity. Among other things, it's boosting precipitation over the North Atlantic and surrounding lands, and it's melting the region's permafrost, glaciers, and ice sheets, including those covering Greenland. As all this freshwater surges into the North Atlantic, it makes the water arriving in the Gulf Stream from the south less salty and less likely to sink.

Scientists have measured a freshening of North Atlantic water over the past forty years—a trend that accelerated in the past decade and has been called the "largest and most dramatic oceanic change ever measured in the era of modern instruments."[59] Recently they've also detected two striking changes in flows of North Atlantic water. In 2001, they discovered a 20 percent decline in the amount of deep water flowing southward through a crucial, kilometer-deep channel between the Faroe and Shetland Islands.[60] And in 2005, in the most worrisome finding yet, they identified a 30 percent decline in the flow of deep water from the Greenland–Iceland Norwegian Sea south along the eastern coast of North America.[61] These signals, together with other evidence, suggest we might be closer to disrupting the Ocean Conveyor than researchers previously assumed.[62] "We are squeezing a trigger in the North Atlantic," says Robert Gagosian, president and director of Woods Hole Oceanographic Institution in Massachusetts. Computer models show, he notes, that a complete shutdown of the Conveyor—in which, for instance, the Gulf Stream breaks up into a series of eddies—would cool the North Atlantic region 3 to 5 degrees Celsius. "That's enough to send mountain glaciers advancing down the Alps. To freeze rivers and harbors and bind North Atlantic shipping lanes in ice. To disrupt the operation of ground and air transportation. To cause energy needs to soar exponentially. To force wholesale changes in agricultural practices and fisheries. To change the way we feed our populations. In short, the world, and world economy, would be drastically different."[63]

A cooling of this magnitude, Gagosian told world business leaders in 2003, would produce winters "twice as cold as the worst winters on record in the eastern United States in the past century." Northern Europe would be as frigid as Alaska and probably unable to feed itself. Moreover, the effects could extend far beyond the North Atlantic: when the Conveyor shut down in the past—for reasons scientists don't fully understand—there were widespread droughts around the planet.[64]

Most experts regard Gagosian's scenario as somewhat extreme. They don't think the Ocean Conveyor will shut down entirely, although they'll admit that a partial shutdown sometime this century is plausible—the Gulf Stream could weaken but not disappear. All climate experts, however, acknowledge that we face something of a paradox: while Earth is, on the whole, getting hotter—very quickly so by historical standards—some regions could become colder, at least for a while, because of radical shifts in the way heat circulates around the planet. Indeed, Europe could turn cooler while the Arctic, including Greenland, continues to melt. "These changes could happen within a decade," Gagosian warns, "and they could persist for hundreds of years. You could see the changes in your lifetime, and your grandchildren's grandchildren will still be confronting them."[65]

Yet Earth's ocean-atmospheric interaction is so complex that it's impossible to know exactly when or how something like this might occur—when the next abrupt shift in equilibrium could happen and where it would take us. Terry Joyce, a physical oceanographer and one of Gagosian's colleagues at Woods Hole, admits he's "in the dark" about how close we are to a new climate regime. "But I know which way we are walking," he says. "We are walking toward the cliff."[66]

Frayed Networks

Even the steady, linear climate change we've seen in the past three decades can do a huge amount of harm, especially because it's happening at exactly the same time that we're damaging land, forest, and ocean ecosystems all over the planet, as I showed in the last chapter.

An ecosystem is a network. The species and organisms making it up are nodes, and the flows of energy, materials, and genetic information between them are links. Through natural selection, species adapt to each other—they "co-evolve," in the language of biologists—and their relationships often become deeply symbiotic. A wonderful example is the symbiosis between flowering plants and the insects that pollinate them.[67] Plant species, writes the ecologist E. O. Wilson, "came to seal obligatory partnerships with their insect counterparts," and "nature is kept productive and flexible by uncounted thousands of such partnerships."[68]

But as we damage ecosystems—by polluting them or destroying their species' habitats, for example—we fray their networks. From temperate grasslands, forests, and wetlands to tropical jungles, coral reefs, and mangroves, the world's ecosystems are losing nodes and links. It's as though we're pulling the threads, one by one, from a complex tapestry. Depending on the pattern of loss, an ecosystem can become more susceptible to shocks—in a word, less resilient.[69] As the ecosystem's species disappear, it can lose keystone species that provide essential services, like pollination.[70] And because ecosystem networks are scale-free—with some keystone species acting as highly connected hubs—if they lose too many hubs, they can fragment into smaller networks or even collapse entirely.[71]

Pollinators deserve special attention because they supply us with much of our food. "One of every three mouthfuls of food we eat, and of the beverages we drink," Wilson notes, is "delivered to us roundabout by . . . pollinators."[72] Yet in many places the number and variety of pollinators—including bees, wasps, flies, and butterflies—seem to be in steep decline. They're apparently succumbing to a barrage of insults, including pesticide overuse, habitat disturbance, pollution, new diseases, and competition from other insect species spread by human beings. So instead of the happy, riotous buzz that used to emanate from our flowering plants, shrubs, and fruit trees in the spring, too often now we hear only silence.

Climate change can push already weakened ecosystems—including systems of pollinators—over the threshold. It disrupts species' habitats, food supplies, and relationships with other species.[73] By 2050, according to one major study, global warming will have started a slide toward extinction for somewhere between 18 and 35 percent of a representative sample of over one thousand plant and animal species worldwide.[74] Says Chris Thomas, the study leader, "The mid-range estimate is that 24 percent of plants and animals will be committed to extinction by 2050. We're not talking about the occasional extinction— we're talking about 1.25 million species. It's a massive number."[75]

Ocean biologists have discovered something that might be even more disturbing: since the early 1980s the productivity of ocean phytoplankton has dropped 6 percent. These microscopic plants play a key role in absorbing carbon dioxide from the atmosphere and are the essential foundation for the ocean's food network. Although many factors could be causing their decline—including changes in the winds that distribute nutrient-laden dust from continents to oceans—global warming seems to be key: the largest declines in productivity have occurred where the ocean's surface has warmed the most.[76]

So over time, a negative synergy of climate change and other environmental stresses may further damage the soils, plant and pollinator ecosystems, forests, and fisheries that are vital to people's health and livelihoods, especially in poor countries.[77] Climate change may also combine with the other tectonic stresses I describe here—population pressure, energy shortages, and economic instability—to produce enormous hardship for societies and people. Again, a story from Haiti shows what we may see, in time, far and wide.

On September 18, 2004, Hurricane Jeanne plowed through Gonaïves, Haiti's third-largest city, sweeping waves of mud off nearby hills that people had stripped of trees. Whole sections of the city were washed away or buried in mud, and almost two thousand people were killed immediately, while hundreds more went missing. The next day their bloated bodies, along with those of pigs and cows, drifted listlessly down the city's flooded streets. Residents had no food or clean water.[78]

Once a prosperous hub of Haiti's cotton trade, Gonaïves has in recent decades seen chronic economic crisis and violence. Today, its government is feeble, and heavily armed rebels and gangs of young men intimidate the population.[79] For many people, Hurricane Jeanne was the last straw. In the days following the storm, bandits plundered convoys of relief trucks, and thugs robbed women of their rations as they left food-distribution centers. Said Medira Jarmis, a forty-five-year-old laborer fleeing the city with eight family members, "There is no way to stay here anymore. There are so many deaths that the place stinks and there is nothing to eat and nothing to drink."[80]

Hurricane Jeanne went on to drench Florida—the fourth major

hurricane to hit the beleaguered state in 2004. Was this extraordinary sequence of storms—and the even worse sequence the following year—a foreshock of a coming climate crisis? We can't yet be sure, although scientific evidence is mounting that warmer oceans are making hurricanes, typhoons, and cyclones more intense around the world, if not more frequent.[81] We can be sure of one thing, though: Jeanne's devastation of Haiti shows what happens when synergistic social and ecological stresses pummel a society relentlessly. Haiti is now locked in a downward, self-reinforcing spiral of ecological, economic, and social breakdown.

Those of us living in rich countries might think that we're immune to this kind of tragedy, far away behind our well-guarded borders—at least we might have thought so, until Katrina ripped open New Orleans and left anarchy in its wake. But the crumbling of the world order we depend on, if and when it happens, is most likely to begin at its margins. It will start largely unseen in the planet's ignored corners. The signals will first appear in remote places like the ice fields of the Arctic and Kilimanjaro, our silent gardens in spring, or the putrid, mud-clogged streets of places like Gonaïves.

NO EQUILIBRIUM

"**W**ELCOME TO THE COUNCIL ON FOREIGN RELATIONS." The room fell silent. The speaker was Leslie Gelb, president of the Council and former columnist and opinion-page editor of *The New York Times*. He stood at a lectern in the Council's elegant Mansion Ballroom, its walls paneled in dark wood and adorned with oil portraits in heavy golden frames. Beyond the floor-to-ceiling windows, evening was gathering along New York's Park Avenue. It had been raining much of the afternoon, but the storm had lost its force and was moving on.

In the audience facing Gelb were 150 of the city's intellectual, economic, and policy elite—academics, politicians, journalists, corporate CEOs, and investment bankers. And behind them was a bank of television cameras, their lights bathing the scene in an icy glare.

It was May 1, 1997, and the occasion was a debate on the "Implications of a Global Economy" between two of America's best-known economic commentators, George Soros and Paul Krugman. Earlier in the day, I had cajoled the Council staff into saving a place for me at the evening's event. It was something not to be missed, everyone was saying.

Just a few months before, George Soros, the legendary currency speculator reputed to have made $1 billion shorting the British pound, had published an incendiary article in *The Atlantic Monthly* titled "The Capitalist Threat." Drawing on the ideas of G. W. F. Hegel, Karl Popper, and Friedrich Hayek, he had proposed that unbridled laissez-faire capitalism was undermining the foundations of democratic society and, ultimately, the prospects for the success of capitalism itself.[1]

Soros, echoing Hegel, argued that civilizations fall from a "morbid intensification of their own first principles." Under the influence of economists and prevailing economic theory, Western societies have turned the first principle that competitive markets allocate resources efficiently—and so tend toward equilibrium—into an unchallengeable truth. Actually, though, markets are fundamentally unstable and often grossly inefficient because objective economic reality is intimately entangled with, and influenced by, the subjective perceptions of buyers and sellers. So, according to Soros, "there are prolonged periods when prices [move] away from any theoretical equilibrium."

Unfortunately, the ideology of laissez-faire capitalism denies the possibility of such instability and inefficiencies and relentlessly opposes government intervention that could improve matters. Capitalist societies are locked into their own self-referential, self-validating beliefs—a situation that makes them less flexible and more vulnerable to breakdown. The antidote, Soros concluded, was an open society founded on the acceptance of human beings' fundamental fallibility—a society in which claims to truth are never considered unchallengeable and core values are subject to constant debate.

The article caused a storm of controversy. This was 1997, after all, and the American economy was enjoying one of the greatest booms in its history. Who was this philosopher financier, this renegade capitalist who had turned against the system that had so amply rewarded him? The evening at the Council offered New York's chattering class a chance to find out. Would Paul Krugman, then Ford professor of economics at the Massachusetts Institute of Technology and perhaps the country's best-known liberal economist, agree with Soros? Or would he put him in his place?

In his introductory comments that evening, Soros made two quick points: he acknowledged that we live in a global economy and that global integration has brought tremendous benefits, including vastly greater freedom of choice and thought because of the flow of ideas across borders. But then he came to the nub of his argument. "The global economy is not without its drawbacks," he said, with deliberate understatement. Its benefits are distributed highly unequally, and there are stark asymmetries of power between capital and labor and between the center of the global economy (in New York and London) and its periphery.[2]

"There are winners and losers, both among countries and among various interest groups inside countries, and there are instabilities built into the system as a whole, and this renders the present situation precarious and vulnerable. What makes it more precarious is that the shortcomings of the system are not properly appreciated."

"Those who are adversely affected," Soros continued, "have no recourse but to opt out and turn protectionist, inviting mutual retaliation, culminating in an eventual breakdown. And breakdown may manifest itself in the financial markets, but it is more likely to find expression in a political or even in a military way."

Then it was Krugman's turn, and he opened with characteristic wit: "I'm not sure, in the end, how much of an argument we have, but let me, like the characters in Monty Python, do my best to provide one." There is something "inhuman about the global economy," he conceded, because one's livelihood can depend on "the whims of people . . . half a world away." But, wryly alluding to Winston Churchill's remark about democracy, he asserted that capitalism is the worst form of economic arrangement, except for all the rest. And it has contributed to stunning progress in people's well-being. "The human race has never had it so good," he said. "I'm talking about the most basic essentials, about things like whether people have enough to eat, and the fact is that capitalism, as it's now practiced, has delivered those essentials to more people than ever had them before."

Critics of global capitalism, Krugman complained, "focus on the millions of people who have been hurt by shifting economic tides, and there are many such people, although not as many as some of the critics imagine, and they ignore the hundreds of millions of people for whom the global economy has made the difference between mere poverty and sheer desperation." Krugman then, in comments he perhaps later regretted, given events that followed, downplayed the risk of global economic instability. "There is a constant froth of instability on the markets, in the same way that there is, for example, constantly changing weather. But that doesn't mean that the underlying system is totally unstable or totally unpredictable. . . . Some instability is the price you pay for a market system."

The debate continued, with audience interaction, for more than an hour. After the applause had faded and people were rising to leave,

I sampled opinion. Everyone I asked thought Krugman had won. Yet I came away with a much different impression. Krugman had won, yes, but his victory was purely technical. He was precise, witty, and well informed. He had all the facts, arguments, and theories at his fingertips, and he used them with great skill. Soros, on the other hand, often seemed muddled. He tried to convey an intuition and, almost by definition, intuition can't be conveyed verbally. Listening to him, I felt that something about his extraordinary experiences with modern capitalism had led him to believe that the global economy had gone awry and that this development posed a transcendent threat to individual citizens and societies alike.

Barely two months later, in July 1997, Thailand devalued its currency, triggering a financial crisis that crippled one country after another in East Asia and then ricocheted through the global economy. The crisis peaked in September 1998, when Russia declared a moratorium on payments on its sovereign debt—a de facto default. Investors and speculators were stunned: they thought that countries as big and important as Russia couldn't default, and they fled to the perceived safety of U.S. Treasury securities. As the value of these securities rose, the New York–based hedge fund Long-Term Capital Management suddenly found itself on the losing side of a series of bets on currency, stock, bond, and interest-rate movements. Rumors swirled that the fund was about to go bust, so United States officials rushed to investigate. To their dismay, they discovered that the fund's gambles threatened the solvency of scores of major banks and investment houses. Its failure could cause panic in, and suck liquidity from, financial markets around the world. On September 23, only hours before the fund collapsed, the New York Federal Reserve cajoled and coerced a consortium of banks and brokerage firms into taking it over.[3]

This crisis was more than Krugman's froth of instability. According to the economic journalist Robert Samuelson, the 1997–98 financial crisis "posed the most serious danger to the international economy since the 1930s."[4] As Soros had predicted, the system had lost its equilibrium.

Heading for the Exits

By 2000, the turbulence had abated, and investors and economic policy makers breathed a sigh of relief. During these years, the ideas of George Soros received at best mixed reviews.[5] Yet I think Soros was right in important ways: he identified extreme global economic instability and the widening gap between rich and poor as critical dangers. Taken together, these dangers are the fifth tectonic stress that threatens our future, because they can devastate lives and sometimes even wreck whole economies. They can also generate resentment, frustration, and anger around the world that could, under the right conditions, tear apart countries—both rich and poor—and shred the fragile institutions and moral consensus that underpin our global society.

I'll consider global economic instability first. Conventional economic theory suggests that capitalist economies will gravitate toward equilibrium—just the way, to use the analogy from the last chapter, a beach ball rolls to the bottom of a hollow between sand dunes—as changing prices for goods and services balance supply with demand. In actual fact, though, like any complex system a capitalist economy can sometimes exhibit unbalanced and capricious behavior.[6] Instead of acting like a smoothly functioning and predictable machine—a windup clock, for example—it can act more like the planet's climate with its synergies, feedbacks, multiple equilibriums, and threshold effects.[7] This is what happened in East Asia in mid-1997, when a self-reinforcing feedback of investment, profit, consumption, and more investment flipped overnight to a vicious circle of falling investment, failing banks, and crashing consumer demand.

The world economy was calmer in the years that followed, but there's little reason to believe that it has become any less prone to nonlinear behavior. In fact, financial crises have become more frequent over the past thirty years, especially those that involve both a sharp devaluation of a country's currency and the loss of most of its banking capital.[8] According to the World Bank, ninety-three countries experienced an astonishing 112 systemic banking crises between the late 1970s and the year 2000. "These crises both were more numerous and expensive, compared with those earlier in history," the Bank observes, "and their costs often devastating in developing countries."[9]

The debate among economists about why financial crises occur or why they're becoming more common is contentious.[10] Even without full agreement among experts, many governments and international financial agencies have put in place policies that they hope will make severe instability less likely. Some poor countries have adopted floating exchange rates in addition to monetary policies that tame inflation. At the same time, agencies like the International Monetary Fund (IMF) and World Bank have laid out best-practice standards for poor countries' banking, securities, and insurance industries. World Bank officials regularly assess and publicize poor countries' laws on corporate governance, accounting, auditing, and insolvency. The assumption is that better information and regulation lower uncertainty and help ensure that decisions by investors and speculators are based on the real strengths and weaknesses of economies. The international financial system will be less volatile, the reasoning goes, if people aren't blindsided by surprises.

No doubt such policies are worthwhile, but they don't really address some deep causes of financial crises. One is the inherent fragility of banking systems. Any bank faces a fundamental mismatch between the time frames of its liabilities and assets: although its customers' deposits can be withdrawn quickly, its loans are usually invested for long periods.[11] Nor do the new policies address the risks arising from the soaring connectivity, speed, and complexity of the international financial system. Today's communication technologies have so increased the number, tightened the coupling, and boosted the pace of transactions within globalized markets that once a destabilizing feedback loop—like a stock market crash or a run on a weak currency—takes hold, it can spiral into a crisis before policy makers can respond, and then it can cascade outward to affect other economies far and wide. True, major financial crises occurred long before the advent of fiber-optic cables, the Internet, or even the telegraph, but today's technologies of instant communication greatly magnify the risk that panicky behavior will sweep financial markets worldwide.[12]

Think of what would happen in a crowded theater if someone yelled, "Fire!"[13] We commonly assume that people panic in such situations because they're afraid of the possible fire, but psychologists have found that this fear by itself isn't enough. Crowds that think they're in great danger usually behave in an orderly and sensible manner—as we saw during the calm evacuation of southern Manhattan after the 9/11 attacks.[14] Panic ensues only when people in the crowd believe two things at the same time: that they're in danger *and* that they have to compete with others in the crowd to escape (in a theater, the exits are chokepoints, and only a few people can escape at any one time). Either of these beliefs isn't a problem by itself, but when they combine synergistically, everyone tries to escape first, and many can get trampled and crushed in the ensuing stampede.[15]

Today, the international financial system resembles a huge, crowded theater that's vulnerable to fire. The people in the theater—currency speculators, bankers, and investors of all stripes—control trillions of dollars of highly liquid capital. Because of new communication technologies, any warnings of danger—of, say, weakness in a particular developing country's economy—travel through the audience in a flash. These technologies also make financial transactions virtually instantaneous, so speculative capital can be moved from one economy or currency to another with the click of a computer mouse, which means that everyone can converge on the financial system's exits simultaneously. But only those who escape first win, and investors and speculators are terrified of being left behind with a worthless stake in an imploded currency or economy. Fear is multiplied, herd behavior takes over, and people have neither time to reflect nor scope for independent action.[16] A financial panic is a classic example of a "positive feedback" with harmful results: fear causes people to do things—like bail out of investments—that make everyone even more afraid.

This is crowd psychology on steroids. Even if there's no fire in the first place, panic can create its own justification by devastating a fundamentally sound economy. By the time the last investor or speculator has left the theater, an economy hit by one of these runs is likely to be a burnt-out shell, with a broken currency, collapsed banks, and ruined industries. In this kind of situation, better information about the country's economic fundamentals—the goal of many reforms since the East Asian crisis—won't help much, because panicking investors and speculators create their own dire economic reality.[17]

It's worth noting that relatively *low* connectivity and speed in one part of East Asia helped stop the 1997–98 crisis from spreading. China wasn't fully integrated into world capital markets. In particular, the Chinese currency, the yuan, wasn't freely convertible into other currencies, so the country stayed largely isolated from the economic turmoil around it.[18] "The giant nation's relative immunity to currency crisis," wrote a less ebullient Paul Krugman three years after his encounter with Soros, "was one of the few things that kept the Asian crisis from turning into a complete nightmare."[19] Somewhat ironically, then, China's policy of insulating its economy from global capitalist pressures helped protect modern capitalism from itself.

Most proponents of greater global economic integration don't dare acknowledge this fact. And clearly it wasn't on the mind of the dot-com millionaire at the Asilomar conference in May 2001 when he declared—as described in chapter 5—that "the more connectivity in our world the better." That day we were all lucky that China had been only loosely connected to the world economy; otherwise, our panel might have been discussing the world's recent economic meltdown.

Those of us living in North America and Europe are receptive to such glib statements about connectivity's benefits because our economies have stayed strong even as crisis has whipsawed distant places like East Asia and Latin America. Since World War II, we've rarely paid a large price for global economic instability. Instead, the costs are "disproportionately borne by the poor," says the Nobel Prize–winner and former World Bank chief economist Joseph Stiglitz.[20] Most of the poor people and societies are distant from us, so they're easy to ignore—just as it's easy to ignore the faraway people hurt by the world's most severe environmental problems. Neither do our economic elites and policy makers spend much time worrying about instability's human consequences. For them, all that really matters is that battered currencies are stabilized and foreign lenders repaid. "On Wall Street," notes Stiglitz, "a crisis is over as soon as financial variables begin to turn around."[21]

But that's not what matters to the people who've lost their jobs or who've been thrown into poverty. For them, there's no recovery until they get new jobs and their economic conditions improve to the precrisis level, if they ever do.

Take Indonesia, the world's fourth most populous country. From the 1960s till the mid-1990s, the country was a golden success story of rapid economic development, with average incomes climbing tenfold in these decades. Then the East Asian crisis hit. As Indonesia's stock market and currency, the rupiah, crumbled under pressure of speculative attacks, the IMF stepped in with its standard package of policies to stabilize the economy and reassure investors and speculators. Insolvent banks were closed, government spending slashed, interest rates sharply increased, and a range of special deals for President Suharto's friends and family members eliminated.[22]

These harsh policies were in many respects sensible and long overdue. But when applied in the middle of a market panic, they sucked spending power out of the economy and helped turn an already critical economic situation into a truly desperate crisis. Indonesia spiraled into a depression, millions of able-bodied Indonesians were thrown out of work, and malnutrition soared—by 2000 nearly half the population couldn't afford adequate food. Meanwhile, insurgencies, ethnic and criminal violence, and (somewhat later) terrorist attacks exploded across the archipelago. The World Bank described the country's economic reversal as "the most dramatic economic collapse anywhere in fifty years."[23]

But didn't Indonesia's economy boom again and its citizens overthrow Suharto's autocracy? By May 2006, Indonesia had indeed found its footing in some ways—even accounting for the impact of the ferocious earthquake and tsunami that hit Sumatra in December 2004. The country's new democracy had taken root, and its stock market had tripled in value in the previous three years. But the effects of the financial crisis nine years before still lingered. The rupiah had recouped only a quarter of its value, and the world's international investors hadn't regained anything like their previous enthusiasm for Indonesia's economy.[24] Most important, unemployment and underemployment were surging. The working age population was growing by 2.5 million people a year, but the economy was generating only half that many jobs.[25] A senior World Bank official warned that the "alarming" lack of work threatened political stability.[26]

As is true for all countries affected by a severe economic shock, Indonesia will never fully recover from the losses suffered during the crisis. The country lost almost a decade of development: by 2006 per capita income had just returned to precrisis levels. "An economy which has a deep recession," says Stiglitz, "may grow faster as it recovers, but it never makes up for lost time."[27]

Income Gap

"Capitalism is, indeed, inherently unstable," says Bill Emmott, the erudite former editor in chief of *The Economist*. Motivated by greed and fear, energized by high-tech globalization, and often misdirected by bad economic policy, capitalism "veers wildly from boom to bust and back again."[28]

This is not by itself a decisive argument against modern capitalism because instability and occasional crisis may simply be the price we pay for the vast wealth that the system generates. However, critics point to an additional and perhaps more serious problem: yes, modern capitalism is astonishingly good at creating wealth, but its bounty isn't distributed fairly; most of it ends up in the hands of a privileged few. At the Council on Foreign Relations, George Soros argued not only that today's version of globalized capitalism is unstable but also that it widens the economic gulf between the world's winners and losers. Is Soros right?

If the debate among economists and other experts over the causes of capitalism's instability is contentious, the debate over capitalism's role in widening or narrowing the income gap between winners and losers—or, more simply, between the world's rich and poor—is nothing short of raucous (yes, even economists can be raucous). The stakes are as high as they get. The more people perceive that today's globalized capitalism is making the already rich vastly richer while simultaneously leaving much of the world's population behind, the lower will be its moral standing as a set of principles for ordering people's social and economic lives.[29] And any social system that loses its moral standing—its "legitimacy," in the jargon of social scientists—is a target for rebellion.

There's abundant evidence to support widely varied positions in the debate. Supporters of today's globalization focus on the soaring aggregate standard of living in countries like Chile, Malaysia, and Taiwan; in the booming commercial zones of China; and in the high-tech cities of southern India. Critics, on the other hand, focus on the chronic economic crisis in Africa and stagnation in much of Latin America.

But there are two things we can say with certainty: never in history have the differences of income and opportunity among us been so great, and these differences are prima facie evidence of a moral failure of almost incomprehensible magnitude. "In a sense, Karl Marx has been proved right," Emmott writes. "The world has become more and more divided between the few rich and the many poor."[30]

The statistics are truly breathtaking. According to a recent report from the World Bank, about 1.1 billion people, or one-fifth of the population of the world's poor countries, live on less than what $1 a day would buy in the United States. About 2.7 billion people, or over half the developing world's population, live on less than $2 a day.[31] According to the UN's Food and Agriculture Organization, more than 800 million people in poor countries face chronic hunger, an increase of nearly 20 million since the 1990s.[32] And according to UNICEF, 1 billion children—or nearly half the children in the world—are severely deprived of nutrition, water, sanitation, health, shelter, education, or information. Over 640 million children lack adequate shelter, and every day four thousand die because of dirty water or poor sanitation.[33]

. At the other end of the spectrum, in 2006 the world has 793 billionaires with a combined wealth of $2.6 trillion—equivalent to 20 percent of the United States' annual gross domestic product (GDP). Between 2003 and 2006, the number of billionaires increased 66 percent, and their total net worth rose 86 percent. If they'd liquidated this wealth in 2006, they could have hired the poorest half of the world's workers—the 1.4 billion workers who earn a few dollars a day—for almost two years.[34] Indeed, an average billionaire could have hired nearly 2 million of these workers. Never before have so few had the ability to command the labor of so many.

Are these disparities getting worse? Yes, they probably are, though the story has a few twists and turns. Orthodox economic theory says that rich and poor countries should eventually converge to similar levels of average income (GDP per capita) because investment should flow from rich countries, where capital is abundant and returns are limited, to poor countries, where capital is scarce and returns are much higher. If convergence is happening, GDP in poor countries should grow faster than it does in rich countries. But in many poor countries today, GDP is actually growing far more slowly than in rich countries. And, if we look back over a century or more, the inequality between the average income in poor countries and the average income in rich countries has widened. This inequality, says former IMF deputy managing director Stanley Fischer, "appears to have been increasing for at least four hundred years."[35]

Until the 1990s, though, assertions like Fischer's couldn't be backed by hard evidence because economists didn't have reliable historical data on people's incomes, especially in poor countries. Then the World Bank economist Lant Pritchett solved the problem by making some reasonable assumptions about the lower limits of average incomes in the past. The results of his analysis, published in a groundbreaking paper in 1997, were astounding. Growth of incomes in poor countries had indeed lagged far behind those in the rich countries: whereas in 1870 the average income in the world's richest country was about nine times greater than that in the world's poorest country, by 1990 it was forty-five times greater. Pritchett concluded, "The magnitude of the change in the absolute gaps in per capita incomes between rich and poor is staggering."[36]

Contrary to orthodox economic theory, then, it seems that rich countries can grow at relatively high rates indefinitely because they have advanced technology and know-how, stable and capable institutions, and highly skilled workforces that let them profitably use their abundant capital. They also have the political and economic clout to make sure the global economic system works in their favor. Many poor countries, on the other hand, stay locked in poverty because they suffer from chronic government failure and corruption, their workforces are poorly educated and unskilled, and the international economy and trading system is biased against them. Responding to these disturbing findings, some economists have put forward a "twin peaks" theory of convergence. They argue that average incomes in rich countries are converging at a high level, while those in poor countries are converging at a lower level.[37]

Of course, economists with a more upbeat assessment of humankind's progress see things very differently. Some argue that the World Bank's widely used data and methodologies chronically overstate poverty, which means that average incomes in poor countries are actually a lot higher than the Bank's statistics indicate.[38] Others admit that inequality between rich and poor countries increased until the 1970s, but they say it has fallen over the past two or three decades as industrialization and globalization have propelled economic growth and raised average incomes in poor countries.[39] Because incomes have gone up, the number of people who are desperately poor has dropped sharply. Between 1990 and 2001, the number living on less than $1 a day declined by about 130 million, and the percentage of poor countries' population below this threshold fell 7 percent, from 28 to 21 percent. The figures were even more impressive in the East Asia and Pacific region: there, the number of people living on less than $1 a day dropped from 470 to 270 million, or from 30 to 15 percent of the region's population.[40] Given that two hundred years ago, about three-fourths of humankind lived below this threshold, such progress cheers enthusiasts of globalized capitalism.[41]

The gains have been almost entirely due to the explosive economic growth in coastal China and more recently in some regions of India. In these two countries incomes have gone up two to four times faster than in rich countries.[42] For this reason, some economists contend that any calculation of global income inequality needs to weight each country by the size of its population. When we give proper weight to the economic success of India and China, they say, we find that global economic inequality is now declining.[43]

In the past few years, the World Bank economist Branko Milanovic has looked closely at just such population-weighted calculations. He points out that almost all of them either ignore economic inequality within countries (by assuming everyone in a given country has the same income) or estimate it indirectly. So Milanovic uses a radically different approach: he exploits the Bank's vast database on household incomes and expenditures—a database that details the economic behavior of individuals and families in countries that together include over 84 percent of the world's population—to measure inequality between the world's individual people, not between its countries. In other words, he treats each person as a member of a single global population, as if national boundaries suddenly vanished. This approach shows that economic inequality among the world's rich and poor people stayed roughly constant—and extremely high—in the 1980s and 1990s.[44] It turns out, Milanovic discovers, that widening inequalities inside China and India have counterbalanced any reduction in global inequality that might have come from these countries' overall growth.[45]

The Dirty Little Secret of Development Economics

For the sake of argument, let's assume that the upbeat economists are right—that average incomes in poor countries are now growing faster than those in rich countries and will continue to do so indefinitely. Surprisingly, it turns out, even though this growth might lower inequality between the average incomes of rich and poor countries, the absolute gap between these incomes will still widen for a very long time.[46] This is the dirty little secret of modern development economics, and it's something that's hardly ever discussed. If it were, the character of the entire debate about the nature, advantages, and disadvantages of modern capitalism and globalization would change.

Consider the assumptions in the 2005 edition of the World Bank's *Global Economic Prospects*. This annual report is widely regarded as the definitive assessment of the world's economy. A close reading of its statistical tables shows that the growth rate of the average income in poor countries was below that of the average income in rich countries in both the 1980s and in the 1990s. No convergence there. Only in the first five years of the new century, according to the report, did income in poor countries grow faster than that in rich countries.[47] Nevertheless, for the decade from 2006 to 2015, the report predicts robust income growth of 3.5 percent in poor countries and 2.4 percent in rich countries.

Now, this may look like convergence, because incomes in poor countries are predicted to grow faster than those in rich countries. But it's not. The gap between poor and rich average incomes will continue to widen: although the average income of rich countries is growing at a slower rate, this rate multiplies a vastly larger income base—$32,000 annually per person in 2006, according to the Bank, compared with $1,500 in poor countries. So the absolute size of the gap between the average incomes of rich and poor countries steadily widens. And it widens not just for a few years or even for a few decades but for *hundreds* of years to come.

It becomes startlingly wide very quickly. By 2050, when the average income in rich countries increases to more than $91,000 (in 2006 dollars), the average in poor countries will be only $7,000, leaving a gap of $84,000. By 2075, easily within the lifetime of today's children, the gap will have widened to almost $150,000, five times larger than today's. The average income in poor countries won't reach the level enjoyed by people in rich countries *now* until almost 2100. But by then people in rich countries will be enjoying nearly $300,000 a year, and the gap between rich and poor will be eightfold larger than today's. It will continue to widen until the year 2256, two hundred and fifty years from now. And the average income in poor countries won't fully catch up to that in rich countries until the year 2291—almost three centuries from now—at the staggering level of $27.7 million a person a year.

Think of a footrace with perverse handicaps. Rich countries are like powerful runners given a twofold advantage: they get to start the race well ahead of everyone else, and the initial speed they run is proportional to their distance from the starting line—in other words, not only do they start far ahead of the line, they're also able to run much faster at first. Poor countries begin much nearer the starting line. They may accelerate faster (that is, they may have a higher growth rate), but because they start so far behind, and because their pace is slower at first, they can't catch up for a very long time.

Some people might say that I'm cooking the books here by assuming that rich countries can sustain a 2.4 percent per capita income growth rate indefinitely.[48] But even in the extremely unlikely event that incomes in rich countries grow just 1 percent annually while those in poor countries continue to grow 3.5 percent indefinitely, the income gap still widens till 2080, and incomes in poor countries won't fully catch up with those in rich countries until 2130, five generations from now.[49] The bottom line, then, is this: not only has the gap between the average incomes of the world's rich and poor widened steadily for a long time— a trend that hasn't changed significantly in recent years—but it will continue to widen for decades, probably for centuries.

Curious Fixation

We're still left with a critically important question: are the two problems Soros highlighted in his Council speech—economic instability and widening gaps between rich and poor—really the fault of the modern form of globalized capitalism?

One school of thought says absolutely not. It says we need more and better globalized capitalism, not less. Instability is a result of poor information, inefficient markets, and weak financial institutions. Income gaps arise because markets in many poor countries aren't truly free, or because tariffs and subsidies in rich countries keep poor countries from getting a fair price when they sell their goods internationally. Also, corruption in poor countries and low investment in infrastructure, health, and education crimp the very growth that would produce convergence with rich countries. In general, these commentators argue, countries that are highly integrated into global markets tend to grow faster because globalization forces them to take tough medicine to make their economies work properly, and then they benefit from huge inflows of investment.[50]

In many respects, these arguments are right: better information, stronger financial institutions, less corruption, and lower subsidies would all help the international economy work better. But the larger thesis—that we need more of today's globalized capitalism to reduce instability and narrow income gaps—runs headlong into some awkward realities. During the decades from 1970 to 2000—the very decades when globalization surged ahead and free markets penetrated every nook and cranny on the planet—economic crises became more common, the income gap between rich and poor stayed wide, and, perhaps most surprisingly, the growth of average income per person declined.[51] Economic globalization did indeed benefit rich countries selling high-value-added goods and services around the world, just as it benefited poor countries—like China and India—that employed cheap labor in large-scale standardized production. But middle-income countries hardly gained at all, including those in Latin America that aggressively privatized state-owned industries and opened their borders to trade and investment.[52] Also, some of the countries that grew the fastest—again including China and India, but also Malaysia and Chile—actively protected their economies using capital controls and trade barriers.[53]

Modern globalized capitalism, then, doesn't remedy the problems of instability and the income gap—which, taken together, make up the fifth tectonic stress that increases the risk of social breakdown. Today's capitalism may even make these problems worse. How? Our societies and economies—as currently set up—need constant economic growth to maintain social and political stability. But sometimes our policies to achieve this growth increase economic instability and widen the income gap.

To follow these links, it's first necessary to recognize growth's role in our modern economies. We take the value of constant growth for granted in our day-to-day discussions of economic matters—in our newspapers and business magazines and in political discussions. That growth is a good thing is an unchallenged, almost sacrosanct assumption. One might even say that we're collectively fixated on maintaining growth. But this is a rather curious fixation because beyond a certain point—a point many of us passed long ago—the higher incomes that growth produces apparently don't make us any happier.

When psychologists have questioned people over the years about how happy they are, they've found that people in rich countries are on average no happier today than people were in the 1970s or even the 1950s.[54] During the intervening decades we've become far richer. In the United States, personal income (in constant 1995 dollars) more than doubled between 1957 and 1998. But over this period the percentage of people who said they were "very happy" actually declined slightly. Notes the American psychologist David Myers, "We are twice as rich and no happier."[55] And when we look at countries around the world, we find that happiness is correlated with income up to about $10,000 to $13,000 per person annually, but beyond this threshold the correlation vanishes.[56]

Money, in economists' terminology, produces "diminishing returns" of happiness. Once our basic material needs are satisfied, it turns out, we don't need more money to be happy, but we do need loving families, supportive social relationships, absorption in satisfying activity, a sense of purpose in our lives, novelty, and security from catastrophic threats to our income and health.[57]

So, if above a relatively modest threshold, greater material wealth doesn't make us much happier, why do those of us who are already well off in rich countries work so hard to get more of it? Psychologists and behavioral economists have offered a range of answers to this question.[58] Some say we're stuck on a "hedonic treadmill": our aspirations tend to exceed our income, and as our income rises, our aspirations rise in lock-step. Others stress that our happiness is partly a result of our relative social status because human beings naturally compare themselves with other people. We're all trying to at least keep up with Mr. Jones next door. If our yardstick of comparison is income, a higher income makes us happier only if it goes up relative to Jones's income. But because Jones is working as hard as we are, nobody gets ahead, and no one feels any happier. We are, essentially, in an unwinnable income race with other people.[59]

These theories may explain why most of us work so hard to get ahead economically, and why all this effort doesn't make us happier, but they don't really address the deeper conundrum that's our central concern here: why do our politicians, policy makers, economists, and public commentators remain so fixed on maintaining economic growth even when higher incomes don't make us happier?

Cultivating Discontent

The best explanation, as I see it, is as follows. By its very nature, capitalism constantly displaces labor, which can erode the economic demand essential to maintaining capitalism's vigor. Economic growth—sustained in part by a culture of consumerism—absorbs displaced labor into new jobs and industries. The process both maintains demand and ensures that the unemployed and underemployed don't coalesce into an angry and destabilizing underclass.[60]

In the brutal competition of a capitalist economy, companies survive in part by keeping their production costs as low as possible. They often do this by inventing or buying new technologies that help them use the inputs to their production process—capital, labor, and resources—more efficiently. One result is usually higher labor productivity: each worker in a firm that makes computers, for example, can produce on average more computers per unit time. In 1999, in the vast Dell computer factory in Austin, Texas, it took two people fourteen minutes to build a personal computer; by late 2004, one person could do the same job in about five minutes—an astonishing 82 percent increase in worker productivity in just five years.[61]

Most economists say that rising productivity is the engine that drives the steadily expanding wealth of capitalist economies and their steadily rising per capita income over the long term.[62] All things being equal, higher productivity makes computers cheaper and boosts the economy's total output of computers. And as computers become cheaper, consumers can have more of them in their lives, or they can spend less of their income on them. If the latter, they'll have money left over to spend on other things, which is the same thing as an increase in income.

But higher labor productivity comes with social costs. When computer makers introduce technologies that make their workers more efficient, they don't necessarily keep all their workers and simply produce more computers. Sometimes, to boost profits, they lay off workers. According to conventional wisdom, this isn't a long-term problem. Companies will invest their higher profits in new industries—making, say, computer games—and the workers who've lost their jobs making computers will find jobs in these new industries and have incomes once again to buy more products.[63] By this logic, technological change plays a dual role: it displaces labor, but it also creates possibilities for new industries with new jobs.[64]

There's ample evidence to support this conventional wisdom— enough evidence, in fact, to sustain the widespread conviction that capitalism will produce ever-expanding material well-being.[65] But the conventional wisdom downplays several critical problems.

First, it implicitly assumes that people have insatiable material desires—that, put simply, people always want to have more stuff. If this assumption is right, people who save money because computers are cheaper will spend it on other things rather than put it in a bank or under a mattress. Companies, in turn, can be confident that they'll sell the additional output that their higher labor productivity generates. But if the assumption is wrong, and people don't have insatiable material desires—and the research on income and happiness suggests that sometimes they don't—they may not spend enough money to spur companies to create new jobs for the workers displaced by rising productivity.[66]

Second, conventional wisdom rarely acknowledges the scope and relentlessness of technological displacement of workers. It's happening in nearly all industries all the time. People who lose their jobs in one industry will usually find that their new jobs in another industry bring little relief from chronic economic insecurity. Until recently, the economy's service sector did offer some refuge: wages for cashiers, restaurant workers, and salesclerks weren't particularly good, but at least labor productivity in the service sector wasn't rising as quickly, so jobs there were more plentiful, if not more secure. Now, though, this sector is seeing rapid productivity increases too.[67] We've all noticed the changes. For instance, when was the last time you spoke to a live telephone operator? Automated operator services using advanced voice-recognition technology have displaced most of these workers.

And third, the conventional wisdom implicitly assumes either that the new jobs available for displaced workers will have roughly the same cognitive and skill requirements as the old ones, or that if the requirements are higher, displaced workers can be retrained to meet them. But as technology has improved, the fraction of jobs in advanced economies involving tasks of high cognitive complexity has risen, while the fraction involving tasks of low complexity has fallen.[68] New technologies of production tend to replace relatively unskilled workers first—because their activities can be more easily automated—while they boost demand for people with the technical expertise and cognitive skills needed to operate, maintain, and repair the new technologies. In offices, for example, information-processing technology has eliminated secretaries, while the accelerated pace, organizational complexity, and technological sophistication of work has increased the need for highly proficient administrative assistants. In factories, robots have replaced assembly-line workers while creating jobs for engineers and technicians who design and maintain robotic systems. Not everyone can be educated or trained to fill such jobs, though; we can't all be computer programmers, financial analysts, or robotic-systems designers. And in any case, the supply of such jobs is limited. So an increasingly high-tech and knowl-edge-based economy heavily rewards a relatively small proportion of workers who have certain cognitive abilities, while everyone else com-petes for the remaining low-skill jobs that pay far less.

If our material desires aren't insatiable, and if significant numbers of workers can't move smoothly to more high-skill jobs, over time our economy's workforce will tend to polarize between, on one hand, a shrinking but ever-richer elite of hyper-cognitively-adept workers who own and run the ferociously productive technologies of our economy but who are unwilling to consume all its products and, on the other, an expanding mass of technologically displaced workers—with, at best, menial jobs—who can't generate enough demand to consume the economy's bounty.[69]

What's the way out? Answer: constant economic growth. Growth creates the new industries and generates the new jobs needed to absorb technologically displaced workers. The American economy, for example, must expand 3 to 5 percent annually—doubling in size every fifteen to twenty-five years—just to keep unemployment from rising.[70] And to get this growth, our leaders and corporations—operating on the implicit assumption that people can be inculcated with insatiable desires and ever-rising expectations—relentlessly encourage us to be hyper-consumers.[71] With our willing and often eager acquiescence, merchants, credit card companies, and banks barrage us with adver-tising (often showing how we're falling behind our neighbor, Mr. Jones), while our economic policy makers ply us with economic incentives—like low interest rates and tax cuts—all to get us to borrow and buy with abandon. Despite the fact that our lives are saturated with stuff, that we've already reached a level of material abundance unimaginable to previous generations, and that more money and possessions add little to our happiness, we must be made to feel chronically discontented with our lot.[72]

In essence, then, the logic underpinning our economies works like this: if we're discontented with what we have, we buy stuff; if we buy enough stuff, the economy grows; if the economy grows enough, tech-nologically displaced workers can find new jobs; and if they find new jobs, there will be enough economic demand to keep the economy humming and to prevent wrenching political conflict. Modern capi-talism's stability—and increasingly the global economy's stability—

requires the cultivation of material discontent, endlessly rising personal consumption, and the steady economic growth this consumption generates. Without economic growth, rich and poor people in our societies would soon confront each other in a fierce zero-sum conflict, and over time the widening gap between these groups would tear our societies apart.[73]

Our economic role in this culture of consumerism is to be little more than walking appetites that serve the function of maintaining our economy's throughput.[74] Our psychological state is comparable to that of drug addicts needing a fix: buying things doesn't really make us happy, except perhaps for a moment after the purchase. But we do it over and over anyway.[75]

Why? There are many reasons.[76] But a central and often overlooked one, I think, is that consumerism helps anesthetize us against the dread produced by empty lives—lives that modern capitalism and consumerism have themselves helped empty of meaning. New technologies create constant economic upheaval, which means that an average person in the American economy, for instance, has to change jobs every four years or so.[77] This chronic economic insecurity cranks up stress in our lives, while the churning of our economy atomizes society as a whole. Both of these trends shred the social fabric of caring, trust, and reciprocity essential to our happiness.[78] So, in place of vital connections between people that come with strong communities, families, and friendships, we substitute the transitory pleasures provided by newly bought things. We also substitute the hyper-stimulation afforded by frenetic movement in cars and planes and by a torrent of incoming information from TV, video games, and the Internet. These diversions further fragment our lives and our social relations. They also rob us of the periods of uninterrupted time we need for the pleasure of focused activities—for pastimes like writing, reading, dancing, gardening, playing a sport or musical instrument, or playing with our children—that bring us real happiness. And because we're chronically dissatisfied, we try to distract ourselves by buying even more stuff.[79]

Psychologists tell us that the flip side of addiction is denial. Our addiction to growth—and to the anesthetizing consumerism that growth makes possible—can be sustained only if we deny growth's often-negative effects. But these effects are real and can be deeply personal.[80] They likely include rich societies' epidemic of eating disorders like obesity, anorexia, and bulimia, and their soaring rates of clinical depression. The ever-rising productivity of our farms and food producers has made food abundant and cheap, so agribusiness naturally barrages us with ads and marketing ploys to get us to buy ever more of their products.[81] (One astonishing result: the number of overweight people in the world—about 1.2 billion, mostly in rich countries—now roughly equals the number of underfed and undernourished, almost all in poor countries.) At the same time, advertisers' constant cultivation of discontent with who we are, what we look like, and where we rank in society has paradoxically helped entrench in our minds the notion that being thin is ideal, with devastating psychological costs for many people, especially girls.[82]

Then there's the related and more general problem of depression. Depending on how it's measured, the illness has become three to ten times more common in the past fifty years. In each successive generation, a larger percentage of people experience at least one severe episode of depression, and this depression strikes people earlier in their lives.[83] Researchers have suggested many explanations, but repeatedly they highlight a cluster of factors: the erosion of community and family, an economic culture that promotes chronic insecurity and extreme individualism, unattainable standards of beauty (especially for women), and an overwhelming proliferation of consumer options—because for a surprising number of people an abundance of options doesn't produce happiness but instead acute stress and feelings of inadequacy.[84]

The Growth Imperative

But hold on. Anti-globalization activists, environmentalists, and others on the ideological left who criticize the consumer mania of rich countries often forget that economic growth is vitally important for a large majority of the world's population. For the roughly 5 billion people living in countries with an annual per capita GDP of less than $13,000, economic growth definitely boosts happiness. For the 2.7 billion people living on less than $2 a day, growth is needed to satisfy the most basic requirements of human dignity. And for the poorest billion people, growth can mean the difference between life and death.

Some version of capitalism is likely the best way to achieve this growth—at least so far no one has invented another economic system that can generate such staggering quantities of wealth so fast. But we can reasonably ask if the costs imposed by today's version of globalized capitalism are too high.[85] As it draws an ever-larger proportion of humankind within its ambit, and as it penetrates ever more deeply into people's lives around the world, it seems to worsen the already severe problems of instability and the income gap—problems that together make up, once again, the fifth tectonic stress that boosts the probability of major social breakdown.

A good label for the combination of economic forces and motivations I described in the previous section is the *growth imperative*—it's imperative that our modern economies grow, if we're going to maintain social peace. Rising productivity steadily reduces the portion of the labor force needed to create a given level of output—a process that, if unaddressed, depresses consumer income and undermines the economy's aggregate demand. Since the 1930s, generations of economic policy makers, especially central bankers, have been acutely aware of the dangers of inadequate demand. The grim lesson of the Great Depression has been seared into their minds: a chronic demand shortfall—and the frightening price deflation that accompanies it— can cripple economies, cause unemployment to skyrocket, and catalyze political extremism. The Depression taught policy makers that mass unemployment is "politically and socially explosive" and must never happen again.[86]

So, somewhat ironically, even the most conservative of today's economic policy makers are closet advocates of the kind of demand-inducing taxation, spending, and monetary policies advocated in the 1930s by the economist John Maynard Keynes—the bête noire of conservatives. If there's a hint of an economic downturn, our politicians eagerly cut taxes and otherwise prime the pump—often, in the process, racking up enormous national budget deficits. Meanwhile, central bankers boost consumption by holding interest rates down to discourage saving. Whether these actions can keep demand high enough over the long run is unclear: today we're seeing symptoms that our rich economies and even the global economy as a whole have—relative to the level of demand—too many steel mills, auto factories, farms, air carriers, hotels, magazine publishers, and the like.[87]

But even if efforts to boost consumption do create sufficient demand, they often make instability worse. When interest rates are too low for too long, especially if inflation appears to be under control, people and companies tend to borrow recklessly to invest in factories, technology, and other productive capital (which further worsens the problem of too much productive capacity) and to speculate on assets like real estate, stocks, and bonds. Returns on capital fall and prices of equities soar far above realistic levels. Such investment and asset bubbles always pop, usually when interest rates suddenly rise (perhaps because government deficits have become unsustainable), and individual and corporate borrowers can't maintain payments on their huge debts. The potential result: an economic crash that leads to a severe credit crunch as investors and speculators withdraw from the marketplace.[88]

The growth imperative worsens instability in other ways too. It tends to cause ever tighter connectivity among economies, financial institutions, corporations, and investors. Because economic policy makers in both rich and poor countries often believe that creating larger, more integrated markets will help sustain growth, they sign free-trade treaties and build institutions like the World Trade Organization. Larger markets, more open trade and investment, and a more efficient global economy do indeed produce huge benefits, especially for rich countries and for some poor countries that specialize in standardized production. But they can also create the kind of close linkages among financial networks that, when a shock happens, allow panic to ricochet through the world economy.

What does the growth imperative mean for the global income gap? Again, in some ways it makes it worse. Because policy makers in rich countries will do everything possible to maintain strong economic growth, and because these countries have such a head start in per capita income, it's extremely difficult for poor countries to catch up—even if they have significantly higher growth rates, as I showed earlier. Modern globalized capitalism encourages countries that are already enormously wealthy to make themselves steadily wealthier, ad infinitum. And a good portion of this enormous wealth underwrites domestic policies, like agricultural subsidies, that simultaneously handicap growth in poor countries.

The growth imperative not only worsens instability and the income gap but also often conflicts with another, increasingly clear, imperative—

to conserve resources and protect the environment.[89] It drives our enormous and relentless appetite for raw materials, like oil, just as it drives our gargantuan output of waste, including carbon dioxide. Policy makers and corporations implore us to buy more cars, bigger houses, and fancier vacations—all for the good of the economy. In the 1990s in the United States, for example, they were thrilled when American car manufacturers succeeded in marketing gas-guzzling sport utility vehicles, because the auto industry is a mainstay of both the domestic and global economies.

Many free-market advocates portray efforts to conserve resources and energy as a kind of eat-your-peas self-abnegation. These people usually believe that we can have endless growth at the same time that we conserve resources and protect the environment. They note that wealthier countries generally produce less of certain kinds of pollution, and they highlight how intangible ideas increasingly drive growth, how competition encourages companies to reduce costs by using resources more efficiently, and how as a result we need less energy and raw material—from oil to steel—to produce a dollar of GDP.[90] So, they say, rich countries can largely "decouple" their economies from resource consumption and, in turn, steadily reduce their impact on Earth's environment.[91] As evidence, they highlight the long-term drop in the amount of energy needed to produce a dollar of GDP: between 1980 and 2001 in the world's rich economies, this "energy intensity" of production fell an impressive 26 percent.[92]

Others make an even more ambitious argument: they say that the claim we face resource and ecological imperatives is sheer nonsense. There are, simply put, no limits to growth because mere physical reality needn't constrain us. Empowered by free markets and modern science, rich economies can create substitutes for key resource inputs that are scarce. "Human ingenuity, energized by sensible policies," *The Economist* magazine writes, "creates resources faster than people use them; people learn to substitute sand (in the form of microchips) for sweat, and fuel cells for petrol engines."[93] And, says the conservative environmental commentator Peter Huber, "Cut down the last redwood for chopsticks, harpoon the last blue whale for sushi, and the additional mouths fed will nourish additional human brains, which will soon invent ways to replace blubber with olestra and pine with plastic. Humanity can survive just fine in a planet-covering crypt of concrete and computers."[94]

These are strong views, but they're not unusual.[95] In debates with geologists about the supply of conventional oil, or with environmentalists about the dangers of climate change, economists in particular say that human beings, if given the right incentives, are smart enough to solve just about any problem that comes their way. We'll see in the next chapter that this view increasingly looks mistaken.

But even the free-market advocates' less ambitious arguments are weak: it may be true that as countries get wealthier and their populations become more concerned about the visible environment, their output of some forms of waste, like sulfur dioxide and lead, declines. It may also be true that new technologies help us make more with less and help us substitute relatively abundant resources for scarce ones.[96] But overall there's little evidence of a general decoupling of economic production from resource consumption.[97] In most rich countries, economic growth has canceled out efficiency gains, so that total resource consumption and waste output have stayed more or less constant for several decades, even though visible air and water pollution has decreased.[98] In many rich countries and especially in the United States, some kinds of consumption and pollution have actually gone way up.[99] Americans use far more paper, plastic, and aluminum than before.[100] And between 1980 and 2004, all those SUVs that helped keep the American economy afloat also helped boost U.S. energy consumption by more than 27 percent and carbon dioxide emissions by 24 percent.[101]

So, as we saw in chapter 4, there's a critical flaw in the argument that total resource consumption and waste output can fall while our economy grows. For this to happen, we need to improve our efficiency faster than our economy grows. But we've learned here that many rich countries need growth of at least 3 percent *every year* just to keep unemployment from rising. And hardly anyone argues that we can improve the efficiency of our use of resources at such a high rate indefinitely.[102]

In any case, much of the apparent decoupling of economic activity from resource and energy use is an illusion. Our economies may have shifted to industries based more on ideas, but that's only because much of our resource-intensive manufacturing has moved beyond our horizon to countries like China and India that then export their wares back to us.[103] In these countries, resource use and pollution are skyrocketing, largely because of the growth of new export-oriented industries. Put simply, economic globalization helps hide the growth imperative's environmental consequences, at least in rich countries.[104] Also, while it may seem that information technologies, from computers to the Internet, are miserly users of resources—because information, after all, is immaterial—they can actually be surprisingly resource hungry. For instance, manufacturing and using a single two-gram thirty-two-megabyte microchip gobbles up more than eight hundred times the chip's weight in chemicals and fossil fuels.[105]

In sum, then, no matter how much we wish otherwise, we can't escape the conflict between our growth, resource, and environmental imperatives. At best, by improving efficiency, conserving resources, and cleaning up pollution, we can diminish this conflict a bit. But we can't come close to eliminating it. That's a real problem, because there's no sign we're about to give up our commitment to growth. Meanwhile, our energy consumption is pushing the limits of supply, and our output of waste, especially of carbon dioxide, is pushing Earth's natural systems beyond their thresholds of resilience.

Clouded Hope

Today's globalized capitalism—which has linked the world together and subordinated people everywhere to its relentless energy as never before—is producing a stew of changes and stresses that's an almost perfect recipe for widespread and even violent resentment of the world's rich by the world's poor. Because capitalism cultivates material discontent, because we naturally compare ourselves with each other, and because comparisons are now so easy through the Web and television, widening gulfs of wealth and opportunity can't but encourage huge numbers of people to be chronically dissatisfied with their lot.

We now live with an inescapable juxtaposition of winners and losers. Everyone everywhere knows who's ahead and who's behind. Every country competes with every other country in a vicious productivity race, as governments strive to make their own nations more efficient, productive, and nimble than the rest. Because rich countries start this race with countless advantages, most poor countries don't have a hope of winning. The poorest can't even get to the starting line. In Latin America, the Middle East, Asia, and sub-Saharan Africa, societies that together encompass billions of people are losing, some irrevocably. Worse, some countries that are key to global geopolitical stability, like Pakistan, are also falling behind. Even in China and India, the gap is widening fast between the minority of people who are participating in the economic boom and the majority, mostly rural farmers, who are still largely on the sidelines.

At the same time, many people in poor societies feel that Western markets and culture are submerging their local cultures and institutions. Combined, these are astonishingly dangerous trends. If our global society is to be resilient and adaptable—and if it's to be peaceful—it must give its citizens roughly equal opportunities to advance their economic and political interests, and it must give them ways of expressing their diversity within a shared human identity. But instead it's giving people ever more unequal opportunity and ever starker differences in their daily lives. Compare, for instance, the day-to-day experiences of a bond trader in the City of London—or even of a suburbanite in Chicago—with those of a slum dweller in São Paulo or Karachi. These people have little common ground on which to build a shared identity—an imagined community or a single, unifying sense of we-ness—that can be the basis for solving our critically serious common problems, from climate change to looming shortages of energy and water.[106] At the same time, globalized capitalism is relentlessly promoting a homogenized, mass culture of consumerism. The result: just when the daily lives of human beings have become vastly different and often vastly unequal in opportunity—probably the most different and unequal in human history—our worldwide consumerist culture is encouraging all of us to adopt universal standards of success and failure.

"Everywhere, expectations have run ahead of opportunities, and resentments have clouded hope," writes the World Commission on the Social Dimension of Globalization.[107] These trends, says the British economist Robert Wade, threaten people in rich countries. "Growing inequality is analogous to global warming," he cautions. "Its effects are diffuse and long-term, and there is always something more pressing to deal with. The question is how much more unequal world income distribution can become before the resulting political instabilities and flows of migrants reach the point of directly harming the well-being of the citizens of the rich world and the stability of their states."[108]

CYCLES WITHIN CYCLES

A S VISITORS CROSS THE THRESHOLD of Room VII in Florence's Museum of the History of Science, the sight stops them in their tracks.

Commanding the room is a device that looks like the orb from a giant's scepter—a glittering, magical sphere that's almost three meters high. It's made up of dozens of circular golden bands, each notched with teeth along one edge, like a clock's cogwheels. The bands with the greatest diameter crisscross the sphere's outside, while layers upon layers of others are nested inside, their cogs intricately meshed. Somehow this weird contraption manages to appear massive yet ethereal, ferocious yet delicate.

As visitors approach from across the room, their eyes are drawn to the obscure recesses of the sphere's heart. What's there? Peering deep inside, past the layers of bands, they're startled to find that at the heart is . . . our planet Earth.

The device is an armillary sphere—a mechanical model of Earth's place in the cosmos. This particular model reflects three ancient assumptions—that the sun, moon, and visible planets (Mercury, Venus, Mars, Jupiter, and Saturn) orbit a stationary Earth, that these seven celestial objects travel in perfectly circular concentric orbits, and that they move along their orbits at a constant rate. Inside the sphere, each celestial object is attached to its own circular band or ring (called an armilla). Turning a crank on the sphere's side engages the cogs, rotates

Antonio Santucci's armillary sphere in the Museum of the History of Science, Florence, Italy

the bands, and carries the attached object around Earth along a path that reproduces the object's movement as seen in the night sky.

At least that was the hope. It was Plato, Aristotle, and most significantly Ptolemy who insisted that the sun, moon, and planets follow steady, circular orbits around a stationary Earth at the center of the universe. But these ancient theorists ran headlong into insistent reality, because the planets' movements, as seen from Earth, are actually any-thing but regular or easily predictable. For instance, on successive nights Mars usually appears to move a little against the night sky's stationary background of stars, but sometimes it stops, and sometimes it bizarrely retraces its steps for a while before resuming its onward journey. How can this behavior, observable with the naked eye, be reconciled with cir-cular orbits around a stationary Earth?

Claudius Ptolemaeus, the last of the great Alexandrian astronomers, was born in Egypt around 85 CE, just five years after the Colosseum opened in Rome. He wrote multivolume books on geography and optics as well as treatments of such tricky geometric problems as the appro-priate angles for a sundial. But his truly monumental and most enduring achievement was the *Almagest*, a thirteen-volume mathematical treatise, likely written around the middle of the second century, on the motions of the sun, moon, and planets. Here Ptolemy forced a marriage between unchallengeable assumptions and recalcitrant reality, and the offspring of this union was one of history's most elaborate exercises in tortured reasoning. He explained the planets' observed motions by way of using epicycles—circular orbits that center on, and so revolve around, the path of other circular orbits. Eventually his scheme had forty of these con-trivances operating around and within each other—cycles within cycles within cycles.[1]

Mechanics, artisans, and tinkers soon began building models that duplicated these baffling celestial affairs. By the late Italian Renaissance, the models had become objects of art and high fashion, much prized by the wealthy and powerful. The armillary sphere in Room VII of Florence's history museum was the largest ever built. It was constructed over five years, from 1588 to 1593, for Ferdinand I de' Medici, the grand duke of Florence, under the supervision of Antonio Santucci, an astronomer and mathematician at the University of Pisa. It may have been the largest armillary sphere, but it was among the last of its kind. Within a few years, Ptolemy's almost absurdly cumbersome view of the cosmos began to disintegrate under a barrage of scientific break-throughs by some of history's greatest minds. Nicolas Copernicus had already—three-quarters of a century earlier—tried to perfect the Ptolemaic system by arguing that one could make much better sense of the wanderings of celestial objects if one assumes that Earth orbits a point near the sun and that the planets in turn orbit Earth.[2] Then, early in the seventeenth century, Johannes Kepler determined that planets move at varying speed along elliptical, not circular, orbits. And finally Galileo Galilei delivered the coup de grâce when he spied Jupiter's moons through his new telescope, effectively demolishing the idea that everything in the cosmos revolves around Earth.

But the old worldview wasn't given up easily. As telescopes delivered a stream of detailed observations of the heavens supporting the heliocentric view that Earth orbits the sun, learned Aristotelians closed ranks to defend the status quo. The greatest opposition came from academic circles. Contrary to popular lore, the Catholic Church, including the pope and especially some Jesuit astronomers, wasn't initially opposed to the heliocentric hypothesis, although some later turned against it. As Arthur Koestler writes in his marvelous history of these events, "The inertia of the human mind and its resistance to innovation are most clearly demonstrated . . . by professionals with a vested interest in tradition and in the monopoly of learning. Innovation is a twofold threat to academic mediocrities: it endangers their oracular authority, and it evokes the deeper fear that the whole, laboriously constructed intellectual edifice might collapse."[3]

Moderates within the Church searched for a compromise that would prevent a crisis. Yet Galileo's stubbornness and pride propelled the uproar to a climax, and in 1633 he was excommunicated. It wasn't until Isaac Newton published his *Principia* in 1687—with its grand synthesis of gravity, mass, inertia, and acceleration—that the scientific issues at stake were finally resolved and our modern view of the solar system began to take firm root.

Santucci's great armillary sphere is a lovely illustration—so tangible in all its elaborate physical glory—of what happens when people who support a particular theory are faced with evidence that doesn't fit the theory's predictions. Quite understandably, they don't want to throw out the theory's core assumptions. Instead, they do their utmost to preserve the assumptions by adjusting other parts of the theory. This was exactly Ptolemy's strategy: he used epicycles to explain recalcitrant reality, more or less, while he protected the metaphysical heart of his cosmology—in this case that Earth is the stationary center of the universe and that the planets' orbits are perfectly circular and uniform. When a few epicycles didn't work, he added more and more, and eventually the whole apparatus became absurdly complex.

All of us, not just ancient astronomers or boneheaded academics, are highly conservative when it comes to our theories of reality. We don't relinquish our core assumptions until the contrary evidence—what philosophers of science call "anomalous data"—is overwhelmingly abundant and relentlessly obvious. And often such conservatism is a good thing: if we threw out our theories wholesale every time a bit of anomalous data came in, we'd create chaos for ourselves. We probably wouldn't survive for long. But sometimes we take this conservatism and denial to extremes, and the result is the kind of cycles-within-cycles intellectual neuroticism that we see in Santucci's great sphere. The contraption speaks of desperation—of a willingness to do just about anything to preserve the sanctioned order of things.

Licensing Denial

What kinds of denial do we employ today? Aided and abetted by politicians, commentators, and so-called experts who are often only too willing to tell us what we want to hear, we use, I believe, a range of strategies to convince ourselves that the problems we face aren't terribly serious, that our future will look more or less like our past, and that the road in front of us—beyond the fog—is straight and clear.

Evidence that doesn't fit this happy vision is accumulating—witness the evidence I've detailed about the tectonic pressures building under the surface of our daily lives. Such pressures seem to be causing the kind of foreshocks that commonly precede larger system breakdowns. Wide swings and sharp surprises in the behavior of our large-scale technological, social, and environmental systems seem to be happening more often. Think of what we've seen just in the past two decades: HIV/AIDS has killed more people than any other pandemic in history, a continent-sized hole in the stratospheric ozone layer has opened over the Antarctic, some of the world's greatest fisheries have collapsed, the international economy has experienced the worst financial crisis since the Depression, the United States has gone to war twice in the Persian Gulf (in large part over oil), and Al Qaeda's terrorism has shaken the world. Just in the past three years we've seen the electrical grid in Italy and eastern North America fail, hurricane Katrina devastate a poorly prepared New Orleans, and angry children of immigrant families riot in cities across France.

We ignore or downplay such evidence in many ways. First, we some-times arrange our lives so we don't see it. We don't watch the news or read newspapers, we don't engage in conversations about dispiriting issues, we avoid going to places where we see things unpleasant, and we allow ourselves to be diverted by consumerism and infotainment. Sometimes such avoidance strategies are easy. For instance, in rich countries we've created in our cities a physical space that is largely divorced from the natural environment—a self-referential human-scale bubble of artificial reality. This bubble allows us to ignore evidence that the climate is changing—that our winters are getting progressively shorter and our summers hotter—because we live in a largely climate-controlled environment.[4] Other changes, however, will be harder to avoid. If oil shortages become serious, we'll see the evidence every time we fill up our gas tank or pay our heating bill.

When avoidance doesn't work, because the anomalous data are simply too intrusive, we can try the strategies of "existential" and "con-sequential" denial that I discussed in chapter 6. Astronomers wedded to the Ptolemaic view of the universe up to Galileo's time were engaged in aggressive existential denial: they were denying the *existence* of a helio-centric order—the reality that Earth orbits the sun. And the inevitable result was that they made their theory relentlessly more complex as they tried to explain away anomalous bits of data—evidence that they could see with their very eyes when they looked at the sky—by adding further epicycles to their explanation of the cosmos.[5]

So existential denial eventually entails big costs: the dominant theory of reality can become so convoluted and arbitrary that it's almost unin-telligible. Then, if someone proposes a theory that explains the things we observe more completely and simply (as Copernicus, Galileo, and Kepler did), the old theory can suddenly look patently absurd in com-parison.[6] Almost overnight, it can be discarded and replaced by the new—a phenomenon described by the philosopher of science Thomas Kuhn, in his brilliant book *The Structure of Scientific Revolutions*.[7] When one cosmological view replaces another, everything around us suddenly looks strangely different. One moment the sun is moving across the sky—it obviously does, because we can see it move—the next we're told that the sun is stationary and that, absurdly, we're flying around it.[8]

This kind of shift may be psychologically wrenching, and it may create upheaval in institutions that are wedded to a particular cosmological worldview. In this instance, though, it's ultimately just a matter of cosmology. The shift's practical, real-world implications are limited: we have to alter our view of our place in the universe, but that's about all. Yet the situation is very different when it comes to shifts in our view of today's pressing problems. Once we move beyond existential denial—once we recognize, for instance, that higher temperatures aren't just the result of the normal cycles of climate but are, instead, strong evidence of global warming and that our lifestyles are making this problem worse—we have to decide what, if anything, we're going to do about it.

And this is where consequential denial kicks in: we don't have to do anything if we can convince ourselves that our problems won't have serious *consequences*. We can, for example, assume that someone will solve our problems. So, rather than worrying about high gasoline prices due to oil shortages, we can insist that our political leaders keep gasoline prices low. To get the job done, they can do just about anything they want—they can risk our national security by allowing us to become too dependent on foreign oil supplies, they can make deals with corrupt regimes, they can go to war to protect supplies, and they can lie to us about how their policies are succeeding. There's just one rock-bottom stipulation: they shouldn't ask us to change our lifestyles or our core values.

We can also deny the consequences of our problems by convincing ourselves that we can deal with them once they get really bad.[9] Yes, we're facing energy shortages, wealth gaps, terrorism and the like, but necessity is the mother of invention, and we're the most inventive species ever, so it will all work out. Also, science and free markets allow modern capitalist democracies to do a better job of creating the right incentives, generating the right knowledge, and implementing the right solutions than any societies in history.[10]

Such arguments are sometimes exactly right. Sometimes it does make sense for us to wait until problems get serious before addressing them; and it's always risky to underestimate human inventiveness and adaptability. But other times these arguments are exactly wrong—they're no more than glib self-gratifying rationalizations that give us license to delay and deny. They exploit the fact that everyone, quite naturally, wants to hear that they don't have to make disruptive changes in their lives.

We may be great problem solvers, but unfortunately we're increasingly creating problems that we aren't effectively solving.[11] The sunny presumption that we can fix our problems once they get bad enough is undercut by the stark fact that some are *already* really bad, and they're *not* getting fixed. Problems like the HIV/AIDS pandemic, the world fisheries crisis, the growing wealth gap between rich and poor, and the possible diffusion of weapons of mass destruction to small groups show that too often we're simply failing to produce the solutions we need when and where we need them. Sometimes, as in the case of HIV/AIDS, our scientific research doesn't advance quickly enough; or, as in the case of the fisheries crisis, our markets don't provide the right incentives. Other times, as in the case of the wealth gap and, to some extent, the diffusion of weapons of mass destruction, our political systems—even the vaunted democracies of the West—can't marshal the necessary collective will.

Unalloyed and simplistic optimism about the future is really just denial in another guise.

Why Don't We Face Reality?

Yet a puzzle remains. Avoiding evidence of our problems, denying their existence, displacing responsibility for dealing with them onto others, or falsely believing we can fix them when they get bad are all, in the end, rotten strategies for survival in our more dangerous world. Why do we adopt these strategies, if they're so bad for us?

Denial has both psychological and social causes. Psychologically, we often choose to ignore things that scare us or that threaten assumptions that give our lives meaning and security. The Ptolemaic system had enormous unconscious appeal because people feared change and craved "stability and permanence in a disintegrating culture," says Arthur Koestler. "A modicum of split-mindedness and double-think was perhaps not too high a price to pay for allaying the fear of the unknown."[12]

We also have great difficulty taking novel threats seriously because we tend to imagine the future through the lens of the past. For instance, the use of civilian airliners against skyscrapers wasn't implausible; it just hadn't happened before the 9/11 attacks, and American policy makers couldn't imagine that kind of threat (although some intelligence analysts did suggest such attacks were possible).[13] Anyone in senior circles in the U.S. government who had pushed the possibility aggressively would have been regarded as a kook.

Also, our brains aren't good at identifying, tracking, and acting on *slow-creep* problems. Problems like global warming develop incrementally over long periods of time; they can become critical in an instant when the underlying stress crosses an unknowable threshold. We don't see the slowly building pressure, as we adapt almost unconsciously to incremental changes in our surroundings.[14] Our attention is instead drawn to things that change quickly and substantially. This problem is made worse when a great deal of uncertainty surrounds the change, because then we have added room for denial. As we saw in chapter 7 with the issue of global warming, if there's at least the possibility of a happy ending, we often choose to emphasize that possibility—however remote—while ignoring the rest of the story.

Finally, and perhaps most fundamentally, our brains are adapted to a world with generally weak connections among events. In the lives of our ancestors tens of thousands of years ago, problems like finding food, preparing shelter, and defending against marauding tribes could usually be separated from each other and addressed sequentially. Our brains, with their limited processing power, could ignore most of the world's complexity because most of it didn't matter to our day-to-day survival.[15] Today, though, our world is vastly different—it's far more tightly connected together, and the problems it generates are often horribly entangled. But our brains are largely the same as our ancestors' brains, so we don't easily see the links and synergies among our problems, and we often can't see how serious our situation is.

Then there are the social causes of denial. Probably the most important is the self-interest of powerful groups—corporations, government agencies, lobbyists, religious institutions, unions, non-governmental organizations, and the like—that have a vested interest in a particular way of doing things or viewing the world. If outside evidence doesn't fit their worldview, these groups can cajole, co-opt, or coerce other people to deny this evidence. Some groups, of course, will be much more effective in the effort than others, owing to their enormous political and economic power. In the United States, for example, some of the world's biggest coal extractors, oil and gas producers, and auto manufacturers have financed an immense and (for a long time) largely successful effort to discredit mainstream climate science and scientists.[16]

This is a fairly conventional explanation of denial, one that points to the influence of powerful interest groups. But it can't fully explain the thoroughgoing pervasiveness and the sheer endurance—in the face of abundant contrary evidence—of much of the denial we see around us today.

It especially can't explain our denial of the often-harmful consequences of economic growth in rich countries. In the last chapter we saw that higher incomes don't make us happier and that the growth imperative likely adds to international economic instability, worsens the global wealth gap, and encourages us to do things, like buy bigger houses and more cars, that hurt Earth's environment. Yet almost all of us are completely convinced that endless growth is a good and natural thing. How can this be so? My answer again highlights the influence of powerful groups: our economic elites encourage us to consume as much as possible, which helps to create jobs for labor that's displaced by technological change—an outcome that helps maintain social and political stability. This answer tells us that growth serves a key social function, but it doesn't really tell us why the conviction that growth is good persists at all levels of our societies—not just among elites but also in the middle and working classes. Nor does it tell us why we stubbornly deny growth's harmful consequences.

We need a more radical explanation of these strange phenomena, and I think it should go something like this.[17] Our economic elites don't just encourage consumerism. Through their influence on the media and on our society's political process, they create, reproduce, and justify a pervasive and interlocking system of rules and institutions—from property rights and capital markets to contract and labor laws—that promotes growth and that, in the process, buttresses their power and privilege. A particular language of capitalism—a "discourse" of economic rationality and competition that penetrates into every nook and cranny of our economies, societies, and lives—helps us understand and abide by these rules and institutions. This language says that people maximize their pleasure from consumption and that they make decisions as if they were calculating machines, constantly weighing costs and benefits to evaluate their choices. Capitalism's language also says that our labor is a commodity to be bought and sold in a competitive marketplace. And it equates our personal identities with our economic roles in that marketplace. Think, for instance, of the intrinsically bizarre way we describe someone to a stranger. We usually say "He (or she) is an x," where "x" is the person's profession.

Taken as a whole, modern capitalism's system of rules, institutions, and language is formidably resistant to change.[18] As we saw in the last chapter, it creates chronic economic insecurity for workers. Companies can choose to invest their profits wherever they want, or not at all, and workers know that their jobs and well-being hinge on their employers' investment decisions. The system also tends to pit workers in different firms against each other, which makes it harder for them to organize to protect their collective interests. And even if people outside our society's economic elites want to make fundamental changes to the economic system, they don't really have the power to do so. Although everyone's political rights are formally the same in capitalist democracy—we all have the right to vote and to organize ourselves politically, for instance—in practice people's abilities to change things vary greatly.[19] The defenders of the economic status quo can marshal an overwhelming onslaught of attacks and ridicule against anyone who dares to publicly challenge capitalism's logic—much the way an organism's immune system launches an attack against an incoming pathogen, with its macrophages and T-cells finding, identifying, and destroying the intruder.

For the vast majority of us who sell our labor in the marketplace, our economic insecurity and relative powerlessness impel us to play by the rules. And in capitalist democracy, playing by the rules means not starting fights over big issues like our society's highly skewed distribution of wealth and power. Instead, it means focusing on achieving short-term material gains—such as bettering our contracts with our employers. Put simply, our economic elites have learned, largely through their struggles with workers in the first half of the twentieth century, to protect their status by creating a system of incentives, and a dynamic of economic growth, that diverts political conflict into manageable, largely nonpolitical channels. As long as the system delivers the goods—defined by capitalist democracy *itself* as a rising material standard of living and enough new jobs to absorb displaced labor—no one is really motivated to challenge its foundations.

We find it far easier to play by the rules if we actually believe in the legitimacy and reasonableness of the larger system that lays down those rules. We become invested in the capitalist worldview. Without it, our modern world wouldn't make much sense at all: we wouldn't know our social and economic roles, and we'd have difficulty connecting and communicating with people.[20] We realize, too, that it's senseless to challenge openly our economic system's overarching logic because we'd be challenging the source of our own paycheck—the goose that laid the golden egg, so to speak. The basic truth of this economic arrangement is crystal clear to everyone: the interests of business prevail over all others.[21]

So our economic system generates pervasive insecurity; this insecurity impels us to play by the rules; our need to play by the rules encourages us to find these rules morally legitimate; and our belief that the rules are legitimate creates a huge obstacle to changing them. For many of us, then, denial is entirely rational.

Buffeted and often terrified by our modern world's relentless flux, many of us desperately want to believe that at least the tenets of capitalism—the right to private property, the benefits of competition, and the imperative of growth—are eternally true. A cadre of experts—central bankers and economists in governments, universities, and think tanks—is only too willing to oblige by substantiating the belief structure. When a bit of anomalous data pops up, like a financial crisis or an energy shortage, these "professionals with a vested interest in the monopoly of learning," as Koestler calls them, rush to add an extra epicycle to the economic theory that underpins capitalist ideology. This keeps the central core of the worldview undisturbed and gives the whole system "scientific" legitimacy. (In reality, though, the experts often know little more about what's going on than the rest of us; and they're certainly little better than an educated layperson at seeing the future.)[22] Just like the people who held fast to Ptolemy's theory of the cosmos, we'd rather accept this extra epicycle—we'd rather have someone tell us a fanciful story—than shift wholesale to some new, untried, and unknown way of organizing our lives. Better the devil we know than one we don't. Indeed, if Koestler is right that insecurity causes people to hold tenaciously to their core assumptions and values, and if capitalism itself is a prime generator of personal insecurity, then capitalism perversely reinforces its own appeal.

We shouldn't expect any challenge to capitalism's tenets to come from the top of our social hierarchy, either. Members of our economic elite rarely have qualms about the prevailing economic worldview because it sustains their status, and because they generally believe that they've achieved that status through their superior intelligence, guts, and drive. The conviction that one's advantages are entirely fair seems virtually a condition of membership in the rarefied upper strata of our societies.

So the tacit arrangement among our elites, our experts, and the rest of us is essentially symbiotic—a mutually gratifying and self-sustaining cycle of denial and delusion. Through our acquiescence in and often active support of modern capitalism, we legitimize our elites' and experts' status and power, while those elites and experts give us an over-arching ideology of permanence, order, and purpose that lends our lives a sense of place and meaning. According to this ideology, economic growth is a panacea for all our social and personal problems. Growth equals health. Unfortunately, though, this psychological attitude is exactly the opposite of the prospective mind I advocated in chapter 1. When we're in denial, we can't think about the various paths that we might take into the future. Nor can we prepare to choose the best path when the opportunity arises. Radically different futures become liter-ally inconceivable—they are "beyond imagining," as the Israeli political scientist Yehezkel Dror describes them—in the same way that the heliocentric cosmos was inconceivable to many people prior to the Copernican revolution.[23]

To survive, let alone prosper, in our new and more dangerous world, however, we need to open our minds to the possibility of fundamental change in our lives.

Diminishing Returns

Standing on the ridge overlooking San Bernardino in November 2003, I marveled at the evidence of recovery all around me. Just weeks after fires had devastated the homes along the ridge, new telephone poles and power lines were going up, and people had already cleared rubble from their lots as they prepared to rebuild. Was this evidence of people's astonishing resilience or of their astonishing capacity for denial—or perhaps both?

As it turned out, I was traveling to find answers to this question. I'd made a quick detour to Los Angeles and San Bernardino on my way to meet the first of two people who could help me understand the implications of our world's tectonic stresses. Leaving the wrecked houses behind me, I returned to my rented car, descended the twisting highway out of the mountains, and drove ninety minutes on the busy freeways across the breadth of Los Angeles to the airport.

Three hours later I was on a flight to Albuquerque, New Mexico, to see someone whose research I'd studied for years, but whom I'd never met in person.

Joseph Tainter can be charitably described as an outlier in the scholarly world—an anthropologist fascinated by ancient Rome who until 2005 worked, of all places, at the Rocky Mountain Research Station of the U.S. Forest Service. Yet his theory about how societies evolve has had immense influence. His slim 1988 book, *The Collapse of Complex Societies*, is in its thirteenth printing in English and has been translated into multiple languages around the world, including Russian, Korean, and Kurdish.[24] It's now a classic among the subculture of academics and specialists who think about how and why societies sometimes fail.

The next morning was the Friday after Thanksgiving and a holiday. All the offices in Albuquerque's downtown were closed, and the streets around the hotel where I was staying were quiet. Many of Albuquerque's residents, I assumed, were storming suburban malls on one of the year's biggest shopping days. Tainter had generously promised me the day to chat, so we met at the hotel's front door at 10:30 a.m. A large man in his mid-fifties with a round face, a graying beard and mustache, Joe struck me immediately as a warm and deeply thoughtful person. But there was something reserved, perhaps shy, and even a bit melancholy in his demeanor. We drove to the Forest Service offices outside the city core. Entering the building through a backdoor and winding through a warren of corridors, we eventually arrived at his office—a small space packed with papers and books, but enlivened by a number of Tainter's beautiful landscape and wildlife photographs, as well as several of David Roberts's magnificent nineteenth-century lithographs of ancient ruins in the Holy Land.

Joe Tainter's theory of how societies evolve and sometimes collapse is sophisticated, but its core argument is quite straightforward. All societies face problems of various kinds over time, he points out. Some of the problems might come from outside the society, such as an attack by another hostile society or a change in the prevailing climate that reduces the rainfall needed to grow food. Others might arise from developments inside the society, like the failure of its financial system. Either way, societies often respond to their problems by increasing their complexity. A society dealing with less rainfall might build elaborate irrigation systems so it uses water more efficiently on its farms, and it might create another layer of state bureaucracy to ensure that everyone abides by water-sharing rules. In the short and medium terms, this greater complexity often produces big benefits—like more food—and most people are better off. But eventually, Tainter argues, things usually don't work out so well. "While initial investments by a society in growing complexity may be a rational solution to perceived needs, that happy state of affairs cannot last."[25]

To support this claim, he makes two very important points. First, complexity costs, and greater complexity costs more. This cost is paid in the currency of energy: from a thermodynamic point of view, as we saw in chapter 2, it takes lots of high-quality energy to keep nature's relentless tendency toward degradation and randomness—toward higher entropy—at bay. Work is needed to build an irrigation system and keep it from falling apart, just as it's needed to create and maintain an irrigation bureaucracy. "Not only is energy flow required to maintain a sociopolitical system," Tainter writes, "but the amount of energy must be sufficient for the complexity of that system."[26]

His second point is probably his most significant single contribution: a society's investment in complexity to solve its problems eventually produces "diminishing marginal returns." This simply means that at some point an additional investment in complexity gives us less benefit than the immediately previous investment. So, for example, if irrigation water is limited, at some point an additional investment in irrigation technology and water-regulation bureaucracy will produce less extra food than earlier investments. Why? Because societies almost always try *first* those solutions that are simplest and give the biggest return for the least cost, leaving for later more complex, costlier, and less effective solutions. So inevitably costs rise and returns drop over time. I've talked about this phenomenon before: drilling companies long ago found the largest and most accessible oil fields. Now, using far more sophisticated and complex technologies, they're drilling deeper in far harsher environments for smaller pools of oil. The result: a relentlessly declining energy return on investment (EROI).

In time, Tainter suggests, the benefits of greater complexity fall to zero and can even become negative—just the way they did with the relentlessly rising complexity of the Ptolemaic system, as represented by Santucci's armillary sphere. When this happens, a society is in real trouble because the population must bear complexity's ever-higher cost, even though this complexity generates no net benefit or improvement in well-being. Over time the society's resilience declines. An expanding portion of its wealth is sucked into the task of maintaining existing complexity, while its reserves to deal with unexpected contingencies fall, making it more susceptible to sudden, severe shocks from the outside. Also, subgroups within the society—like restless ethnic groups in peripheral territories—can decide that the balance of costs and benefits no longer favors them and so try to break away (this happened with some barbarian groups that inhabited the farthest reaches of the western Roman empire in its later days). In such a situation, the return on investment in complexity deteriorates "at first gradually, then with accelerated force," Tainter writes. "At this point, a complex society reaches the phase where it becomes increasingly vulnerable to collapse."[27]

As we talked in his office, two things especially puzzled me. First, I found it odd that societies would continue to increase their complexity beyond the point that it benefits them. "Why do societies tend to go beyond the optimum with complexity," I asked. "Why can't they just stop when we get to the point of diminishing marginal returns?"

"Because the problems don't stop," Tainter replied. "The answer is that the universe is perverse, and the challenges never end. But there's another reason, too, that I've expressed more clearly in my recent articles: complexity tends to be adopted as a short-term solution. People don't think about long-term consequences when they adopt more complex behaviors or institutions, but ultimately these things do endure for the long term. They become entrenched, and their long-term costs can be very high.

"Our responses to the September 11, 2001, attacks are a perfect example," he continued. "What were our first responses? We increased the complexity of behavior. We increased the scope of government. We created new agencies and merged old ones. And no one really thought about the long-term cost of it all. These were simply things that had to be done."

So, I thought, some societies are like juggernauts. As they try to deal with their immediate problems, they develop inexorable momentum toward greater complexity, and this complexity brings with it long-term unintended costs. But that raised the second aspect that puzzled me: Tainter's theory focuses almost exclusively on how societies respond to their problems, not on the changing nature of the problems themselves. Are the problems our societies face today—global warming, rising energy costs, mega-terrorism, emergent diseases, and the like—getting harder to solve?

Tainter agreed that our societies are faced with more problems simultaneously and that the pace at which these problems unfold seems to be increasing. In his terms, we are dealing with a growing number of "concatenating" problems—that is, problems that are chained together and that reinforce each other in entirely unexpected ways. "When you look at concatenating problems," he said, "you're looking at increasing complexity and costs just to keep things stable. In these circumstances, problem solving often just maintains the status quo—costs go up, but the benefits remain level. So it's likely that some problems simply won't be addressed, at least not to everyone's satisfaction. People won't understand why they aren't being addressed, which means they'll become dissatisfied, and the government will progressively lose its moral authority or legitimacy. You're also likely to see declining living standards as an increasing proportion of the GDP is allocated to solving these problems."

Our conversation eventually concluded. As we rose to leave, I stopped to admire the Roberts lithographs on his office's walls. One in particular caught my attention—a depiction of a mighty stone gate at the entrance to what looked like an ancient temple. The gate was rectangular, not arched, and it had elaborate carvings on its sides and top. The most striking feature was something unintended by its designers and builders: the central keystone in the horizontal beam across the top was displaced downward—so far downward, in fact, that it appeared to be hanging precariously in midair. I asked Tainter about the scene and remarked on the keystone.

"That's the Temple of Bacchus. It's in Baalbek, one of the greatest collections of Roman ruins in the entire Mediterranean."

"And where's Baalbek?" I asked, feeling a little embarrassed that I didn't know.

"In Lebanon, at the north end of the Bekaa Valley. I'd love to see these ruins, but I doubt if I ever will."

"Why not?" I asked.

"Baalbek is controlled by the Hezbollah, and Americans aren't welcome."

The gate to the Temple of Bacchus in Baalbek

Panarchy

Five months later, in early April, I was back in the southern United States. This time, though, my destination was a spot on Florida's Gulf Coast.

Leaving the Tampa airport, I drove north on Interstate 75 through central Florida. The flat landscape of Sumter County, one of the poorest parts of the state, rolled by for countless kilometers—its scrub forest relieved only occasionally by vegetable and cattle farms. Eventually I crossed into Marion County and turned left at Ocala. The land transformed into lush, rolling pastureland, home to hundreds of posh horse farms that have made the county a center of thoroughbred breeding—their paddocks and riding rings marked off by neat white fences. Then, turning left again, I followed a dead-straight two-lane road across a vast plantation of pine trees on my way to the coast.

I was going to see Crawford "Buzz" Holling, one of the world's great ecologists. I'd followed Holling's research for almost a decade, but I'd met him for the first time only a year earlier at a workshop in Georgia. Today we'd arranged to have lunch in his hometown of Cedar Key, a popular vacation spot, artists' colony, and refuge for retirees. After meeting at Pat's Red Luck Café on the town's touristy waterfront, we sat down to eat deep-fried grouper, home fries, and coleslaw—and talk about resilience.

Buzz Holling is a kind and gracious man, with a shock of white hair and a warm smile. Born in Toronto and educated at the University of Toronto and the University of British Columbia, he worked for many years as a research scientist for the government of Canada, where he pioneered the study of budworm infestations in the great spruce forests of New Brunswick. Later, as an academic researcher and eventually as director of the International Institute for Applied Systems Analysis in Austria, he created powerful mathematical models to explain the ecological phenomena he saw in the field. Using these models, he achieved major breakthroughs in understanding what makes complex systems of all kinds—from ecosystems to economic markets—adaptive and resilient.

Since the early 1970s, Holling's research has attracted attention in disciplines ranging from anthropology to economics. His papers have been distributed like samizdat through the Internet, and Holling himself has become something of a guru for an astonishing number of very smart people studying complex adaptive systems. Some of these researchers have coalesced into an international scientific community called the Resilience Alliance, with over a dozen participating institutions around the world. Although Holling is now retired from his last academic position at the University of Florida, he's still terrifically vigorous and focused on furthering the Resilience Alliance's work.

Holling and his colleagues call their ideas "panarchy theory"—after Pan, the ancient Greek god of nature.[28] Together with Joe Tainter's ideas on complexity and social collapse, this theory helps us see our world's tectonic stresses as part of long-term global process of change and adaptation. It also illustrates the way catastrophe caused by such stresses could produce a surge of creativity leading to the renewal of our global civilization.

Panarchy theory had its origins in Holling's meticulous observation of the ecology of forests. He noticed that healthy forests all have an *adaptive cycle* of growth, collapse, regeneration, and again growth. During the early part of the cycle's growth phase, the number of species and of individual plants and animals quickly increases, as organisms arrive to exploit all available ecological niches. The total biomass of these plants and animals grows, as does their accumulated residue of decay—for instance, the forest's trees get bigger, and as these trees and other plants and animals die, they rot to form an ever-thickening layer of humus in the soil. Also, the flows of energy, materials, and genetic information between the forest's organisms become steadily more numerous and complex. If we think of the ecosystem as a network, both the number of nodes in the network and the density of links between the nodes rise.

During this early phase of growth, the forest ecosystem is steadily accumulating capital. As its total mass grows, so does its quantity of nutrients, along with the amount of information in the genes of its increasingly varied plants and animals. Its organisms are also accumulating mutations in their genes that could be beneficial at some point in the future. And all these changes represent what Holling calls greater "potential" for novel and unexpected developments in the forest's future.[29]

As the forest's growth continues, its components become more linked together—the ecosystem's "connectedness" goes up—and as this happens it evolves more ways of regulating itself and maintaining its stability. The forest develops, for example, a larger number of organisms that "fix" nitrogen—converting the element from its inert form in the air to forms that plants and animals can use—in the specific amounts and in the specific places needed. It becomes home to more worms, beetles, and bacteria that break down the complex organic molecules of rotting plants into useful nutrients. And it produces more negative feedback loops among its various components that keep temperature, rainfall, and chemical concentrations within a range best suited to life in the forest.

Over time as the forest matures and passes into the late part of its growth phase, the mechanisms of self-regulation become highly diverse and finely tuned. Species and organisms are progressively more specialized and efficient in using the energy and nutrients available in their niche. Indeed, the whole forest becomes extremely efficient—in a sense, it effectively adapts to maximize the production of biomass from the flows of sunlight, water, and nutrients it gets from its environment. In the process, redundancies in the forest's ecological network—like multiple nitrogen fixers—are pruned away. New plants and animals find fewer niches to exploit, so the steady increase in diversity of species and organisms slows and may even decline.

This growth phase can't go on indefinitely. Holling implies—very much as Tainter argues in his theory—that the forest's ever-greater connectedness and efficiency eventually produce diminishing returns by reducing its capacity to cope with severe outside shocks. Essentially, the ecosystem becomes less resilient. The forest's interdependent trees, worms, beetles, and the like become so well adapted to a specific range of circumstances—and so well organized as an efficient and productive system—that when a shock pushes the forest far outside that range, it can't cope. Also, the forest's high connectedness helps any shock travel faster across the ecosystem. And finally, the forest's high efficiency makes it harder for it to realize its rising potential for novelty. For instance, the extra nutrients that the forest ecosystem has accumulated aren't easily available to new species and ecosystem processes because they're fully expropriated and controlled by existing plants and animals. Overall, then, the forest ecosystem becomes rigid and brittle. It becomes, as Holling says, "an accident waiting to happen."[30]

So in the late part of the growth phase of any living system like a forest, three things are happening simultaneously: the system's potential for novelty is increasing, its connectedness and self-regulation are also increasing, but its overall resilience is falling. At this point in the life of a forest, a sudden event such as a windstorm, wildfire, insect outbreak, or drought can trigger the collapse of the whole ecosystem. The results, of course, can be dramatic—large tracts of beautiful forest can be obliterated. The ecosystem loses species and biomass and in the process much of its connectedness and self-regulation.

But the effects on the ecosystem's overall health may be very positive. A wildfire in a mature forest creates open spaces that allow new species to establish themselves and propagate; it destroys infestations of disease and insects; and it converts vegetation and accumulated debris into nutrients that can be used by plants and animals that reestablish themselves after the fire. The organisms that survive become much less dependent on specific, long-established relationships with each other. Most important, collapse also liberates the ecosystem's enormous potential for creativity and allows for novel and unpredictable recombination of its elements. It's as if somebody threw the forest's remaining plants, animals, nutrients, energy flows, and genetic information into a gigantic mixing bowl and stirred. Once-marginal species can now capture and exploit newly released nutrients, and genetic mutations that were a bane to survival can now be a boon.

And because the system is suddenly far less interconnected and rigid, it's far more resilient to sudden shock. This is a perfect setting for the forest's plants and animals to experiment with new behaviors and relationships—a pollinator species like a bee or wasp will try gathering nectar from a type of flower it hadn't previously visited, or a carnivore might try killing and eating a different kind of prey. If such experiments fail, the damage is less likely to cascade across the entire system.

In these ways the forest ecosystem reorganizes and regenerates itself, quite possibly in a very new form. Put simply, the catastrophe of collapse allows for the birth of something new. And this cycle of growth, collapse, reorganization, and rebirth allows the forest to adapt over the long term to a constantly changing environment. "The adaptive cycle," Holling writes, "embraces two opposites: growth and stability on one hand, change and variety on the other."[31] It's at once conserving and creative—a characteristic of all highly adaptive systems.

Holling and his colleagues use a three-dimensional image to represent the relationship between a system's rising potential and connectedness and its declining resilience. The shape looks like a distorted figure eight or infinity symbol floating in space.[32] In the foreground is the growth phase—a curve that moves upward as the system's potential and connectedness increase. At the same time, the curve moves forward in three-dimensional space—toward the observer—as the system's resilience declines.

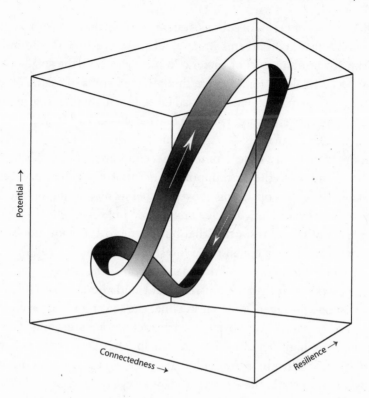

The adaptive cycle can be represented in three dimensions.

Holling and his colleagues call this part of the adaptive cycle the "front loop." It represents a process of incrementally rising complexity. At the top of this curve, the system collapses. Things then happen fast as the system descends into the "back loop," where it undergoes a rapid process of reorganization before beginning once more the slow process of growth.

There's one more essential part to Holling's theory. He argues that no given adaptive cycle exists in isolation. Rather, it's usually sandwiched between higher and lower adaptive cycles. For instance, above the forest's cycle is the larger and slower-moving cycle of the regional ecosystem, and above that, in turn, is the even slower cycle of global biogeochemical processes, where planetary flows of materials and elements—like carbon—can be measured in time spans of years, decades, or even millennia. Below the forest's adaptive cycle, on the other hand, are the smaller and faster cycles of sub-ecosystems that encompass, for instance, particular hillsides or streams. In fact, adaptive cycles can be found all the way down to the level of bacteria in the soil, where the smallest and fastest cycles of all are found. Here things happen on a tiny scale of millimeters or even microns, and they can take place in minutes or even seconds. So the entire hierarchy of adaptive cycles—what Holling and his colleagues call a panarchy—spans a scale in space from soil bacteria to the entire planet and a scale in time from seconds to geologic epochs.

This brings us to the most important point of all for our purposes in this book: the cycles operating above and below play an important role in the forest's own adaptive cycle. The higher and slower-moving cycles provide stability and resources that buffer the forest from shocks and help it recover from collapse. A forest may be hit by wildfire, for example, but as long as the climate pattern across the larger region that encompasses the forest remains constant and the rainfall adequate, the forest should regenerate. Meanwhile, the lower and faster-moving cycles are a source of novelty, experimentation, and information. Together, the higher and lower cycles help keep the forest's collapse, when it occurs, from being truly catastrophic. But for this healthy arrangement to work, these various adaptive cycles must be at different points along that figure-eight loop. In particular, they mustn't all peak at the top of their growth phases simultaneously. If they do—if they are "aligned at the same phase of vulnerability," to use Holling's phrase— they will together produce a much more devastating collapse, and recovery will take far longer, if it happens at all.[33] Should a wildfire hit a forest at the same time as the regional climate cycle enters a drought phase, the forest might never regenerate.

Panarchy theory helps us understand how complex systems of all kinds, including social systems, evolve and adapt. Of course, it shares similarities with other theories of adaptation and change. Its core idea—that systems naturally grow, become more brittle, collapse, and then renew themselves in an endless cycle—recurs repeatedly in literature, philosophy, religion, and studies of human history, as well as in the natural and social sciences.[34] But Holling has done much more than just restate this old idea. He has made it far more precise, powerful, and useful by distinguishing between potential, connectivity, and resilience; by identifying variations in the system's pace of change as it moves through its cycle; and by describing the roles of adjacent cycles in the grand hierarchy of cycles.

As Holling and I sat on the veranda of the Red Luck Café on a blissful April afternoon, watching brown pelicans glide through the air along the waterfront, I felt fortunate to have a chance to ask questions directly of someone whose ideas had influenced me so much. Holling embodies something truly rare—the kind of wisdom that comes when an enormously creative, perceptive, and courageous mind spends a half-century studying a phenomenon and distilling its essential patterns. So, taking full advantage of the opportunity, I encouraged him to expand on many aspects of panarchy theory, filling gaps in my understanding and giving me nuance and perspective that only he could provide. As we came to the end of our conversation, I asked him a question that had been on my mind since our meeting a year before. In Georgia, I remembered, he'd been adamant that humanity is at grave risk.

"Why do you feel the world is verging on some kind of systemic crisis?"

"There are three reasons," he answered. "First, over the years my understanding of the adaptive cycle has improved, and I've also come to better understand how multiple adaptive cycles can be nested together—from small to large—to create a panarchy. I now believe that this theory tells us something quite general about the way complex systems, not just ecological systems, change over time. And collapse is usually part of the story.

"Second, I think rapidly rising connectivity within global systems—both economic and technological—increases the risk of deep collapse. That's a collapse that cascades across adaptive cycles—a kind of pancaking implosion of the entire system as higher-level adaptive cycles collapse, which causes progressive collapse at lower levels."

"A bit like the implosion of the World Trade Center towers," I offered, "where the weight of the upper floors smashed through the lower floors like a pile driver."

"Yes, but in a highly connected panarchy, the collapse doesn't have to start at the top. It can be triggered at the microlevel or the macrolevel or somewhere in between. It's the tight interlinking of the adaptive cycles across the whole system—from the individual right up to the level of the global economy and even Earth's biosphere—that's particularly dangerous because it increases the likelihood that many of the cycles will become synchronized and peak together. And if this happens, they'll reinforce each other's collapse."

"The third reason," he continued, "is the rise of mega-terrorism—the increasing risk of attacks that will kill huge numbers of people and produce major disruptions in world systems. I'm not sure why mega-terrorism has become more likely now. I suppose it's partly a result of technological changes and the rise of particularly virulent kinds of fundamentalism. But I do know that in a tightly connected world where vulnerabilities are aligned, such attacks could trigger deep collapse—and that's particularly worrisome.

"This is a moment of great volatility and instability in the world system. We need urgently to do what we can to avoid deep collapse. We also need to figure out how to exploit the opportunity provided by crisis and collapse when they occur, because some kind of systemic breakdown is now almost certain."

Overextending the Growth Phase

On my drive back to Tampa, I thought about panarchy theory and its connection to my visit to the ridge overlooking San Bernardino five months before. The devastation along the ridge showed what happens when people don't let the natural cycle of collapse and regeneration unfold at its own pace and in its own way. The longer people sustain a social, economic, or ecological system in its growth phase, the sharper, harder, and more destructive its ultimate breakdown will be. The residents of the ridge, like all people living in the interface zones between cities and forests across California, had controlled and regulated the forest around them. They'd been especially diligent about preventing wildfires. So the forest's potential—specifically, its potential energy—continued to grow as underbrush, dead branches, twigs, and duff accumulated into a thick and explosively combustible layer across the landscape. Then the unthinkable happened: a combination of drought and insects killed all the trees. At that point only a random spark was needed to create a conflagration that swept away the dead forest and all the beautiful homes nestled within it.

People's intervention badly distorted the adaptive cycle of this integrated system of humans and forest. Under natural conditions, the ridge's forest might have seen a series of small and localized fires over a period of decades. Instead, people's increasingly aggressive efforts to prevent wildfire even as inflammable debris piled up around them overextended the cycle's front loop—that is, its phase of growth of potential and connectedness. So the system's eventual collapse was vastly worse than it need have been. Put simply, efforts to regulate the system—efforts undertaken apparently without much understanding of how they affected the system's resilience—produced the conditions for catastrophe.

Holling's panarchy theory thus gives us a lens through which we can better understand an astonishing range of phenomena—not just the fires of San Bernardino.[35] It especially helps us make sense of the many issues raised so far in this book.

We can see the danger of the tectonic stresses in a new light if we think of humankind—including all our interactions with each other and with nature and all the flows of materials, energy, and information through our societies and technologies—as one immense social-ecological system. As this grand system we've created and live within moves up the growth phase of its adaptive cycle, it's accumulating potential in the form of people's skills and economic wealth. It's also becoming more connected, regulated, and efficient—and ultimately less resilient. And finally, it's becoming steadily more complex, which means it's moving further and further from thermodynamic equilibrium. We need ever-larger inputs of high-quality energy to maintain this complexity. In the meantime, internal tectonic stresses—including worsening scarcity of our best source of high-quality energy, conventional oil—are building slowly but steadily.[36]

So we're overextending the growth phase of our global adaptive cycle. We'll reach the top of this cycle when we're no longer able to regulate or control the stresses building deep inside the global system. Then we'll get earthquakelike events that will cause the system's breakdown and simplification as it moves closer to thermodynamic equilibrium.

Panarchy theory also helps us better understand another critically important phenomenon: the denial that prevents us from seeing the dangers we face. Our explanations of the world around us—whether of Earth's place in the cosmos or of the workings of our economy—move through their own adaptive cycles. When a favorite explanation encounters contradictory evidence, we make an ad hoc adjustment to it to account for this evidence—just like Ptolemy added epicycles to his explanation of the planets' movements. In the process, our explanation moves through something akin to a growth phase: it becomes progressively more complex, cumbersome, and rigid; it loses resilience; and it's ripe for collapse should another, better, theory come along.

We often invest enormous mental energy to maintain a perspective on the world that's at variance with reality—that's far from intellectual equilibrium, so to speak. But today bits of anomalous evidence—from data on the melting of Greenland's ice cap to reports of steadily falling discovery of new oil fields—are piling up around us like the combustible materials around the houses in the San Bernardino mountains.

DISINTEGRATION

I T ALMOST LOOKS as though the Romans could bend rock, I thought. I was standing on one of the sides of the channel of an ancient Roman aqueduct. The channel, now dry, had once carried millions of litres of water. As I studied its design, I again marveled at Roman engineering's singular blend of pragmatism, elegance, and genius. But there was something peculiar about the material that formed the channel's interior wall—the side closest to the once-flowing water. It wasn't masonry or concrete. It was rock. And it didn't look like any rock I'd seen Romans use before. It didn't consist of individual pieces carved and chiseled to fit tightly together. Instead, it was a seamless whole, and it curved smoothly around a bend as the aqueduct turned a corner. Most peculiar of all, the material was striated, with countless lines running lengthwise along the curved wall's surface, almost as if it were petrified wood that had been steamed and bent for the purpose. How could this be?

Some months after my visit to the Forum in Rome, I was continuing my search to understand the links between energy and society—and ultimately to understand how civilizations fail and recover. And this search had brought me to the region of Languedoc in southern France to visit one of the ancient world's great architectural marvels, the Pont du Gard. Erected to carry an aqueduct across the Gard River, it is perhaps the finest example of Roman bridge engineering and, quite simply, one of the most beautiful structures humans have ever built.

The Pont du Gard, just northeast of Nîmes, France

That day in the fall of 2003, the bridge looked particularly grand, as the late-afternoon sun highlighted it against the dark oak, olive, and arbutus forests of the surrounding valley.

But I wasn't visiting the Pont du Gard to marvel at another Roman structure made of rock. I was there because a few days before I'd visited the Institut d'Art et d'Archéologie on the campus of the University of Paris in Nanterre, a Paris suburb. In a small, spartan cafeteria, I'd talked to Sander van der Leeuw, an archaeologist who studies how people and the natural environment interacted in ancient Mediterranean civilizations. He encouraged me to go to the Pont du Gard, and my later review of both his research and the voluminous studies of his colleagues helped me understand the connection between the bent stone and our circumstances today—especially our rapidly changing energy situation.[1]

During the empire's heyday, Roman cities and towns used huge amounts of water for public baths, fountains, artisans' workshops, and the dwellings of the rich, so engineers laid down aqueducts throughout the empire—in the process building countless bridges, siphons, and tunnels.[2] The Pont du Gard's three tiers of voussoir arches stand almost fifty meters high, and the original bridge extended nearly five hundred meters across the narrow but deep valley below. Erected in about five years sometime in the middle of the first century CE (not long before the Colosseum was built), and made of fifty thousand metric tons of stone, it was a key part of a system that carried water from the Eure and Plantery springs near the present-day town of Uzès to the Roman provincial settlement of Nemausus, which has since become the city of Nîmes. After decades of testing and refinement, the aqueduct eventually carried some forty thousand cubic meters of water a day, enough to fill sixteen Olympic swimming pools. Yet over its entire fifty-kilometer length—as it twisted and turned across very rough terrain—it descended only thirteen meters, or barely twenty-five centimeters per kilometer. The best engineers today would have trouble equaling this construction feat.[3]

The water from the springs at the origin of the aqueduct carries a heavy load of minerals, especially calcium carbonate. As this water coursed through the aqueduct over five centuries, some of the dissolved material was deposited on the channel's sides. Each year another layer was laid down, and over time the material accumulated into the thick deposits of limestone on the channel's interior walls.[4] So it was indeed rock that I could see along the channel's sides—very similar, in fact, to the travertine used in the Colosseum—and it did indeed bend smoothly around corners, having accreted on the sides one microscopic particle at a time. Along some sections of the Pont du Gard, these deposits are almost half a meter thick on each wall, an accumulation that must have greatly restricted the water's flow.[5]

These hundreds of thin layers of limestone—the striations that I'd observed—are like tree rings, because they give us an indelible record of the aqueduct's long history. And archaeologists of the Roman period have studied them carefully. They've learned that when the Romans were properly maintaining the aqueduct, including the various reservoirs along its length, the water it carried ran clean and clear, and the material deposited on the walls formed a light gray layer of firm, pure limestone.[6] When the Romans neglected maintenance, debris washed into the reservoirs, and plant roots penetrated the channel itself. The aqueduct's water then carried a heavy load of dirt and organic material, and the deposits on the channel's sides became brown and soft—almost like a sponge.

Limestone deposits of the Nîmes aqueduct provide a record of its past. (The interior surface of the aqueduct channel is immediately to the right of the sunglasses, which are resting on the striated deposits; the channel itself is at the far right.)

The interior of the aqueduct, in other words, tells a story about year-by-year changes in the capacity and competence of Roman administration—and about the disintegration of Roman power—in southern Gaul in the waning days of the western empire.[7]

Looking carefully at the bent rock, I could see clearly the difference between earlier and later deposits. Reaching down, I felt the brown material that had been deposited most recently. It crumbled under my fingers.

Checkerboard Landscape

The story of that crumbling rock begins with the relationship between Romans and nature. It turns out that they dominated the natural environment of the Mediterranean basin and surrounding regions, just as we dominate Earth's biosphere today.[8] Romans deforested, drained, irrigated, terraced, and overgrazed hundreds of thousands of square kilometers of land.[9] "Indeed, one of the most striking parallels between the

Roman exploitation of much of Western Europe and the comparatively recent exploitation of North America," write van der Leeuw and
his colleague Bert de Vries, "is the checkerboard of roads and drainage
ditches that divides both landscapes into square miles."[10]

A checkerboard landscape—what an evocative image. It gives us a
feeling for the monumental investment the Romans made in organizing
their territory. And how remarkable that this pattern has persisted
through the millennia. But why did the Romans make this investment?
The thermodynamic theory I outlined in chapter 2 suggests an answer:
societies can sustain themselves—especially their complex hubs, like
their towns and cities—only with copious inputs of high-quality energy,
and to get this energy they must dominate and organize the territories
that supply it.

The Romans vigorously organized their territory to aid both commerce and tax collection. The late empire's main tax was imposed on
agriculture—it was essentially a direct extraction of energy from the
countryside. Farmers paid their taxes mainly in silver coin or in foodstuffs that tax collectors converted to coin, and this coin was then often
exchanged for gold coin.[11] The gold was essentially distilled solar
energy. It took a staggering amount of labor—powered by food energy
from the sun—to dig the gold ore from the ground, and then it took
additional energy—in the form of wood energy also from the sun—to
smelt it. The gold could easily be carried long distances and used to buy
grain or human and animal labor far from the field where the farmer's
crops first captured the sun's rays. But the Romans couldn't effectively
administer agricultural taxes if they didn't know where the farmland was,
who owned it, and whether the soil was good.[12] So when they took control of a new region in Western Europe, they would dispatch a professional class of surveyors—the "agrimensores"—to map the land and
carve it into square-mile blocks.[13]

Mapping the landscape wasn't enough, though. The conquerors
needed to control it too, which the Romans did in the most efficient
way possible. Instead of setting up an entirely new administrative
system, they exploited a new territory's existing social and technological infrastructure, especially its cities and towns.[14] Julius Caesar and
other great Roman generals in fact found it hard to hold on to territories that didn't have enough cities and towns.[15] When they captured a
territory, they immediately garrisoned its urban nodes with Roman

troops. Then they co-opted the local urban elites to help manage—and in particular to tax—the surrounding countryside by granting them various types of association with the empire, sometimes including citizenship.[16] Finally, they integrated the new regions into the larger Mediterranean economy by improving a region's existing roads and building new links to the empire's superb interregional system of highways. In time the newly conquered region was suffused with the empire's common institutions—its language, writing, coinage, and administrative and judicial systems.

The empire's roads served many purposes. Not only were they corridors for moving troops rapidly, but they also served as the empire's fiber-optic cables, carrying the vital information—mail, contracts, administrators' reports, tax assessments, government orders, military intelligence, and maps—that allowed it to function and solve its problems.[17] Flows of information and flows of energy reinforced each other: information was needed to organize the extraction of high-quality energy from the countryside, but high-quality energy was also needed to produce, transfer, and store huge quantities of information.

Roads also let Romans move their products.[18] Along the empire's highways and byways, unsettled land was cleared, drained, and irrigated, and new estate farms were set up to make foodstuffs for export—especially wine and olive oil. These farms and their processing and storage facilities were often industrial in size and organization: in southern France, for instance, archaeologists have discovered the remnants of huge oil and wine cellars, including one that could have held 400,000 modern-day bottles of wine. Factory-size pottery workshops also made tens of thousands of the large jars, or amphorae, used as transportation containers.[19]

In this way, the empire became an early version of today's globalized world—a highly interconnected economy from Britain to Egypt, in which individual regions could specialize in producing things they made best. This helped boost the economy's overall resilience: variations in local geography, climate, and soil across the Mediterranean basin allowed farmers to cultivate diverse crops, so a shortfall in one area could often be addressed by imports from other areas where harvests were better.[20] Economic integration spurred economic growth, especially in the provinces. Partly for these reasons, during the first century CE southern Gaul became, in the words of one expert, "the most productive, most industrialized, and indeed most civilized province in the empire."[21]

All these practices—regimentation of the landscape, co-optation of urban elites, expansion of road networks, and integration of regional economies—helped the Romans extract the energy needed to sustain their empire's growing complexity. "For societies powered by solar energy," Tainter and his colleagues note, the main way to increase wealth is "to control more of the earth's surface where solar energy falls."[22] As van der Leeuw and de Vries put it, the Roman empire is best understood as a gigantic system that "slowly structures an increasing area and a growing number of people into a coherent whole. In return, it draws energy (human and animal) and raw materials (food, minerals and water) from the area thus colonized."[23] Wealth was transferred from the provinces back to Italy, mainly to Rome, as taxes, provincial tributes, and customs duties, and as rents from both imperial properties and provincial estates owned by the Roman aristocracy.[24] And much like our modern global economy, the empire had its own growth imperative. "[It] was dependent on continued expansion for survival," continue van der Leeuw and de Vries. "More and more resources—including raw energy in the form of slaves— were brought from an increasingly distant periphery to the center and transformed into a growing range of artifacts and other objects."[25]

As Rome moved into the second century CE, the system became increasingly unstable. For one thing, just as Tainter argues, the cost of a larger and more complex empire eventually rose faster than the benefits it produced.[26] The empire lacked natural frontiers to its north beyond the Rhine and Danube and to the east in what is now called the Middle East. So up to the time of Emperor Trajan's death in 117 CE, Rome's generals often protected each new territory they seized by capturing adjoining territories farther afield.[27] As these territories were incorporated into the empire, Rome's frontiers became longer, as did its lines of communication from the center of power. Also, advanced knowledge about administration, military organization, and weapons flowed to the people in the new territories, boosting their power relative to Rome. So Rome spent ever more resources keeping in line ever more distant and often restless territories and preventing incursions by barbarians—such as Celtic, Germanic, and Gothic tribes—across ever more remote frontiers.[28] Unfortunately, the croplands in the cold climates of Northern Europe produced little surplus wealth or energy to provide for their own control and defense, let alone to sustain the empire's rising core complexity.[29]

The Roman empire maintained a permanent army big enough to

respond to a range of contingencies—the first state to do so until modern times.[30] Considering the empire's expanse, it wasn't an excessive force, averaging around thirty legions through much of the second century, or a total of about 300,000 to 350,000 men. Luckily, until the reign of Marcus Aurelius (161–180), the frontiers were relatively stable. All the same, the army remained stretched dangerously thin along the empire's ten-thousand-kilometer perimeter.[31] And even this limited force imposed a huge burden on Rome because it had to be provisioned, armed, and paid.[32] Simply feeding it required something in the neighborhood of 200,000 metric tons of grain each year.[33] Also, between the middle of the first century CE and the early part of the third, a soldier's pay nearly tripled, to encourage recruitment and ensure the legions' loyalty. "Army pay was by far the largest item in the budget," writes the historian A. H. M. Jones, and the effect of the increases on the empire's finances was, in his view, "catastrophic."[34]

There were other problems, too. Widespread deforestation not only raised the price of fuel wood—a principal source of energy—but also dried the landscape in many places, left hillsides exposed and vulnerable to erosion, and increased flows of silt that plugged irrigation systems. Intensive, industrial-scale use of farmland damaged cropland and lowered grain yields.[35] Also, as the empire grew in size and complexity, the amount of information needed to keep it orderly and coherent grew exponentially, as did the difficulty of managing and moving the information across the empire's territory.[36] And the Mediterranean economy's tighter integration turned out to be a mixed blessing: at the same time that it helped to increase the economy's overall resilience by diversifying the empire's sources of key products, the connectivity made highly specialized local producers— often surviving on razor-thin profits—more vulnerable to distant shocks.[37]

Imperial costs continued to rise. A larger empire required more complex administration, so bureaucracy and officialdom ballooned. And not only did 200,000 people remain on the grain dole in the city of Rome, but to maintain the state's legitimacy as slums overflowed with people migrating from the countryside, emperors supplemented the dole with free oil, pork, and wine.[38] They also boosted spending on extravagant spectacles and public works. The cost of building and maintaining public baths, in particular, must have been huge: by the fourth century, Rome alone had one thousand baths, and many were unbelievably ostentatious.[39]

Energy Subsidy

For a long time, Rome escaped its mounting problems by spending the treasure—especially the accumulated gold—of conquered territories.[40] When Rome captured the king of Macedonia's treasury in 167 BCE, it was able to lift taxes on Italy. Gold from the kingdoms of Pergamon and Syria bankrolled huge jumps in Rome's budget in 130 and 63 BCE. And when Caesar conquered Gaul, so much gold flooded the Mediterranean economy that its price fell 36 percent. "By the last two centuries BCE," writes Tainter, "Rome's victories may have become nearly costless, in an economic sense, as conquered nations footed the bill for further expansion."[41]

The captured gold—much of it held in the Temple of Saturn in the Forum—was essentially a massive energy subsidy of the empire's rising complexity and reach.[42] And because this gold represented far more energy than Rome had invested to get it, the empire enjoyed a highly favorable energy return on investment—or EROI. But after the subjugation of Egypt by Octavian (later Augustus) in 30 BCE and Trajan's seizure of Dacian gold and Transylvanian gold mines in 106 CE, Rome ran out of rich kingdoms to plunder, and its EROI shifted dramatically downward. Abruptly Rome found itself overextended: the empire's now vast territory, much of it captured and till then administered using the proceeds of conquest, had to be run using the solar energy from its annual food output, a flow barely enough to cover the empire's normal needs.[43] Emperors and their administrators soon found they had no buffer—no surge capacity—to cope with nasty surprises. The situation became acute in the years following 165 CE, when Marcus Aurelius faced converging challenges, including repeated poor harvests, savage barbarian attacks that penetrated as far as Italy itself, and a devastating plague that killed one-quarter to one-third of the population.

Rome's emperors had few options to cope with financial deficit. Raising taxes or creating new ones was seldom politically and administratively feasible, even when inflation eroded the value of revenues.[44] Instead, emperors sold off state property, confiscated the aristocracy's assets, and turned to the age-old but self-defeating strategy of debasing the currency. Between 50 and 200 CE, the silver content in Roman coinage dropped almost 50 percent, and by the year 269 it had plummeted to almost zero.[45]

Rome was thus locked into a food-based energy system that left little room for maneuver. Without the scientific skills, institutions, and culture to make the jump to a new kind of energy system—like one based on fossil fuels—the empire was trapped in a thermodynamic crisis that exacerbated its underlying brittleness.[46] This story has been illustrated, in a truly fascinating way, by archaeologists working in the Rhône valley, not far from the Pont du Gard.[47]

By the time Julius Caesar finished pacifying all Gaul in 51 BCE, much of the region's substantial population—estimated at nearly 6 million—was concentrated in towns in the Rhône delta and the lower Rhône valley, from the present-day city of Marseilles west to Montpellier, and northward through Arles and on to Orange. Because this area was relatively urbanized, the Romans colonized it first. Then they moved north up the Rhône valley to less inhabited territory. To guide their colonization, in the last decades BCE the emperor laid out a master plan for roads, settlements, farms, drainage canals, and ditches. We're lucky today to have a remarkable record of this plan: a tax and property map engraved around 77 CE on a marble tablet, now preserved in a museum in Orange. This map shows an astonishing ten thousand square kilometers of the valley divided into a neat grid of fifty-hectare plots. Much of the unsettled land in the north, especially a zone called the Tricastin, was parceled out as retirement sinecures to ex-legionnaires. Better to keep former soldiers happy on the farm than have them rebelling in the provinces.

It's clear from this map, from archeological surveys of nearly a thousand settlements in the region, and from analysis of aerial photographs that the Romans organized the Rhône valley—in particular the Tricastin—to grow agricultural products for export to the regional and Mediterranean economies. It's also clear that the surveyors who laid out the grid of ditches and property boundaries had an urban fetish for straight lines and right angles not suited to the land's natural features. Their drainage grid, in fact, cut directly across the natural flow of rivers and streams at an angle of forty-five degrees, so storms often washed out the ditches or filled them with dirt and rubble.[48] Because the components of the system were highly interdependent, the failure of a single section of ditch could cascade to cause a far larger breakdown.[49]

In Buzz Holling's terms, the Romans' agricultural and irrigation system in this region was intricate, tightly connected, highly regulated—

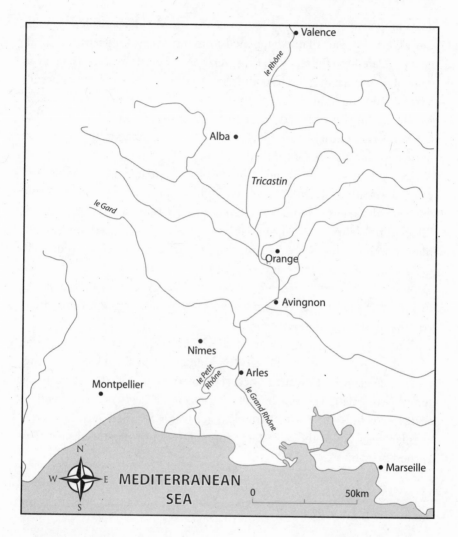

The Rhône valley

and brittle. It worked fine as long as there was enough labor—or energy—to maintain the entire system. But the advent of peace in the empire in the first century CE slowed the flow of ex-legionnaires to the more distant parts of the Rhône valley. More important, by the second century CE, agricultural colonization in the region came to a standstill, and the Romans began abandoning hundreds of farms, first in the southern part of the Rhône valley and delta and later in the north. This rural crisis occurred throughout Gaul and across much of the empire, continuing and deepening into the third and fourth centuries. We don't really know its causes, but the fact that it happened in many places suggests that local factors related to climate, soil, or government weren't key. More likely a general downturn in the interconnected Mediterranean economy hurt all farms producing foodstuffs for export.[50] Also, the great plague of the late second century and general civil instability in the third century probably left farms short of labor. Finally, as I'll show in a moment, the rising tax burden on agriculture forced many farmers to leave their land.

In the Tricastin, the abandonment of farms and decline in population were devastating. There weren't enough people to repair the region's intricate drainage system, so it fell apart. Erosion worsened, ditches filled with debris, and more farms—especially marginal ones with poorer soil and fewer road connections to markets—were abandoned. Over time, similar events unfolded across much of the western empire. Economic integration among regions declined and information flows dwindled. Towns and cities went their own way and turned inward, their command of the countryside fading.[51]

The deposits on the inside of the Nîmes aqueduct tell this story. During these centuries, the city of Nemausus lost control over the aqueduct's route. Powerful local landowners along the structure's length—normally responsible for keeping the aqueduct maintained—instead cut holes into the channel to get water for their farms. And as regular cleaning of the aqueduct stopped, its once-crystal-clear water turned brown.

Ruthless Extraction

For over a millennium in Western culture, Rome's collapse has been an emblem of social catastrophe, one often used as a cudgel in political debate. When people don't approve of a particular social, political, or economic trend, they'll often assert that it caused Rome's demise. So explanations have proliferated. In 1984 the German historian Alexander Demandt listed more than two hundred different explanations for Rome's fall that he found in the historical literature since 1600—from epidemics, plutocracy, and the absence of character to vainglory.[52]

Perhaps it's rash, then, to add another one to the list. Still, recent work by archaeologists, economic historians, and complexity theorists gives fresh insight into what happened. And their story, which has immense relevance to our situation today, comes down to this.

Because energy is a society's master resource, when Rome exhausted its energy subsidies from its conquests—when it had to move, in other words, from high-EROI to low-EROI sources of energy—it faced a critical transition. And, at least in the Western part of the empire, it didn't make this transition successfully.[53] It couldn't sustain the cost and complexity of its far-flung army, ballooning civil service, hungry and restless cities, elaborate information flows, and intricate irrigation systems, like the one in the Tricastin. Not that it didn't try. Rome's prodigious effort to save itself by putting in place a system to aggressively manage its energy problem was simultaneously one of history's greatest triumphs and tragedies. It was a triumph because, for a while at least, the effort reversed what seemed like the empire's inexorable decline; but it was ultimately a tragedy because it didn't address the empire's underlying problem—complexity too great for a food-based energy system—and was thus bound to fail.

After years of warfare, plague, and weak harvests, Marcus Aurelius left the state treasury almost empty when he died in 180 CE. For nearly a century after his death, turmoil, civil war, and barbarian attacks racked the empire, in no small part because of a chronic fiscal crisis. As barbarian tribes coalesced into powerful federations to the north, Rome's control of its frontier territories disintegrated and many were abandoned.[54] Travel and trade became unsafe, and literacy and record keeping plummeted; commerce declined. Although tax revenues were static or declining, government costs continued to go up as emperors tried to secure their power by expanding the dole, increasing the size of the army, boosting soldiers' pay, and holding more games and spectacles.[55] The empire verged on collapse.

The downward spiral stopped only when the emperors Aurelian (270–275) and Diocletian (284–305) revamped Rome's finances, administration, army, and tax system. Diocletian, in particular, introduced complex and harsh measures to extract more energy from the land. Two centuries of debasing the currency had so eroded money's value that Rome now insisted that it receive the bulk of its taxes not in money but in kind—in grain, meat, and the like. But because surpluses of foodstuffs couldn't be stored easily, Diocletian realized that Rome needed to match its annual intake of such goods to its requirements. The result was

Rome's first rudimentary state budget. At the beginning of each financial year on September 1, officials calculated the government's needs for the coming year and divided the total by the number of units of land in the empire. In this way, the tax rate could be adjusted every year according to Rome's needs.[56] Meanwhile, other officials conducted new censuses and resurveyed the countryside to identify and measure every scrap of potentially productive land. "The tax rate was established from a master list of the empire's resources," write Tainter and his colleagues, "broken down province by province, city by city, field by field, household by household. Never before had the state so thoroughly penetrated its citizens' lives."[57] The historian Chris Wickham concurs: "Taxation dominated the economy and was the economic foundation for the state. Nothing in the late Roman economic system escaped the state's embraces."[58]

From the point of view of the landowner, the tax system was cruelly rigid. Taxes stayed the same even if the land was poor or the harvest failed. Villages, towns, and cities were responsible for taxes levied on their surrounding populations, so if one farmer didn't pay his share, others had to make up the difference. Even uncultivated land was taxed, and if the owner couldn't be found, the land was compulsorily assigned to someone who could farm it and pay the taxes.[59] Widows and children inherited tax obligations. The result, writes the historian Edward Luttwak, was "a perfected system of taxation-in-kind, which ruthlessly extracted the food, fodder, clothing, arms, and money needed for imperial defense from an empire which became one vast logistic base."[60]

Ruthless extraction was essential, because complexity and costs continued to climb. Diocletian stabilized the frontiers by building more roads and fortresses along the empire's boundaries. He enlarged the army to around 600,000 soldiers, or somewhere between fifty-six and sixty-eight legions—a force that required conscription of nearly 100,000 new soldiers every year.[61] To reduce the likelihood of revolt, he divided provinces and split the imperial administration into four parts, multiplying a bureaucracy that became increasingly segmented into specialized units. To supply the state and army's essential needs, he and his successors made many occupations—like milling, baking, weaving, dyeing, and shipping and transport—compulsory and hereditary. And because the debasing of currency continued to produce inflation, Diocletian promulgated in 301 his famous Edict on Prices, stipulating the price of hundreds of goods and services.

All these innovations brought relative prosperity and calm to the empire for much of the fourth century.[62] But the cost was extreme and unsustainable. The tax rate was now set annually, which meant emperors could finally raise taxes easily, and before long they did indeed raise them—a lot. Between 324 and 364, taxes apparently doubled, and the result was a sudden deepening of the rural crisis.[63] From Africa to Gaul to Asia, vast numbers of farmers, especially those with small plots or marginal cropland, abandoned their farms. Some sought the protection of large landowners, while others moved to cities. Those who remained on the land had steadily less food for their families and no longer the wherewithal to protect their cropland from long-term damage.[64] Malnutrition and outright starvation appear to have caused rural death rates to surge beyond birthrates, so the empire's population never recovered from the plagues of the second and third centuries.[65] And as peasant populations fell, finding enough conscripts for the army became difficult. Still, with famers chafing under the burden of rising taxes, the army was an increasingly vital source of the emperor's power and legitimacy.[66]

By the fifth century in the West, the empire was literally burning through its capital—its productive farmland and its peasantry. Peasants deserted their lands, so power and wealth were increasingly concentrated in the hands of large landowners, who then used their influence to evade taxes.[67] By 400, fewer than a dozen senatorial families owned most of Gaul and Italy, and they largely controlled the reins of power in the West.[68] As state finances deteriorated, public services like roads, bridges, aqueducts, and the postal service broke down. The army's quality declined, and information flows were attenuated, which sapped the empire's ability to identify and solve its problems. As the empire grew weaker, it started to disintegrate from the periphery toward the core. Barbarians saw opportunities to attack along the West's long and poorly defended frontier, and an apathetic and overtaxed peasantry often did little to resist. The loss of territories to barbarians further undermined the empire's finances, debilitated the army, and invited more attacks.[69]

The western empire was again caught in a downward spiral toward collapse and simplification, and this time there was no rescue. Rome's strategy, Tainter writes, had been "to respond to a near-fatal challenge in the third century by increasing the size, complexity, power, and costliness of the primary problem-solving system—the government and its army." The imperial government's aim hadn't been further conquest but just maintenance of the status quo. In the end, as costs outstripped revenues, the western empire "could no longer afford the problem of its own existence."[70]

Holland Times Ten

The western Roman empire couldn't make the transition from high-EROI to low-EROI sources of energy. Today, our societies are headed toward a similar transition as oil becomes harder to find. Sometime in the 1960s the United States crossed a critical threshold when its EROI for domestic petroleum extraction started to fall, and it's likely that since then just about every other oil-producing region in the world has crossed the same threshold (often it takes a while for data to show clearly that the threshold has been crossed).[71] Very few people—certainly not our society's leaders—grasp the significance of this change, yet it's of epochal importance. It marks the beginning of a shift from our modern industrial civilization to some other kind of civilization.

We can't yet say what form this new civilization will take, but we can be fairly certain that compared with our experience over the century and a half since the industrial revolution, energy will become far more costly as nonconventional and renewable sources replace cheap oil. The price rise won't be steady and linear: we'll see sharp spikes and dips as the global economy tries to adjust. Even an average increase in real energy costs of just 2.5 percent each year—a rate we've consistently exceeded in recent years—will compound into a tenfold increase in a century.

Can we get through this transition wisely and safely? Not if we refuse to understand its implications and simply continue what we're doing now. In Buzz Holling's terms, we're busily extending the growth phase of the adaptive cycle of our planetary economic, ecological, and social system. In the process, this planetary system is becoming steadily more complex, connected, efficient, and regulated. Eventually it will become less resilient; it may, in fact, have already started to lose resilience.

A number of factors drive these changes. First, the desperate need of companies, economies, and societies to maximize performance and productivity forces them to steadily boost their organizational and technological complexity, their internal efficiency and regulation, and their speed of production and transport of materials, energy, and information.

Also, as the world economy expands relative to the size of Earth's resource base and biosphere, we have to use resources and energy far more efficiently and manage our interactions with nature with ever greater care—and this means progressively more elaborate technologies, procedures, regulations, and institutions. Based on current trends, global output of goods and services will quadruple from $60 to $240 trillion (in 2005 dollars) by 2050.[72] If we're going to keep such a gargantuan economy humming—and if we're going to avoid simultaneously wrecking the planet's environment—we'll need everything from high-tech energy and water conservation programs to huge bureaucracies that find and punish the people and companies that emit too much carbon dioxide. And finally, as our EROI declines in coming decades, we'll need far more sophisticated technologies and organizations to scavenge small pockets of oil from all over the world and to pull together lower-quality energy from a myriad of solar, wind, and geothermal generating plants. We've seen this particular outcome before: Diocletian, remember, had to impose an immensely elaborate and intrusive tax regime to gather tiny amounts of solar energy from farms spread across the Roman empire.[73]

In short, in coming decades our resource and environmental problems will become progressively harder to solve; our companies, organizations, and societies will therefore have to become steadily more complex to produce good solutions; and the solutions they produce—whether technological or institutional—will have to be more complex too.[74]

Today's Holland gives us a hint of what this future might be like.[75] One of the world's most crowded countries, Holland has a heavily industrialized, energy-intensive, high-consumption economy, and its people must constantly fight back the sea to survive on their small patch of territory—much of it indeed reclaimed from the sea.[76] Over the centuries, the Dutch have responded by putting in place astonishingly complex systems of technology and social regulation. These have included block-by-block urban residential committees to prevent flooding, detailed laws to maximize efficient use of land, and of course an intricate system of dikes, canals, and pumping stations.[77] As Holland has become progressively wealthier, more crowded, and more hemmed in by resource and environmental pressures, the regulations and technologies have become steadily more intricate and costly.

But if we end up with a global society and economy like Holland's, would that really be so bad? After all, the Dutch live very well. Sadly, even the enormous complexity of today's Holland won't be remotely adequate for the host of planetary challenges we're going to have to address soon, like climate change and worsening shortages of high-quality energy. We'll have to create a global society that I've come to call "Holland times ten," with vastly more sophisticated, pervasive, and expensive rules and regulatory institutions than anything the Dutch live with today. Do we really want such a future for ourselves and our children?

And even if we do, can we really create it? First of all, Holland is in some ways an inadequate example. It's a small, ethnically homogeneous society with relatively low economic inequality, a deeply rooted culture of collaboration, and a citizenry that's receptive to social policies intended to change people's behaviors.[78] These are hardly features of our world as a whole. Also, today's Holland maintains its comfortable lifestyle by importing energy, food, and natural resources from far beyond its boundaries, and by expelling much of its wastes, such as its carbon dioxide, outside its boundaries too—Holland's carbon dioxide ends up traveling in the atmosphere around the planet.[79] Humanity as a whole, though, can't get its resources or expel its pollution beyond Earth's boundaries.

More important, as our global social-ecological system moves through the growth phase of its adaptive cycle—toward a Holland-times-ten future—it's losing resilience because of processes I've outlined in this book. Capitalism's constant pressure on companies to maximize efficiency tightens links between producers and suppliers; reduces slack, buffering, and redundancy; and so makes cascading failures more likely and damaging. As well, capitalism's pressure on people to be more productive and efficient drives them to acquire hyperspecialized skills and knowledge, which means they become less autonomous, more dependent on other specialized people and technologies, and ultimately more vulnerable to shocks (remember how most of us were so ill equipped to deal with the 2003 blackout). Meanwhile, worsening damage to the local and regional natural environment in many poor countries is fraying ecological networks and undermining economies and political stability. And finally pressure is increasing *within* both rich and poor societies too—from tectonic stresses like demographic imbalance, growth of megacities, and widening income gaps.

All these factors are creating the overload condition I talked about in chapter 5—just at the moment when we're entering an epochal shift from high-EROI to low-EROI sources of energy. Because it takes energy to create and maintain complexity and order, and because energy will become steadily more expensive, we'll find it steadily harder to implement complex solutions to our complex problems.

Indeed, in a world of far higher energy costs, a Holland-times-ten global system is likely impossible. Even today's globalized economy won't be viable, because it takes too much energy to keep it running. As energy prices rise, we'll first see cutbacks on long-distance travel and trade.[80] Instead of becoming increasingly "flat" as barriers to commerce and economic integration disappear—as some commentators, such as the *New York Times* columnist Thomas Friedman, suggest—the world will become more regionalized and even hierarchical because manufacturing, commerce, and political power will shift to countries with relatively good access to energy.[81] Eventually those of us in rich countries will have to change many things in our societies and daily lives—not just the machines we use to produce and consume energy but also the work we do, our entertainment and leisure activities, how much we travel in cars and airplanes, our financial systems, the design of our cities, and the ways we produce our food (because our current agricultural practices consume a huge amount of energy).

The growth phase we're in may seem like a natural and permanent state of affairs—and our world's rising complexity, connectedness, efficiency, and regulation may seem relentless and unstoppable—but ultimately it isn't sustainable. Still, we find it impossible to get off this upward escalator because our chronic state of denial about the seriousness of our situation—aided and abetted by powerful special interests that benefit from the status quo—keeps us from really seeing what's happening or really considering other paths our world might follow. Radically different futures are beyond imagining. So we stay trapped on a path that takes us toward major breakdown.

The longer a system is "locked in" to its growth phase, says Buzz Holling, "the greater its vulnerability and the bigger and more dramatic its collapse will be." If the growth phase goes on for too long, "deep collapse"—something like synchronous failure—eventually occurs. Collapse in this case is so catastrophic and cascades across so many physical and social boundaries that the system's ability to regenerate itself is lost.[82] The forest-fire described in chapter 9 shows how this happens: if too much tinder-dry debris has accumulated, the fire becomes too hot, which destroys the seeds that could be the source of the forest's rebirth.

Holling thinks the world is reaching "a stage of vulnerability that could trigger a rare and major 'pulse' of social transformation." Humankind has experienced only three or four such pulses during its entire evolution, including the transition from hunter-gatherer communities to agricultural settlement, the industrial revolution, and the recent global communications revolution. Today another pulse is about to begin. "The immense destruction that a new pulse signals is both frightening and creative," he writes. "The only way to approach such a period, in which uncertainty is very large and one cannot predict what the future holds, is not to predict, but to experiment and act inventively and exuberantly via diverse adventures in living."[83] We'll see shortly that exuberant experimentation is essential to social resilience.

Motivation, Opportunity, and Framing

A buildup of pressure from multiple stresses, a rapid-fire series of shocks, and a weakening of resilience can combine to push a society over the edge, as indicated in chapter 5. From the third century on, these processes pummeled the Roman empire: contemporary writers noted that everything seemed to be going wrong everywhere at the same time—that Rome had to cope with a seemingly never-ending barrage of barbarian attacks, assassinations, internal unrest, troop revolts, famine, plague, and financial crisis.[84] Rocked by these foreshocks, the empire started to crack apart.

As we go through our own EROI transition in coming decades—in a world far along the growth phase of its own adaptive cycle and under enormous pressure from multiple stresses—what can we expect to happen?

In coming years, I believe, foreshocks are likely to become larger and more frequent.[85] Some could take the form of threshold events—like climate flips, large jumps in energy prices, boundary-crossing outbreaks of new infectious disease, or international financial crises. In poor countries where environmental, population, and economic stresses are already severe and social capacity to manage them remains low, we'll probably see a steady increase in outbreaks of civil violence—including riots, insurgency, guerrilla war, ethnic cleansing, and terrorism. Severe spikes in energy costs, especially, could trigger violence in megacities in poor countries, while a combination of higher energy costs and shifts in climate could cause food shortages that spur guerrilla war in rural areas already suffering from serious cropland and water scarcity. Such events could culminate in larger social earthquakes, as multiple shocks topple regimes in zones of geopolitical importance, like the Middle East, South Asia, and East Asia. If this turmoil is unchecked, world order could disintegrate in stages—from the poorest countries at its periphery to the richest countries at its core, much as happened in the western Roman empire. Even early on, though, instability is likely to penetrate into rich countries as terrorism and urban unrest surge where there are large, ghettoized concentrations of underemployed and unemployed young men.[86]

This scenario may be scarily plausible, but it's not inevitable. For example, more stresses acting on our societies don't automatically translate into more instability and violence. Since the 1960s, social scientists have made remarkable strides in understanding the causes of civil violence.[87] They've found that two factors dominate: *motivation* and *opportunity*. One or other isn't enough by itself; lots of both are needed to generate upheaval.

People usually won't participate in group violence unless they're powerfully motivated to do so, because the cost of participating might be very high—they could die, and so could their loved ones. In most cases people will join only if they feel that a particular group or governing regime is treating them extremely unfairly. Generally, too, they must believe that violence offers a genuine opportunity to improve their lot. This means they must believe that the balance of power favors them—that in a violent struggle with the prevailing group or regime they have enough power to eventually get their way.

Because so much rests on these two apparently simple factors, we might think it's pretty easy to predict when and where civil violence will occur. Alas, both motivation and opportunity are a messy mixture of objective and subjective factors. For instance, whether people believe their situation is unfair depends not only on objective facts—their income relative to others, their political and economic freedoms, and the like—but also on their subjective notion of what's just and unjust. Also, whether they believe they have an opportunity to succeed in their uprising depends not only on their access to the objective instruments of power—like money, explosives, and guns—but also on subjective factors like group identity.

A potentially rebellious group will feel more powerful if it believes that other people and groups share its identity—that they are part of its "we" group, so to speak—and could therefore be allies.[88]

The tectonic stresses I've considered in this book negatively affect people's objective situation. Energy shortages and environmental damage dislocate people's lives and can make them poorer, while widening demographic and income gaps allow people to make invidious comparisons between themselves and others. But something else— something subjective—must be added to the mix if these changes are to motivate people to become violent: leaders must *frame* these changes in a particular way for their followers. This means they must tell a story that connects the dots of the available evidence in a way that convinces people that they're being treated unfairly and can do something about it. This story is, essentially, a mini social theory that draws on people's shared values, myths, and symbols to give them a satisfying explanation of their unsatisfying situation. It tells them that their situation is unjust, it identifies who's to blame, and it explains what they can do to ensure that they'll triumph over the injustice they're experiencing.[89]

Power Shift

Dislocated lives, worsened poverty, and wider income gaps affect the motivation to participate in violence by providing fodder for extremist leaders. They create in people general feelings of frustration and anger that leaders can shape and focus—through clever framing—into powerful resentments against governments or specific groups. Income gaps are especially good fodder because people care far more about their relative than their absolute status. People on the losing side of the gaps— or who strongly identify with those who are—can be made to feel profoundly humiliated.[90]

Unfortunately, as I showed in chapter 8, the income gap between the world's rich and poor will widen quickly in coming decades: if current trends continue, the gap will more than double in the next forty years. And the world's connectivity makes the reality of the widening gap inescapable. Our increasingly global society is intimately linked together by fiber-optic cables, air travel, and trade, and suffused with information from TV, videos, and the Web. Millions know what they're missing, whether it's something right next door or on the other side of the planet.[91] And millions—especially in Africa, the Middle East, and Latin America—feel they're being left behind: they see no clear route to success in the modern global economy or even a real role for themselves. Many of these people also live under semi-authoritarian regimes that don't allow for peaceful expression of discontent, so they're left with no political options other than passivity or revolt.

As the income gap widens, our global society is becoming increasingly polarized between rich and poor classes. Political theorists going back to Aristotle have argued that a large and successful middle class is important to peace in any society because it moderates conflict between rich and poor and tempers political extremism.[92] But the world's middle class is in precipitous decline. According to the World Bank economist Branko Milanovic, over 77 percent of the world's people are now poor (with a per capita income below the Brazilian average), while about 16 percent are rich (with incomes above the Portuguese average), which leaves less than 7 percent in the middle.[93] Relative to rich countries, almost all the middle-income countries in 1960 dropped into the ranks of the poor by 2000, and the club of rich countries became largely Western.[94] The poorest countries will almost certainly stay where they are—at the bottom of the world's economic ladder. "Unless there is a remarkable discontinuity with the patterns of development that have lasted during the past half-century (and possibly longer)," Milanovic writes, "the likelihood of escaping from the bottom rung is almost negligible." This is not a prescription for a stable world order.

Many Western journalists, commentators, and policy makers heatedly reject the argument that poverty and income gaps help cause civil violence. To them this argument is little more than a thinly disguised effort to blame the world's dominant economic system—globalized capitalism—for phenomena like international terrorism or instability in poor countries. And they point out that many terrorists—including the members of Al Qaeda who carried out the 9/11 attacks—are well educated and come from relatively comfortable backgrounds.

It's true that extremely poor people are often quite fatalistic about their poverty, so they generally don't feel really angry about it, which weakens their motivation to join in civil violence of any kind. They're also so busy surviving that they don't have much time to think about whether their situation is unfair or to organize themselves to make it better. And because they usually have limited educations, they have limited knowledge of things beyond their immediate situation. The people who are most likely to participate in violent groups have more time and resources to reflect on their circumstances, more information to know that things could be better, and more opportunity to become radicalized by finding and exchanging ideas with others who share their views.

Yet we shouldn't jump to the conclusion that poverty and income gaps are irrelevant to the incidence of violence. First of all, social scientists have found that poor societies are much more prone to civil violence.[95] Even though we don't yet fully understand how these two factors are linked, we can be pretty sure they're linked somehow. Also, it's a mistake to assume that because many of the participants in certain types of civil violence—such as terrorism—aren't desperately poor, poverty can't motivate them. People who are relatively educated and wealthy can still identify strongly with people who are deprived—less fortunate members of their ethnic or religious group, for instance— and be very motivated by this deprivation. They can feel what one expert has called "vicarious humiliation."[96] Finally, it's important to take account of all the participants in acts of civil violence: although the leaders of insurgent or terrorist groups are often relatively educated and well off, it turns out that the people who do most of the work, take most of the risks, and actually launch attacks are in many cases much poorer.[97]

Dislocated lives, worsening poverty, and income gaps can affect the opportunity for violence too. In a poor country, they can undermine the economy and in turn the financial and the military strength of the governing regime, which can weaken support for the regime among powerful elites. They can also erode the regime's moral authority or "legitimacy" in the eyes of the general public: if people are becoming poorer or feel they're falling behind everyone else, they usually conclude that their government is doing a bad job. As the regime loses elite and popular support, potentially rebellious groups may think their chances for a successful uprising are getting better—especially if they receive help from discontented elite factions.[98]

But the factor that has the biggest impact on the likelihood of violence in our world today is the second of this book's two multipliers— what some experts have called the "power shift."[99] Technological change has generally boosted the capabilities of small groups and individuals relative to large institutions and governments. Sometimes this power shift makes our lives better, as when the Internet helps citizens engage in democratic debate about contentious public issues. But sometimes it makes our lives worse, because one kind of power that's diffusing to small groups is an extraordinary capacity to destroy.[100] Put simply, technological change is allowing fewer people to kill larger numbers of people more quickly than ever before.

This trend is particularly visible in poor countries that have been flooded with small arms and light weapons, including assault rifles, rocket-propelled grenade launchers, and small but deadly land mines. In a long list of poor countries, these weapons have given militias, ethnic groups, political factions, and gangs the opportunity to wreak havoc. Some countries have become trapped in perpetual cycles of attack and counterattack, and in places like Somalia, Sierra Leone, Liberia, and eastern Congo the result has been a virtual collapse of government authority. Organized crime backed by heavily armed militias has been quick to fill the vacuum, and criminal activity—from the production of phony passports to trade in illegal diamonds—is an important source of revenue for insurgents and terrorists operating elsewhere. Both Al Qaeda and Hezbollah, for instance, have exploited the West African diamond trade to finance their operations.[101]

New information technologies are also boosting violent groups' power relative to governments. Satellite phones, the Internet, and portable computers allow groups—including transnational terrorist networks—to share information on weapons and recruiting tactics, arrange surreptitious fund transfers across borders, and plan attacks around the planet. Such groups can now communicate almost invisibly using the enormous power available in everyday computers—a common laptop today has as much computational punch as all the computers available to the U.S. Defense Department in the mid-'60s—and state-of-the-art encryption software that can be downloaded as freeware off the Web.

Developments like this are certainly worrisome, but they're not even close to the greatest concern. Violent small groups will achieve their ultimate technological leverage when they start using weapons of mass destruction, including chemical, biological, radiological, and nuclear weapons. Most specialists believe it's just a matter of time before they do so, because the world can't bottle up forever the technologies or knowledge needed to make the devices.

I earlier mentioned the risk that terrorists could get hold of enough nuclear material, especially highly enriched uranium (HEU), to make an atomic bomb. Given the huge amount of HEU sitting in insecure stockpiles around the world, and the tiny fraction needed to make a crude but very effective bomb (with roughly the explosive power of the one that obliterated Hiroshima), some experts say the probability of a nuclear attack by terrorists in the next decade is greater than 50 percent.[102] They have been especially alarmed by one shocking discovery: Western intelligence agencies uncovered a sophisticated and illicit international network, originating in Pakistan, set up expressly to transfer nuclear-weapons technologies and materials to such rogue countries as North Korea, Iran, and Libya.[103] Many experts now believe that, based on this intelligence, nuclear materials and technologies are leaking into international criminal syndicates, from where they may be transferred— for a price—to terrorist groups.

Other possibilities are on the horizon. For example, molecular biologists have lately learned how to create viruses using genetic codes available on the Internet and DNA from companies that sell custom-made genetic material.[104] The codes of devastating diseases, including smallpox and the 1918 Spanish flu, are now publicly available, and the machines that allow biologists to mutate the genes of common viruses to make them more lethal, or even to construct viruses from scratch, can be found in laboratories around the world.[105]

Because of these technological trends, small groups of people will soon be able to humble entire nations. We haven't even begun to grasp the implications of this possibility for our economies and democracies. What, really, would it mean for the liberty, democracy, and prosperity of Western societies if one or more nuclear bombs exploded in the West's great cities? We already know one thing, though: as enormous destructive power diffuses down the social hierarchy from governments to groups and even to individuals, protecting our societies becomes exponentially harder. We have to pay attention to—and potentially control—smaller and smaller details of our security environment.

The intelligence and defense agencies of countries used to focus their resources mainly on tracking, assessing, and responding to threats from large agglomerations of military force, like armies massed along borders, naval fleets at sea, and missiles in their silos. Today such agencies have to pay much more attention than before to small groups and even single persons. A meeting of a half-dozen people in a restaurant in Peshawar, Pakistan, may seal the fate of Washington, D.C.; and if a few people can destroy a city, every cluster of people is potentially a mortal threat. In other words, as destructive power diffuses downward, Western intelligence agencies will argue that they need a progressively higher-resolution picture of what's going on around us, and they need this picture not just of things happening in our own countries but also of events all over the world.

The task of gathering, processing, and interpreting this information is almost incomprehensibly demanding. And even if we can identify the genuine threats our countries face around the world, we almost certainly can't afford a police and military force large, complex, and sophisticated enough to respond to them all. The problem is particularly acute for the United States, the world's remaining superpower and a target of anger practically everywhere. For all intents and purposes, the country now has an infinitely long defensive perimeter: it must now defend itself and its allies against attacks that could come at any time in any corner of the planet, from the Golden Gate Bridge at home to the nightclubs of Bali in Indonesia and the tourist hotels of Mombasa in Kenya. It doesn't have the option of ignoring this threat, because successful attacks have a huge symbolic effect. They demonstrate the weakness of American military and political authority, which exposure, of course, simply invites more attacks.

Rome, remember, found itself fatally overextended. Given its energy and financial resources and manpower, its frontiers were too distant and too long, and it eventually succumbed to relentless barbarian attacks. Soon the United States—faced with declining EROIs and the steadily increasing destructive power of our modern-day barbarians—may also be stretched beyond its breaking point.[106]

A Shattered Sphere

We're awed by Rome's buildings because they were built to inspire awe. The empire derived its authority partly from its fabulous engineering—and from its everyday display of precision and power with mere rock. Engineering ideology helped justify the construction of the Pont du Gard, just as it justified the Colosseum. That the aqueduct was built in a remote corner of southern Gaul only reinforced its underlying message of timeless accomplishment: we have the power and the will, the Romans were saying, to do the work of gods wherever we choose.[107] But today another, very different and even contradictory, message—a message about the ultimate limits of human power—hides inside the aqueduct, where most people don't see it. When Romans no longer properly maintained the aqueduct and its water turned brown, the layers of limestone along its interior became soft. So when I touched this limestone and it crumbled under my fingers, I touched the story of Rome's disintegration.

Some people would say that it's absurd to draw a parallel between Rome's fate and the prospects for today's world. Far from seeing the disintegration of world order, they'd say, we're actually seeing its progressive reinforcement and increasing coherence, as global trade advances and as countries (with some notable exceptions) sign treaties to manage global problems from genocide to climate change. And, they would go on, despite high-profile conflicts in places like Iraq and Afghanistan—and an upsurge in Islamic terrorism—violence has, in reality, steadily declined around the world for a decade, as war between Cold War proxies has ended, democracy has taken root, and the United Nations and other organizations have intervened to stem conflict in poor countries.[108]

This is a dangerously simplistic view. Not only does it downplay the reality that dozens of societies around the world are still experiencing vicious civil conflict, but it also extrapolates a few positive trends linearly into the future. However, complex systems don't advance in a straight line. They develop with progress and setbacks, thresholds and flips, and sudden shifts in direction. In human societies, periods of relative calm that might seem permanent are sometimes interrupted by extraordinary turmoil, when deep stresses combine to release their energy with sudden force.

Today, just as in the late Roman empire, deep stresses are rising and system resilience is declining. Just as was true then, too, the coherence of world order depends critically on the economic, political, cultural, and military might of a single superpower.[109] The foundation of this might is access to abundant energy. "More than in any other modern nation," writes the eminent energy specialist Vaclav Smil, "the United States has acquired its power and influence largely through its extraordinarily high use of energy."[110] And America survives, as the ancient city of Rome did, on lifelines of energy from distant regions.

As American leaders grasp how the shift from a high-EROI to a low-EROI world jeopardizes their country's dominance, they'll do exactly what Rome's leaders did: they'll use every means—including, when necessary, force—to organize and control the world's territory to permit the extraction of energy. In the process, they'll run headlong into other energy-hungry societies trying to do the same thing—in particular, India and China, two rising giants without remotely enough energy at home to satisfy their ravenous appetites.

The likely result: widespread conflict over energy resources, especially since many of the world's large remaining reserves of oil are in geopolitically unstable regions, like the Middle East and Central Asia.[111] Some analysts dispute oil's importance as a factor in the 1991 and 2003 wars against Iraq, but no one can dispute the fact that the Middle East is a major focus of American foreign and military policy because it has a lot of oil. America's need for overseas energy forces it to ingratiate itself with corrupt, authoritarian, and often unstable regimes in regions rife with ancient conflicts. The people of these countries frequently despise their governments and, because the U.S. supports the governments, they despise the U.S. too. In September 2001, we all learned how dangerous such hatred can be.

So in the next decades, while the U.S. desperately tries to extract energy resources around the planet and likely gets into fights with other countries as a result, it will also have to fight another battle spawned partly by its thirst for energy—an endless battle along an infinite frontier against a largely unseen and protean enemy.[112] Modern networks of terrorists and insurgents often aren't scale-free. They don't have static, highly connected hubs—like a cluster of leaders at the top of an organizational hierarchy—that can be identified, targeted, and destroyed with potentially devastating effect on the overall network.[113] Instead, their connections are loose, limited in number, and constantly changing, depending on immediate tactical needs.[114] The people in such networks are bound together mainly by radical ideologies that give them only the most general framing and injunctions for action. So Al Qaeda, in the wake of heavy pressure from the United States and its allies, has mutated from a coherent global organization into a motivating symbol or potent idea—that of Al Qaedaism and its resistance to Western oppression of Muslims.[115] This means that the U.S. "war on terrorism" can't be war in any conventional sense. It's more like a worldwide guerrilla conflict, in which the enemy shifts its tactics as necessary, chooses where to strike at will, and then disappears into a vast crowd of passive supporters, all deeply resentful of the U.S., its allies, and their policies and power.[116]

In the spring of 2003, at the beginning of my journey for this book, I sat in the Roman Forum and reflected on America's military triumph in Iraq. At the time, the U.S. vice president, secretary of defense, and their advisers were supremely confident that America could achieve its military and political interests in both the country and region. This confidence seemed justified: after all, the American victory over conventional Iraqi forces had been swift and utterly decisive. But the guerrilla war that followed has proved bloody and intractable, and America has gradually learned a rude lesson in the limits of modern military power.[117]

The lesson has far broader implications. It's one bit of anomalous data that, together with many others, contradicts the explanation offered by conservative elites, especially in the United States, of how things are supposed to work in our world. The assumptions that energy will be endlessly abundant, that economic growth can continue forever, and that American military dominance is unchallengeable are essential parts of today's armillary sphere—that smoothly functioning machine of cycles within cycles, of intermeshed bands, that gives us an explanation of our world's order. Yet despite the miracles of the market, energy constraints are worsening. Global warming fundamentally challenges capitalism's growth imperative. And too many people in the world's distant corners aren't awed by America's military power any more. The armillary sphere's bands are starting to break. At some point the sphere will shatter, as will its reassuring predictability and order. When this happens, we need to be ready.

CATAGENESIS

THE BLACKENED FOREST and wrecked houses on the ridge on each side of me, devastated by the terrible wildfire a month earlier, weren't nearly as sinister as the dust storm that was slowly blocking my view of San Bernardino as it sucked precious topsoil off nearby hillsides and distributed it far and wide. I knew the storm covered a huge area of Southern California: earlier that day, while I was driving up the Interstate thirty kilometers to the west, the same gritty miasma enveloped the highway and brought traffic to a screeching halt.

The fires of San Bernardino told me about the dangers of negative synergy, and they showed me how our efforts to keep the world around us stable can lead to catastrophe. The aftermath of the fires also said something about what tomorrow might bring. I was watching the wind randomize the land's wealth.

Could this be an image of our future, I wondered—a future in which disintegration and entropy triumph over order? Will our incaution, hubris, and stupidity transform things we should hold dear—our natural environment, engineering marvels, political institutions, and our culture and great art—into dust?

No matter whether our future brings disintegration, renewal, or something in between, we can be sure that the road in front of us won't be straight and that our grandchildren's world will look starkly different from ours today. The famous claim that we have reached the end of history—that capitalist democracy will become the universal and final form of social order—will look truly silly in retrospect.[1] We in the West, anesthetized by materialism and egged on by a self-satisfied intelligentsia, may have convinced ourselves that our way of life is the apex of economic and political achievement and that, within its bosom, we can sustain some kind of endless plateau of hedonistic satisfaction. In reality, though, history isn't close to ending. It has barely begun.

True, in the past half century Western societies have been astonishingly creative, adaptive, and prosperous. But they've also been astonishingly lucky. They avoided, as much by chance as by design, a catastrophic nuclear war with the Soviet Union; they've had access to a vast pool of petroleum in the Middle East, even as reserves dwindled at home; and they've enjoyed and exploited a stable and productive natural environment. This luck has given us ample room for denial—so far.

Luck can't last forever. In coming decades, oil is almost certain to become far scarcer, Earth's climate will likely become warmer and its weather more unstable, and chances are good that terrorists will import weapons of mass destruction into one or more of our major cities. All the same, we shouldn't be fatalistic: to some extent we can choose among the plausible futures in front of us. In the past, we usually haven't done this very well because we haven't really understood that our challenge isn't to preserve the status quo but rather to adapt to, thrive in, and shape for the better a world of constant change.

We typically use a two-stage strategy to preserve some semblance of the status quo, as I indicated in chapter 1: first, we deny our problems, and then we reluctantly try to manage them. When we deny our problems, we can't possibly choose our future. Sometimes neglected problems do fix themselves, but we can't expect such an outcome with many of the problems—like global warming—we're facing now. Yet the current wisdom is that we should pander to denial. Don't depress yourself or other people, we're told, with negative and scary messages. But there's one big problem: we're not going to crack through the hard shell of chronic denial by downplaying the dangers we face.

Our second strategy for keeping things more or less the same—a strategy we reluctantly adopt only when denial is clearly not working—is management. The goal here is to keep our problems from becoming so bad that we have to significantly change our lifestyles. Managing our problems is often formidably difficult, because they're developing inside exceedingly complex technological, social, or natural systems—

like Earth's climate or the international economy. Often, we don't really know how such systems work, nor can we predict how they'll behave, as they sometimes flip between modes of behavior that are radically different.[2] Also, when we manage challenges like our tightening energy supply or the persistent instabilities in the global financial system, we usually have to make our technologies, procedures, and institutions progressively more complicated and often (in the process) less resilient. And finally, when it comes to our most vexing and serious problems, our efforts at management often fail completely because vested interests block or divert policies that genuinely address the problems' underlying causes.

Consider our response to global warming. If we're going to manage our greenhouse-gas emissions, we will have to make a host of astonishingly difficult transitions in the next couple of decades.[3] Technologically, we'll need to revamp everything from our cars to our methods of making concrete and growing rice. Institutionally, we'll need to set up a vast superstructure of scientific, monitoring, financing, and enforcement mechanisms that spans the globe and affects every person and organization's life. It will probably be the most elaborate and intrusive system of global institutions humankind has ever developed, so we might reasonably be wary of implementing it. All the same, if managing our greenhouse-gas emissions is a key goal for our societies—and it's entirely reasonable that it should be—we need to begin the transitions right away: our emissions are now soaring, they will affect our environment for centuries (so every year we delay we're locking in more environmental damage in the future), and it looks as if Earth's climate is already changing fast.[4]

Yet in every economy in the world, the price of hydrocarbon-based energy fails to incorporate even a tiny fraction of global warming's long-term cost to our children and grandchildren, which means that entrepreneurs don't have enough incentive to innovate to reduce our energy appetite. Also, just about every suggestion for substantial change in the way we should produce and consume goods and services—change that could lead to a real drop in carbon emissions—runs headlong into a wall of opposition from industries, politicians, and consumers with vested interests in the status quo. These groups have a potent argument to back their resistance to change: any policy adopted must not reduce economic growth, they say, because of growth's multiple benefits. It maintains our quality of life (as currently defined), absorbs workers who have lost their jobs because of technological change and, ultimately, ensures the political stability of our societies.[5]

So even if the technological and institutional transitions we need to make are possible, opposition from powerful groups and individuals has ensured that our *rate* of progress in dealing with the issue is far too slow to forestall a dangerous buildup of greenhouse gases. When it comes to the problem of global warming, our efforts at management aren't working. Instead, we're just papering over the problem and, at best, talking a lot about it.

We need a new approach to the great challenges we are confronting. Efforts at management are often important, even essential, but sometimes they aren't going to give us a satisfactory solution. The alternative approach I advocate requires us to adopt what I've termed a prospective mind. We need to be comfortable with constant change, radical surprise, and even breakdown, because these are now inevitable features of our world, and we must constantly anticipate a wide variety of futures. With a prospective mind we'll be better able to turn surprise and breakdown, when they happen, to our advantage. In other words, we'll be better able to achieve what I call catagenesis—the creative renewal of our technologies, institutions, and societies in the aftermath of breakdown.

A Watch List

"People will rarely acknowledge that an accustomed way of life is unsustainable except in the face of prolonged, devastating failure." So Joseph Tainter and his colleagues write.[6] And thus far in human history, this principle seems to have been almost universally true. People and societies don't easily give up on denial or admit that management isn't working. The two familiar strategies must fail dramatically before the wrenching psychological transition to a new approach can be justified. Anomalous evidence has to be so obvious and overwhelming that it can't be ignored.

Indeed, we often need to experience an abrupt and harsh threshold event, breakdown, or surprise before we're willing to accept that we can't continue the way we're going. In our own lives, for instance, we might have to get really sick before we change unhealthful habits like smoking or overeating. The same is true in international and global affairs. In 1938, people around the world paid little attention to foreshocks like the

Anschluss, the occupation of the Sudetenland, or Kristallnacht. Instead, it took the earthquake of Hitler's invasion of Poland in 1939 to make them wake up to the reality of Nazi aggression and to start doing something about it. More recently, it took the attacks on New York's World Trade Center to bring home the dangers of international terrorism, and the bankruptcies of giant firms like ENRON and WorldCom to raise awareness of widespread corporate corruption in the U.S.

In the same way, the majority of people are unlikely to pay much attention to the tectonic stresses we've discussed in this book—population imbalances, energy shortages, environmental damage, global warming, and widening gaps between rich and poor—until something truly dramatic happens. In coming years foreshocks produced by these stresses will become more frequent and severe. Just as people did in 1938, most of us will screen foreshocks out of our consciousness. Eventually, though, full-blown social earthquakes—like Hitler's invasion of Poland—will occur, and we won't be able to ignore them.

How and where might such earthquakes occur? We can't possibly know for sure, but we can identify some plausible scenarios. I believe three countries—Saudi Arabia, Pakistan, and China—and one region—Europe—deserve particular attention: each is especially vulnerable to some combination of the tectonic stresses we've discussed, and in any of them a social earthquake could reverberate around the world. National security and intelligence officials would call this a *watch list* of places where things might go truly haywire.[7]

Saudi Arabia is on my watch list because even a partial disruption of its oil output would have an immense impact on the world economy. The country is the world's largest producer of oil (at about 12 percent of world output, just ahead of Russia) and has by far the largest stated reserves (somewhere around a quarter of the world's total). As production peaks in non-OPEC countries, Saudi Arabia will become a supplier of last resort to the world economy. But no one outside the kingdom really knows how much oil the country has, and it could have much less than it claims (oddly, the country has pumped a total of 46 billion barrels of oil in the past seventeen years—at a current rate of about 3.3 billion barrels a year—without any decrease in its stated reserve figure of about 260 billion barrels). The world is likely to get no warning before Saudi output peaks—an event that credible authorities suggest could happen soon.[8] And when the peak has passed, the output decline is likely to be rapid, perhaps more than 8 percent a year.[9] Falling output will have a severe impact on a Saudi economy that remains (despite efforts at diversification) highly dependent on oil revenues.

Saudi Arabia isn't well positioned to cope with this kind of economic shock. In fact, Saudi society is vulnerable to a potentially explosive mix of demographic, economic, political, and social stresses. Fertility rates remain high, and the United Nations estimates that the country's population will balloon almost 40 percent to 34 million people in the next fifteen years.[10] Young adults (fifteen to twenty-nine years old) make up almost half of the adult population, and despite the country's oil wealth about a quarter of them are unemployed.[11] The kingdom's budget has been in deficit almost every year since the mid-1980s, as the government has poured vast sums into the Saudi military and underwritten the profligate lifestyles of nearly fifty thousand princes and their relatives. Corruption is endemic, the most basic human rights are abused, and there is barely a glimmer of democratic process.[12]

In the face of simmering and sometimes violent discontent, the ruling al-Saud family struggles to sustain its legitimacy and power. In particular, it provides political and legal support for, and funnels money to, the country's ultraconservative Wahhabi clerics, who then use the country's religious schools to propagate their brand of austere anti-Western Islamist ideology. By buying Saudi oil, the West thus helps fund a system that spreads hatred of the West. Many of the hundreds of thousands of young men influenced by this education system despise the ruling regime and dismiss it as a pawn of the United States. To top it off, the country has no constitutional process for succession, so it faces a potentially crippling struggle among royal leaders as the long-entrenched ruling gerontocracy, mostly in their seventies and eighties, inevitably dies in the next few years.

With the recent run-up in oil prices, Saudi revenues have soared, improving the country's economic situation, lowering unemployment, and allowing authorities to buy time.[13] Still, the country remains an incubator of Islamic terrorism both at home and abroad (of the nineteen 9/11 attackers, as many commentators have noted, fifteen were Saudi). Some analysts worry that terrorists will attack the hubs of the Saudi oil network, such as its key oil ports and processing facilities.[14] Others believe that popular unrest could bring to power a more extreme fundamentalist regime that would stop selling oil to the West.[15]

One has to wonder how the West, and the United States in particular, managed to get itself into a position where its core economic and security interests are intimately entangled with such a troubled society. Heavy dependence on Saudi oil makes Western economies far less resilient and the global economic and political system of which the West is a significant part far more brittle. It's easy to imagine how events in the desert kingdom could hurt people in just about every corner of the world.

Second on my watch list is Pakistan, because events there could allow terrorists to get hold of nuclear weapons. Pakistan is a country also besieged by multiple stresses including, like Saudi Arabia, severe demographic strain. The country's population will grow by almost a third—from today's 157 million to around 210 million—in the next fifteen years.[16] Again, young adults make up nearly half of the adult population. In the countryside, population pressure has combined with poverty, corruption, and bad land-use practices to produce severe deforestation, water scarcity, and damage to cropland. Partly as a result, the country won't be able to grow enough food to feed itself in coming decades.[17] The environmental and economic crisis in the countryside spurs people to move en masse into cities that have been doubling in size every twenty-five years. Although the country's economy is growing quickly today, largely because of a booming textile industry, remittances from overseas Pakistanis, and money brought home by wealthy Pakistanis who have recently returned after terrorism fears made them less welcome overseas, there's scant evidence that this growth is trickling down to the poor.[18] Illiteracy still exceeds 60 percent, and after military spending is subtracted from the national budget there's little government money left for education, clean water, housing, and health care for the poor.[19]

One of Pakistan's defining features is its warrior class—a military elite that regularly overthrows corrupt elected governments and that collaborates with landowning, bureaucratic, and judicial elites to run the country. This civil-military oligarchy holds the masses in contempt and is deeply distrustful of India. It cynically exploits Islam and the threat of Indian aggression to promote the state's legitimacy. Internally, it uses radical Islamic political parties to undermine the country's moderate opposition, while externally it has used jihadis—Islamic holy warriors—as a tool of foreign policy in both Afghanistan and the disputed territory of Kashmir, as well as in India more generally.[20] The predictable result: anti-Western and anti-modern Islamic groups have become entrenched in the society, especially in its poorest quarters. Pakistan's network of twelve thousand Islamic seminaries, or madrassas, is now the population's main source of education, while radical Islamic political parties have become the principal avenue of political expression, for the country's swelling tide of marginalized and unemployed young men. Radical Islam has also insinuated itself into the country's wealthy elite and the ranks of senior officers in the military and intelligence service.[21]

Pakistan is considered a linchpin in the fight against international terrorism, yet the regime's commitment to the fight seems halfhearted because of popular animus toward the West, and toward the United States in particular.[22] Meanwhile, Pakistan's nuclear weapons program has produced over a metric ton of highly enriched uranium (enough for about twenty atomic bombs of the size and type that destroyed Hiroshima), and it has been caught shipping bomb technology to some of the world's most deplorable regimes.[23] Commentators have warned that the assassination of President Pervez Musharraf could cause the country to disintegrate in a spasm of civil violence, creating abundant opportunities for the theft or clandestine sale of its nuclear materials and technology. This "failed state" scenario may be unrealistic, because the army dominates the society and has a strong interest in order. Still, Pakistan has a history of turmoil following on the heels of periods of rapid economic growth, and it's exceptionally vulnerable to economic shocks, especially since it imports over 80 percent of its oil.[24] A sudden shortage of oil would devastate the economy and state finances and cause an explosion of unrest in cities deeply divided by ethnicity and religion. Radical officers could exploit this disorder to seize power and then share the country's nuclear secrets with sympathizers far and wide.

Saudi Arabia and Pakistan are pretty obvious entries on practically everyone's watch list, as it's easy to see how disruption there could ripple through country after country. Some commentators would also identify China as a threat, but only because its economic and political might is growing so fast. They extrapolate current trends into the future, so they assume that China will become the next rival to the United States for world economic and even military dominance. Its emergence as a superpower, they predict, will bring great instability.[25]

But the real worry isn't China's potential strength but its weaknesses. Like Saudi Arabia and Pakistan, the country faces a host of rapidly worsening stresses that threaten to cause an economic and social break-down.[26] While leaders in Saudi Arabia and especially in Pakistan are dithering in response, China's leaders are trying to manage their problems with ingenuity, political ruthlessness, and enormous resources. All the same, it's not clear their efforts at management will succeed.

China faces four fundamental challenges: first, it has a huge population (now just over 1.3 billion), about half of which is still desperately poor; second, a significant portion of its labor force remains employed in grossly inefficient state-owned industries; third, the governing regime has dubious political legitimacy, especially now that old-style Communist doctrine has lost its appeal; and fourth, the country's domestic supplies of natural resources—especially water, cropland, and energy—are increasingly degraded or limited. Since the late 1970s, China's leaders have dealt with the first three challenges by unleashing the forces of capitalism. In many ways this strategy has been a spectacular success: 250 million people have been catapulted out of destitution, and economic growth has become the key source of the Communist regime's legitimacy. As long as most Chinese can see year-by-year improvements in their material circumstances, the large majority are willing to tolerate, and even support, the state's authoritarianism.

But the limits of this strategy are now brutally clear. Economic growth has been concentrated in China's cities, especially along its central and southern coast, while the country's interior rural zones have fallen further and further behind.[27] Average rural incomes are now less than a third of those in cities.[28] To make matters worse, the social safety net has vanished in rural areas: country folk have no pensions, and routine medical costs drive many into penury.[29] In under thirty years, China has gone from being one of the world's most egalitarian societies (albeit one where nearly everyone was very poor) to one of its most unequal, and this disparity is driving one of the largest migrations ever seen, as a "floating population" of well over 100 million people, mostly dirt-poor farmers, leaves the countryside to work in cities for months or even years at a time.[30]

Years of breakneck growth, coupled with the regime's general indifference to the environment and conservation, have made the fourth of China's challenges—natural resource scarcity—truly critical.[31] Indeed, resource scarcity threatens to severely undermine China's recent economic progress.[32] The whole northern half of the country now suffers from severe water shortages. Because of water withdrawals, the mighty

Yellow River is dry at its mouth for up to seven months every year, and the water table under the northern plain, the source of much of China's wheat, is falling by 1.5 meters a year and that under Beijing by 1.2 meters a year.[33] Deforestation in the south contributes to horrific annual floods along major rivers, and in the north and west it helps deserts expand by ten thousand square kilometers a year. Partly as a result the frequency of sandstorms—sometimes storms so bad that they cover Beijing in suffocating grit—has quintupled since the 1960s.[34] China has also lost so much cropland to erosion, salinization, and metastasizing cities that the country's leaders worry about its ability to feed itself.[35] In addition, the country is now extremely vulnerable to energy shocks. Domestic oil output is almost flat, in part because the largest oil field in Daqing in the northeast has passed its peak and is declining by 5 percent annually. As recently as 1993, China could meet its oil needs from its own fields; now it imports half its oil, and in fifteen years it will import three-quarters.[36]

These economic and resource stresses are tearing apart the country's social fabric. According to official figures, there were 240 mass protests *every day* in 2005, an increase of almost 20 percent from the previous year.[37] Rural communities especially are rebelling against poverty, environmental damage, official neglect and corruption, and rampant land seizures by speculators for factories, housing complexes, shopping malls, and golf courses.[38] Young and educated men in the cities are also starting to violently protest their limited job opportunities.[39] A deep split has developed in the country's political elite between those who want to push ahead with rapid modernization and those who want to return to more traditional socialist economics to protect workers.

China is caught in a seemingly impossible dilemma. On one hand, the growth imperative is a stark fact of life: the country must expand its economy by at least 7 percent a year to absorb workers from the countryside and from defunct state-owned industries. Such growth is central to maintaining civil peace. On the other hand, the same growth is creating great social and environmental stresses and dangerous imbalances in the economy, including wild overinvestment in certain sectors, like real estate, and a huge accumulation of bad debt in state banks.[40] To generate enough cash flow and foreign investment to service this debt, the country's leaders have encouraged an export boom—especially to the United States—by keeping the Chinese currency cheap.

And here's where China's predicament is most immediately relevant to the rest of the world, because the most plausible scenario for a new international economic crisis involves a rupture of the financial relationship between China and the U.S. This relationship is a kind of "twin bubbles" symbiosis driven by growth imperatives on both sides of the Pacific. The Chinese currently sell 40 percent of their exports to the American market. Then, by buying Treasury Bills, corporate bonds, and short-term securities, they lend back to the U.S. some of the dollars they receive for their exports. This helps finance America's huge budget and trade deficits, which stimulates domestic demand while at the same time propping up the dollar (relative to the Chinese currency) and holding down U.S. interest rates. Low interest rates, in turn, discourage Americans from saving. Instead, Americans speculate on real estate and borrow on rising household equity, which liberates cash that they then use to buy more Chinese goods.[41]

How long can this symbiosis continue? Probably for a good while, given that each side has an interest in maintaining the current equilibrium. But the equilibrium crucially depends on the ability of the U.S. government and consumers to take on ever more debt. Sooner or later, many analysts believe, the American deficits will become so large that they'll frighten potential creditors.[42] Even the Chinese may decide not to lend because they'll fear that a fall in the dollar will devastate the value of their U.S. investments.[43] Or the U.S. housing bubble will burst, sapping consumers of spending power and throwing the American economy into a recession.

A sharp downturn in the U.S. economy would be a body blow to China, especially since the country now has a trade deficit with the rest of the world. It would immediately make the banks' problem of nonperforming loans—somewhere between a third and half of all domestic debt—far less tractable. It could even lead to a run on the banks.[44] And the shock waves wouldn't be limited to China. The historian Niall Ferguson notes that a Chinese currency or banking crisis could have "earth-shaking ramifications."[45] If the country's growth rate fell significantly below 7 percent, the regime would likely shore up its legitimacy and popular support by stoking nationalist fury against Taiwan and Japan. Even today, popular resentment of Japan simmers just below the surface of Chinese society, while in Japan right-wing nationalist sentiment against China has been gaining ground.[46] Should times turn difficult, many of the ingredients are in place for a bitter conflict between these old enemies.[47]

A heavy burden of debt—whether borne by a country, company, or individual—boosts vulnerability to external shock. It lowers resilience. In many other ways too—by degrading its environment, drawing down its energy resources, and creating appalling disparities between its rich and poor—China is stretching the limits of its elasticity, perhaps close to the breaking point. "The extremity of China's condition," writes the Harvard ecologist Edward O. Wilson, "makes it vulnerable to the wild cards of history. A war, internal political turmoil, extended droughts, or crop disease can kick the economy into a downspin."[48]

Some Western countries and regions are susceptible to social earthquakes too and therefore deserve a place on our watch list. Europe's challenges are especially daunting. Just like Saudi Arabia, Pakistan, and China, the continent will have to cope with several tectonic stresses simultaneously.

The first is the stark difference in population growth between Europe and its neighboring countries from North Africa to West Asia. Today, Europe's population is roughly the same size as that of its neighbors, but by 2050 it will be only a third the size. This growing imbalance is ratcheting up migration pressure from surrounding countries and swelling the continent's immigrant underclass, now numbering about 20 million. Many of the immigrants live in dreary urban and suburban enclaves— isolated "internal colonies," some analysts call them—that are already cauldrons for an explosive brew of high unemployment, failed assimilation, resentment, and strong group cohesion.[49] The second stress is the decline in Europe's already limited oil and gas reserves (Britain's oil output from its North Sea fields, for example, is falling by nearly 8 percent annually and its gas output by 12 percent).[50] The decline is making the continent more vulnerable to the vicissitudes of outside supply. Even today, domestic sources of energy, including hydro and nuclear power, cover barely half the continent's needs; in fifteen years the amount will drop to about a third. Much of Europe's oil will come from the Middle East, especially Saudi Arabia, and almost all of its natural gas will come from suppliers like Algeria and Russia, both of dubious reliability.[51] The third stress is climate change. Europe's geographical position makes it particularly exposed to a change in ocean heat circulation. Scientists have little idea when such an event might occur, but there's now evidence that it's starting to happen, and even a partial disruption of the North Atlantic conveyor could hurt European agriculture by making the continent cooler and drier and more susceptible to extreme storms, droughts, and floods. This would matter to people in Europe of course, but it would affect people around the world too, because the continent's farms supply almost 40 percent of internationally traded food.[52]

If energy and climate shocks hit Europe simultaneously, everything there from gasoline to bread would skyrocket in price. As economic activity slowed, unemployment would surge, and the result would likely be dreadful violence along the continent's ethnic fault lines.

Moments of Contingency

Events in Saudi Arabia, Pakistan, China, Europe, or in a hundred other places could trigger a social earthquake that shakes everyone in the world. Here's just one scenario.

Al Qaeda detonates a radiological device in the Abqaiq oil-processing facility in Saudi Arabia, taking 5 percent of world production off line for at least six months and provoking civil turmoil throughout the kingdom. Then Iran, exploiting a critically tight oil market, cuts its own oil exports in half as a protest against United Nations Security Council sanctions over its nuclear program. As oil prices spike to $150 a barrel and investors flee stock markets around the world, Europe, Japan, and the United States tip into a steep recession. Riots break out in cities in poor countries because of skyrocketing fuel prices. Pakistan, its economy reeling, is convulsed by protests. Hard-line Islamic army officers launch a coup, and President Musharraf is assassinated. In China, the sudden worldwide slump causes widespread layoffs, while unrest breaks out in rural and urban zones, and several major banks fail. To maintain its popular legitimacy at home, China plays the nationalism card, threatening both Taiwan and Japan over disputed oil fields in the East China Sea. Meanwhile, Western countries begin laying plans for a massive invasion of the Persian Gulf to secure oil supplies.

Such a string of events could signal that our global system has passed the apex of its growth phase—the upper cusp of its adaptive cycle—and that a larger, perhaps uncontrollable, process of breakdown is beginning. If so, we'll have entered what I call a *moment of contingency*—a junction in our movement through time that amplifies both danger and opportunity. A moment of contingency is a moment of choice, like the fork in the pathway encountered by the traveler in Robert Frost's poem "The Road Not Taken." It's an instant when small things matter a lot, because they can give us the nudge that sends us down one road rather than another. Frost's traveler had an advantage: he could see ahead to where a road "bent in the undergrowth." But in the aftermath of a globe-shaking social earthquake like the one I've just described, we'll have trouble seeing anything at all. The comfortable old road behind us will have vanished, but the new ones in front of us will be barely visible in a fog of fear and uncertainty.

In moments of contingency, nothing is definite, and everything is tentative. Choices made by societies, groups, and individuals may be less constrained than previously, but the consequences of choices are far more opaque. Social reality loosens its grip on us. It becomes more fluid. Long-standing relations of authority between people, groups, and institutions weaken, while deeply ingrained patterns of social behavior lose purpose and meaning. Actions and futures that were once unthinkable—because they were too wonderful or too horrible—are suddenly possible. In moments of contingency, surprise and bewilderment create mental polarities: anticipation alternates with fear, and hope with despair. And these polarities evoke the best and worst attributes of human character—courage and cowardice, generosity and greed, kindness and malice, and integrity and deceit.

Moments of contingency are thus easily exploited for good or ill. Fear, hope, and greed are unleashed at the same time that social reality becomes fluid. This means that people's motivation to change their circumstances soars just as their opportunities to accomplish change multiply. Whether the outcome of this powerful confluence is turmoil or renewal hinges—in large measure—on how the situation is framed.

People will want reassurance. They will want an explanation of the disorder that has engulfed them—an explanation that makes their world seem, once more, coherent and predictable, if not safe. Ruthless leaders can satisfy these desires and build their political power by prying open existing cleavages between ethnic and religious groups, classes, races, nations, or cultures.[53] First they define what it means to be a good person and in so doing identify the members of the *we* group. Then they define and identify the bad people who are members of the *they* group.

These are enemies such as immigrants, Jews, Muslims, Westerners, the rich, the poor, or the nonwhite, who are the perceived cause of all problems and who can serve as an easy focus of fear and anger.

Particularly receptive to such stereotypes are people who already feel humiliated or victimized; so too are those who feel alienated or marginalized and who believe, as a result, that they have no stake in society and no peaceful means to express their unhappiness. These people are not necessarily, or even mainly, the destitute; rather, they're people who see a rapidly widening gulf between what they're getting and what they think they rightfully deserve. In a world where environmental stress has dislocated the lives of hundreds of millions of people, where billions are on the losing side of an ever wider and ever more visible wealth gap, where globalized capitalism's relentless pressure upends communities and traditions, and where most people feel they have no real political voice, the raw material for this kind of radicalization is available practically everywhere—even before a severe social earthquake hits, like the one I've described above.

Indeed, large numbers of people are already primed to see the world in terms of a Manichean division into good and evil—those who have adopted some form of religious fundamentalism, whether Christian, Islamic, Jewish, Hindu, or Buddhist.[54] In both rich and poor societies, many people find that the certitudes that give their lives meaning and purpose are under siege: change seems to happen faster every year, economic insecurity is pervasive, scientific advances assault traditional beliefs, and many once-powerful secular dogmas, like pan-Arabism and Marxism, have imploded. In this stormy world, fundamentalist creeds can seem to provide a firm anchor.[55] All such creeds claim privileged access to absolute truth, and all establish what's right and wrong, provide strict rules of behavior, and identify friends and enemies. And because that truth comes from revelation not research, creeds justify the suspension of reason and deliberation—a kind of psychological denial that may be a balm for the bewildered but that's a truly inept response to an ever more complex reality.[56] Some extreme forms of fundamentalism even encourage their followers to look forward with joy to the wholesale obliteration of both society and nature. For these fanatics, Christian dispensationalists in the United States among them, today's corrupt world must be destroyed before a new, pure, and truly righteous one can emerge.

When a social earthquake erupts—when the established order starts to crack and crumble—much depends on what happens in the period immediately following the first shock. In this moment of contingency—this moment of choice—the worst personalities and passions too often prevail. When "the centre cannot hold," wrote W. B. Yeats in 1919 in "The Second Coming," "the best lack all conviction, while the worst are full of passionate intensity."[57] If fanatics of one stripe or another succeed in framing the choice, and if they and their followers have already stocked up on resources, identified their enemies, struck deals with potential allies, and worked out plans for exploiting the moment when it arrives, the rest of us will find we really have no choice at all. Then we'll all travel together toward a future of turmoil, violence, and disintegration. We won't be able to retrace our steps, because momentum will build and animosities will deepen as a cycle of action and reaction, attack and retaliation sets in. The power shift, in fact, makes this kind of cycle more likely: as people's capacity to kill increases, any single attack can hurt more people and provoke more people to seek revenge; yet the very same power shift means it takes smaller numbers of people seeking revenge to keep the cycle going.

If terrorists were to use weapons of mass destruction against major population centers, they'd destroy not only people and property. They could also destroy our tolerance and capacity to see alternatives to retaliation, and they could reinforce our most rigid and negative stereotypes. If these things happened, we'd lose our ability to see differences between individuals in groups we classify as *them*. And we would start to see whole sections of humanity as the enemy.

If we're unprepared, such attacks would also erode the political will and the economic, political, and social institutions—like free media, liberal democracy, and open scientific research—that we need to deal with the violence's deep causes, including wealth gaps, demographic imbalances, or environmental damage. Should radiological bombs go off in our major cities, we wouldn't think about these causes. Instead, we'd commit our remaining resources to the immediate tasks of self-preservation. We'd seal our borders and leave our cities. If we were rich, we would barricade ourselves in our communities. And we'd round up and torture suspicious people, while launching preemptive strikes against those we feared. The end result could well be an unraveling of social complexity—a retreat back to local concerns and to simpler, more authoritarian (even brutal) institutions, rules of behavior, and economic and social relationships.[58]

Resilience

Events don't have to turn out this way, because we really do have some ability to choose our future. But we have to recognize what kinds of forces we're up against, we have to have courage, and we have to be smart—not only at the time of the social earthquake and the moment of contingency that follows but also well in advance. Specifically, if we're going to have the best chance of following a different and positive path, we must take four actions. First, we must reduce as much as we can the force of the underlying tectonic stresses in order to lower the risk of synchronous failure—that is, of catastrophic collapse that cascades across boundaries between technological, social, and ecological systems. Second, we need to cultivate a prospective mind so we can cope better with surprise. Third, we must boost the overall resilience of critical systems like our energy and food supply networks. And fourth, we need to prepare to turn breakdown to our advantage when it happens—because it will.

The first action, reducing the force of underlying stresses, is the most obvious, largely because it resembles a conventional management approach to dealing with our problems. Experts of all types have generated a considerable quantity of good ideas about how we can reduce the force of the tectonic stresses I've identified in this book—population imbalances, energy shortages, environmental damage, climate change, and income gaps. Yet too often the experts operate only within the silos of their disciplines and professional communities. Demographers don't talk to energy specialists, agronomists don't speak to economists, and climate scientists don't talk to epidemiologists. Instead, experts usually target the problem they understand, and because they don't think much about how to integrate their ideas with the ideas of experts focusing on related problems, the policies they propose are too narrowly focused.

This highly compartmentalized approach doesn't work in a world of converging and synergistic stresses. We must bring experts together across disciplinary barriers, just as we must bring governments together across cultural, ideological, and political barriers.[59] And we also need to realize that there's no magic bullet: there's no single technical solution, institutional response, or policy that will neatly resolve all our challenges in one fell swoop. More than ever in humanity's history, we have to be aggressively proactive on multiple fronts at the same time. For example, we need to simultaneously increase our support for family planning in countries that still have high fertility rates; boost efforts to conserve the planet's soils, forests, fisheries, and fresh water; implement a global transition to new and cleaner energy sources; sharply reduce carbon emissions; pour resources into reducing disease in poor countries; work to rebuild societies shattered by conflict; and reform the international financial system so that it doesn't wreck economies in reaction to the corrupt or incompetent policies of their elites. Perhaps more urgently than anything else, we need to reduce the risk of weapons of mass destruction falling into the hands of small groups, especially by securing and destroying as much as possible of the world's huge stock of highly enriched uranium.

These are all really hard yet vital tasks. Alas, humankind's track record when it comes to proactive policy, especially in response to slow-creep problems, doesn't inspire much confidence that we will succeed in these tasks. Today, most of us are simply too deep in denial, and our political and economic systems are too hobbled by powerful vested interests for real change to happen in the absence of a sharp push or shock from outside. With colossal effort by the relatively small numbers of people today engaged in trying to do something about these problems, and perhaps with a good deal of luck, we might divert or somewhat weaken the tectonic stresses. But we're unlikely to weaken them enough to reduce significantly the danger we face, so we'd better get ready for social earthquakes.

This is where cultivating a prospective mind comes in—the second action we need to take. We can't possibly flourish in a future filled with sharp nonlinearities and threshold effects—and, somewhat paradoxically, we can't hope to preserve at least some of what we hold dear—unless we're comfortable with change, surprise, and the essential transience of things, and unless we're open to radically new ways of thinking about our world and about the way we should lead our lives. We need to exercise our imaginations so that we can challenge the unchallengeable and conceive the inconceivable. Hunkering down, denying what's happening around us, and refusing to countenance anything more than incremental adjustments to our course are just about the worst things we can do. These behaviors increase our rigidity and dangerously extend the growth phase of our adaptive cycle. When a social earthquake eventually occurs, we'll have no new concepts, ideas, or plans to help us cope and no alternative ways of seeing our future.[60] Without alternatives, there will be no constraint on fear, and we'll be especially vulnerable to the kind of amplification effect we experienced within our psychological networks in the immediate aftermath of the 9/11 attacks—when a torrent of sensational commentary and raw emotion coursed through the Internet and across the airwaves and only terrified us more.

A prospective mind knows that scientific knowledge is the best tool to determine the boundary between plausible and implausible futures. But precise prediction is impossible because our complex and nonlinear world is full of unknown unknowns—things we don't know we don't know.[61] Some of these unknown unknowns—perhaps an overlooked interaction between components of an electrical grid or an unseen genetic mutation in a virus in southern China—can hurt us badly. So a prospective mind also stresses the value of prudence—a long-neglected and even derided quality of mind and behavior.

A prudent way to cope with invisible but inevitable dangers is to take the third action I identified above, which is to build *resilience* into all systems critical to our well-being. A resilient system can absorb large disturbances without changing its fundamental nature.[62] Roman engineers intuitively understood this idea. When they designed and built an arch, for example, they weren't acquainted with—as engineers now are—all the precise laws and formulas governing the structure's load, tension, compression, bending, and shear. In many ways, they were operating in a world of unknown unknowns. So they didn't try to maximize the efficiency of their structures by using the least amount possible of rock, brick, and mortar. To do so would have made these structures subject to sudden collapse. Instead, they compensated for their lack of precise knowledge by consciously designing all their structures with large margins of error. They appreciably overbuilt their arches and the like, and their prudence helped the structures better withstand unexpected shocks like geological earthquakes.[63]

Likewise in an increasingly uncertain and dangerous world, we should sometimes give up extra efficiency and productivity in order to gain resilience—especially to improve our ability to prevent foreshocks from triggering synchronous failure. We can do this in many ways. One involves loosening some of the coupling inside our economies and societies and among our technologies. For instance, we can promote the distributed supply of vital goods like energy and food: the more power we produce with solar panels on our rooftops, the less vulnerable we'll be to power disruptions far away.[64] And in our technology-supercharged international financial markets, if we lower the incentive to "head for the exits" when there's a hint of trouble, investors will have more time to think before they panic.[65] We can gain resilience, too, by increasing the buffering capacity or slack in our economies. Industries can rely less on just-in-time production—a particular obsession of the past two decades—and instead build up inventories of feedstocks and parts so they can keep running even when supplies of essential inputs are temporarily interrupted.[66] And finally, since malicious groups and individuals will probably someday target the highly connected hubs of our scale-free energy, food, information, transportation, and financial networks, we can identify these hubs and either redesign our systems to remove them entirely or replicate them to create redundancy.

This is exactly what many financial firms in lower Manhattan did during the 1990s. Forewarned by the 1993 bombing of the World Trade Center, they made contingency plans by setting up redundant data, information, and computer facilities in more remote locations. Some firms even made advanced arrangements with companies specializing in providing emergency relocation facilities in New Jersey and elsewhere. As a result they better withstood the effects of the 9/11 attack. Though the NASDAQ headquarters was demolished in the attack, for instance, the exchange's data centers in Connecticut and Maryland remained connected to trading companies through two separate lines that passed through twenty switching centers. NASDAQ officials later claimed that their system was so resilient that they could have restarted trading only a few hours after the attack.[67]

Resilience is an emergent property of a system—it's not a result of any one of the system's parts but of the synergy between all its parts. So as a rough and ready rule, boosting the ability of each part to take care of itself in a crisis boosts overall resilience. We can apply this rule in our personal lives and families. We should have battery-powered radios and flashlights for the next time the power goes out; food, water, and cash on hand; and plans on where we'll rendezvous with our loved ones should we get separated. This is not new advice: in rich countries, all government disaster agencies recommend that we have enough supplies at home to survive for several days in the event of infrastructure failure. We can apply this general principle, too, in our communities and neighborhoods. The 1906 San Francisco fire spread across the city with devastating effect largely

The 1906 San Francisco Fire

because the water-supply system was tightly coupled. The earthquake had shattered the system's iron mains in three hundred places, and in most places firemen couldn't get water. To keep this from ever happening again, the city later built 175 stand-alone cisterns under intersections across the city—each holding 300,000 litres of water or more—so that individual blocks would have their own water supplies to fight fires. Now if the mains break in a disastrous earthquake, firemen can pop open a hatch in the middle of an intersection, drop a submersible pump into the reservoir beneath, and get on with fighting fires.[68]

When it comes to connectivity in its networks, a resilient system is a bit like Goldilocks's favorite bowl of porridge: it has neither too much nor too little connectivity.[69] In a resilient system, individual nodes—like people, companies, communities, and even whole countries—are able to draw on support and resources from elsewhere, but they're also self-sufficient enough to provide for their essential needs in an emergency. Yet in our drive to hyper-connect and globalize all the world's economic and technological networks, we've forgotten the last half of this injunction. And we've forgotten, too, that resilience is a "public good"—something in whose benefits everyone shares, whether or not they pay for it. As with any public good, whether national defense or fire protection, if the government doesn't intervene, everyone tends to wait for someone else to pay. In order to create resilient economies, for instance, companies won't voluntarily shift away from just-in-time production, because any company that does so by itself will be crushed by competitors who don't have to carry the cost of extra inventory.

So in the end it's up to governments of all kinds—municipal, national, and international—to develop incentives and enforce rules that encourage everyone to do what they can to boost the overall resilience of our critical systems. It's also up to governments to employ the necessary resources—money, time, and human capital—to identify and address specific weak points in these systems. We've learned from Buzz Holling that our world's capacity to avoid "deep collapse"—or synchronous failure—depends on resilience throughout the system. In practical terms today, this means we must focus our attention on boosting the resilience of the world's weakest societies—those with horribly damaged environments, endemic poverty, inadequate skills and education, and weak and corrupt governments.[70] If we don't, our entire global social-ecological system will become steadily more vulnerable to the diseases, terrorism, and financial crises that emerge from its least resilient components.

Of course, many of these recommendations fly in the face of the ideology of today's globalized capitalism. In its most dogmatic formulation, this ideology says that larger scale, faster growth, less government, and more efficiency, connectivity, and speed are always better. Slack is always waste. So resilience—even as an idea, let alone as a goal of public policy—isn't found anywhere on the agendas of our societies' leaders. At the annual G8 meetings, the leaders of the United States, Canada, England, France, Germany, Italy, Japan, and Russia talk endlessly about managing global systems, but they never talk about making them more resilient. No member of Buzz Holling's extraordinary international research consortium, the Resilience Alliance, has ever attended the World Economic Forum—that annual self-congratulatory tête-à-tête of world leaders in Davos, Switzerland. And because our leaders hardly ever think about resilience, we keep doing things that make our lives progressively *less* resilient—we pile on more debt, build tract housing over our finest cropland, develop addictions to distant sources of energy, become so specialized that we can't take care of ourselves when everyday technologies fail, and fill every nook and cranny of our days with so much junk information and pointless running around that we don't have time to reflect on what we're doing or where we're going.

Open Source

The great 1906 earthquake and the resulting fire cut like a scythe through San Francisco. The disaster's direct cost totaled between $350 and $500 million in 1906 dollars—about $6 billion in today's dollars—and a staggering 1.3 to 1.8 percent of the United States GNP that year. Its economic shock spread quickly from the city to the larger U.S. and world economies.[71] While aid poured into San Francisco from across America, stock markets plummeted in New York and London. And because British insurance companies underwrote about half of the fire policies held by the city's residents and businesses, huge claims were soon presented in London. At first the insurers tried to avoid payment by arguing that the earthquake, not the fire, caused the damage. But after a public outcry and threats of lawsuits, they relented and agreed to meet their obligations.

In those days, most advanced economies adhered to the gold standard. Countries tied their money supplies to their reserves of gold and guaranteed that they'd redeem their currencies in gold at a fixed rate. So the practical result of the insurers' payments—when the payments finally started flowing in the late summer and fall of 1906—was a massive flood of gold out of London into San Francisco. In fact, so much gold left England that the country's money supply suddenly contracted, creating a liquidity crisis and threatening to push the economy into a deflationary spiral. In response, the Bank of England nearly doubled its core interest rate in just over a month and cracked down on purchases of U.S. debt by English banks. These emergency measures worked: by the end of 1906, gold was flowing into England again.

But the earthquake's aftershocks continued. Higher British interest rates and the abrupt squeeze on U.S. debt by England, France, and Germany hammered the American economy, causing one of the most rapid economic downturns in the country's history. Gold hemorrhaged overseas, and suddenly it was the United States' turn to deal with contracting liquidity. The stage was set for a full-blown economic crisis, and, sure enough, when a New York bank failed in October 1907, financial panic erupted. Across the country, people rushed to withdraw their money from banks. A group of bankers led by J. P. Morgan lent millions to prop up banks, trust companies, and the stock market, but to little effect.[72] Only when the U.S. Treasury lent much larger amounts to keep banks afloat did the crisis ease.

The 1907 panic was a bucket of cold water in the face of American capitalists and lawmakers. They couldn't deny that the country's private banking system was chronically vulnerable to liquidity crises. Even the intervention of the likes of J. P. Morgan hadn't restored stability. Clearly worried, Congress created a commission under Senator Nelson Aldrich to investigate the panic's causes and recommend solutions. From 1908 to 1912, Aldrich and his colleagues looked into banking systems around the world, and in 1913 their recommendations led to the creation of a U.S. central bank. The Federal Reserve System, or Fed, as it's now commonly known, would keep banks afloat in the event of a financial panic and, with its ability to influence the economy's money supply, became a pillar of American economic power in the twentieth century.

So a devastating earthquake caused a financial crisis that catalyzed a deep reform of key institutions. Here we see catagenesis at work:

catastrophe is followed by creativity and eventually renewal. And because of an odd coincidence, this particular story has a personal twist.

In November 1910, Aldrich called a handful of senior bankers to a secret meeting. Using assumed names, the bankers boarded his private rail car as it sat on a siding in Hoboken, New Jersey, and were taken to Jekyll Island—an exclusive gentlemen's club on the Georgia coast frequented by America's richest men and their families. J. P. Morgan, still bruised by the 1907 debacle, was a club member, and he likely arranged for the group to use the facilities, although he didn't join the group himself. Over several weeks in the club's elegant lodge, the men drafted the outlines of the Federal Reserve System.[73]

Today the lodge has been converted into a hotel that trades on the building's old Southern charm. And in the spring of 2003, in the very room in which the Aldrich group deliberated, I attended a Resilience Alliance workshop and met Buzz Holling for the first time. Even though I didn't know then about the room's strange connection with the San Francisco earthquake, as I listened to the conversation among the workshop's participants, I began to appreciate more fully how all highly adaptive systems go through cycles of breakdown and regeneration. Breakdown is greatly disruptive to parts of the system, but it needn't be catastrophic overall, and it can produce exactly the conditions required for a burst of creativity, reorganization, and renewal.

Sadly, though, history shows that most human civilizations overextend the growth phase of their adaptive cycle, so they eventually suffer deep collapse. "A long view of human history reveals not regular change but spasmodic, catastrophic disruptions followed by long periods of reinvention and development," writes Holling. "In contrast to the sudden collapses of biological panarchies [such as forest ecosystems], there are long periods of ruinous reversal, followed by slow recovery and the restoration of lost potential."[74]

I realized at the meeting on Jekyll Island that this is, in fact, the fundamental challenge humankind faces: we need to allow for breakdown in the natural function of our societies in a way that doesn't produce catastrophic collapse but instead leads to healthy renewal. This idea isn't quite as radical as it first sounds. Cycles of breakdown and renewal are normal in modern capitalist economies. Companies go bankrupt, and new ones emerge in their place; established economic sectors disappear, to be replaced by industries driven by new technologies; and recessions shift capital from inefficient firms to productive ones, while helping to purge the excesses of earlier boom times.[75] Joseph Schumpeter, one of the twentieth century's greatest economists, famously called these processes a "perennial gale of creative destruction" that's spurred, in part, by the relentless innovation of entrepreneurs.[76] But elsewhere in our societies, rigidity is the rule rather than the exception. Powerful habits, beliefs, and vested interests hold sway, so things like underlying structures of wealth and power and entrenched patterns of social and consumer behavior don't really change.

And, while lots of people cite Schumpeter's memorable phrase, modern capitalist economies aren't always the paragons of adaptivity they're claimed to be. Since the 1960s, better management of the economies of rich countries has reduced short-term economic volatility. Recessions have become less severe as central banks and governments—scared of the political aftershocks of economic downturns—have learned how to maintain demand without high inflation.[77] But this better short-term management may have just made our economies more prone to larger crises later; some economists believe it encourages a buildup of inefficient and unprofitable investment in real estate, factories, and the like.[78] Meanwhile, in the larger global economy, as we saw in chapter 8, financial crises, especially banking crises, have become more frequent, and they've often hurt poor countries very badly. Such crises are breakdowns of a sort, and sometimes they even lead to vital reforms—like the flowering of democracy in Indonesia that followed the East Asian financial crisis. On balance, though, they're often too severe to be helpful, and they only make people who are already desperately poor even more miserable.

So somehow we have to find the middle ground between dangerous rigidity and catastrophic collapse. In our organizations, social and political systems, and individual lives, we need to create the possibility for what computer programmers and disaster planners call "graceful" failure. When a system fails gracefully, damage is limited, and options for recovery are preserved.[79] Also, the part of the system that has been damaged recovers by drawing resources and information from undamaged parts. Holling explained how a collapsed ecosystem regenerates itself by drawing support from panarchy cycles that operate both above and below—or, put differently, on larger and smaller scales—than the ecosystem itself. For instance, a forest that has burned regrows when large-scale cycles of water and nutrients help to germinate tiny seeds left behind in the soil. The recovery of San Francisco after the earthquake and fire is a good example of this process too: the disaster was limited to one geographical zone, and while people in the city worked to rebuild their individual households and businesses, relief arrived from the larger systems of American society and governments and from insurance firms as far away as London and Germany. At one point, so much gold had arrived that the city didn't have enough vault space to store it all.

Breakdown is probably something that human social systems must go through to adapt successfully to changing conditions over the long term.[80] But if we want to have any control over our direction in break-down's aftermath, we must keep breakdown constrained. Reducing as much as we can the force of underlying tectonic stresses helps, as does making our societies more resilient. We have to do other things too, and advance planning for breakdown is undoubtedly the most important. This is the fourth action we must take if we're going to follow a positive path into the future.

We can't know exactly what breakdown will look like, and we don't know when it will happen, but we can still start figuring out now how we'll respond. In vigorous, wide-ranging, yet disciplined conversation among ourselves, we can develop scenarios of what kinds of breakdown could occur.[81] In this conversation, we shouldn't be afraid to think "out-side the box"—to try to imagine the unimaginable—because in a non-linear world under great pressure, we're certain to make wrong predictions if we just extrapolate from current trends. Then we need to lay down plans and organize ourselves so that we're prepared to take advantage of the opportunities that various types of breakdown might offer to build a better world. For instance, depending on the scenario, we might plan to aggressively disseminate information through the Internet, mass media, and various social networks to frame the rapidly changing situation in a humane and constructive way. Or we might plan non-violent disruption of efforts by extremists to organize themselves. Or we might organize coordinated mass civil disobedience of the kind we've seen recently in democratic popular protests in Serbia and the Ukraine.[82] In general, we can be sure that when breakdown happens we'll be much better off if we have contingency plans ready to go.[83]

In preparing for breakdown, we need to keep one thing in the fore-front of our minds: people who aren't extremists face a huge disadvan-tage in any kind of political struggle with extremists. To use the jargon of social scientists, nonextremists have a formidable "collective action problem." They're rarely organized in coherent groups and thus find it hard to act in a coordinated way. They differ widely in their values and perspectives, and they vary in their strength of commitment to political and social causes. Extremists, on the other hand, are often organized in coherent and well-coordinated groups that have clear goals, distinct identities, and strong internal bonds that have grown around a shared radical ideology. As a result, they can mobilize resources and power effectively. Also, since extremists usually believe that their ends justify any means, they're willing to be violent and ruthless to get what they want. Faced with such adversaries, people who aren't extremist must work hard to build bonds of trust and understanding among themselves and to lay down action plans for a wide range of possible futures.[84]

In our communities, towns, and cities, we can use small-scale experiments to see what kinds of technologies, organizations, and procedures work best under different breakdown scenarios.[85] How, for instance, will we move ourselves around, feed ourselves, and generate energy if our conventional ways of doing these things have been greatly disrupted? And how are we going to keep extremists from manipulating people who are suddenly scared and angry? We can experiment with ideas, too, because ideas powerfully shape our relationship with the natural world and with people around us. In a moment of contingency, the struggle over how we should frame our options and our future is really a battle of ideas. By experimenting with new ideas about politics, economics, and values, we'll be better advocates of a coherent vision of the future and a plausible way of getting there—we'll be stronger, more confident, and less afraid.

In countries that are already very rich, we especially need to figure out if there are feasible alternatives to our hidebound commitment to economic growth, because it's becoming increasingly clear that endless material growth is incompatible with the long-term viability of Earth's environment. What might a "steady-state" economy—an economy that maintains a roughly constant output of goods and services—look like? What economic and ethical values might it be based on? Could it incorporate some (albeit radically transformed) version of market-based capitalism, and would it be compatible with political and personal liberty?[86] And how would we deal with the political and social conflicts that would inevitably arise if there were no growth? Right now conventional wisdom says that a steady-state society would be, at best, a miserable place to live and, at worst, brutally oppressive. Perhaps that's just because so far we've simply found such a radically different future too hard to imagine.[87]

Thinking about alternatives to the growth imperative means thinking about alternatives to conventional economics—an elaborate apparatus of assumptions, theories, and empirical research that reinforces the legitimacy of globalized capitalism and the power of the world's capitalist elites. At the heart of this view is the assumption that the economy is separate from nature and operates much like a machine. The machine's behavior is linear, predictable, and reversible, so it can be managed by a planet-wide class of technocrats—including central bankers and government officials—trained in the arcane science of economics. An alternative theory would recognize that the economy is intimately connected with nature and its energy flows. This larger economic-ecological system often doesn't act like a machine at all. Instead, its behavior is path dependent, marked by threshold effects, and often neither predictable nor controllable. An alternative view would also recognize there are no good substitutes for some of the most precious things nature gives us, like biodiversity and a benign climate. Because we can't adequately replace these things with something else once they're gone, we need to create ways of giving them explicit economic value so people will have an incentive to protect them.[88] Such an alternative view, if developed in detail, would help everyone understand that conventional economics is not unchallengeable truth but rather a particularly potent ideology—a blend of scientific finding, analytical gymnastics, value judgments, and self-congratulation.

Conventional economics is the dominant intellectual rationalization of today's world order. As we've overextended the growth phase of our global adaptive cycle, this rationalization has become relentlessly more complex and rigid and progressively less tenable. Breakdown will, all at once, discredit this rationalization and create intellectual space for new ideas to flourish. But this space will be brutally competitive. We can boost the chances that humane alternatives will thrive by working them out in detail and disseminating them as widely as possible beforehand.[89]

Advance planning means we need to develop a wide range of scenarios and experiment with technologies, organizations, and ideas. We'll do better at these tasks, and we'll also do better in the confusing aftermath of breakdown, if we use a decentralized approach to solving our problems, because traditional centralized and top-down approaches aren't nimble enough, and they stifle creativity. Scientists have found that complex systems that are highly adaptive—like markets and even the immune system of mammals—tend to share certain characteristics. First of all, the individual elements that make up the systems—such as companies in a market economy or T-cells and macrophages in an immune system—are extraordinarily diverse. Second, the power to make decisions and solve problems isn't centralized in one place or thing; instead, it's distributed across the system's elements. The elements are then linked in a loose network that allows them to exchange information about what works and what doesn't. Often in a market economy, for example, several companies will be working at the same time to solve different parts of a shared problem, and important information about solutions will flow between them. Third and finally, highly adaptive systems are unstable enough to create unexpected innovations but orderly enough to learn from their failures and successes. Systems with these three characteristics stimulate constant experimentation, and they generate a variety of problem-solving strategies.[90]

We're all very familiar with just such a system—the Internet, and its subsystem, the World Wide Web. In one respect, humanity is extraordinarily lucky: just when it faces some of the biggest challenges in its history, it has developed a technology that could be the foundation for extremely rapid problem solving on a planetary scale, for radically new forms of democratic decision making, and most fundamentally for the conversation we must have among ourselves to prepare for breakdown. So far, though, we've barely tapped this potential. The Internet and Web—rather than becoming powerful instruments of problem solving, adaptation, and social inclusion—have simply turned into venues for a screaming cacophony of electronic narcissism.

This situation may be changing. Recently, we've seen an explosion of distributed and collaborative problem solving on the Web using various "open-source" approaches. These efforts include the development of the free computer operating system Linux and the Wikipedia online encyclopedia. In both cases a large number of volunteers work on components of a larger task, and they make the products of their work freely available for use by others and for correction and improvement. Over time, the individual components and the larger project that they're part of—an operating system or an encyclopedia—steadily improve in quality.

So far, though, open-source approaches have been applied to solving technical problems like the creation of complex software or large databases. Now we urgently need research to see if we can use this kind of problem-solving approach—and the culture of voluntarism that underpins it—to address the ferociously hard social, political, and environmental problems discussed in this book. It's not at all clear that open-source methods can do the job. That's because, at root, our global problems are not really technical ones: they're political problems fraught with conflicts over values, interests, and power; surrounded by scientific uncertainty; and burdened with deep moral implications. In almost all cases, people disagree about goals and about standards for measuring the success of solutions. Often they don't even agree on what good solutions look like. What, for instance, is a good solution for the problem of global warming? Is it simpler lives and less material and energy consumption? More renewable energy? How about more nuclear power?

But even if open-source methods can't give us clear and final solutions to problems that are ultimately rooted in politics, they're still a powerful way to develop scenarios, experiment with ideas, and lay plans in advance of breakdown. And most important, they can help us build worldwide communities of like-minded people who, in the course of working together on tasks, become bound together by trust and by shared values and understandings.[91] Such communities would then be better able to act with common purpose in a moment of contingency and to seize the opportunity for catagenesis.[92]

BAALBEK:

THE LAST ROCK

"TARGET: SAUDI OIL INDUSTRY," the headline read.
It was the evening of Monday, May 31, 2004. My short flight
from Amman, Jordan, to Beirut, Lebanon, had reached cruising
altitude, and I'd just opened my copy of the day's *International Herald
Tribune*. The Saudi story capped the front page.

Twenty-two people had died as government commandos stormed a
luxury residential compound in the city of Khobar in oil-rich eastern
Saudi Arabia. The commandos were trying to free dozens of hostages—
including Western business executives—held by militant gunmen. They
arrived too late for some. "The gunmen were going through the compound and either shooting people or letting them go," said one Western
diplomat. "They shot non-Muslims, and they let the Muslims go."

I flipped through the paper. Four American soldiers had been killed
by a land mine as the southern part of Afghanistan erupted in violence.
An assassination of a cleric had touched off Sunni-Shiite riots in
Karachi. American forces were again fighting a radical Shiite militia in
Najaf, Iraq. And in southeastern Haiti, tens of thousands of villagers
were huddled together in the open after floods destroyed their homes.
In only thirty-six hours, nearly two meters of rain had poured onto the
region's deforested hillsides, unleashing a torrent of mud, rock, and
debris that buried or drowned nearly two thousand people. It was,
according to the news story, one of the worst natural disasters in
Caribbean history.[1]

Just the everyday crises of a turbulent world? Or were these events the kind of early foreshocks we should expect from a planet under extreme stress? Of course, no one could possibly know for sure. But I was beginning to see some answers to the questions I posed at the start of my investigations.

I was on my way to visit the great Roman ruins in Baalbek, a town in the Bekaa Valley of eastern Lebanon. Many archaeologists and historians of Roman antiquity believe that the construction of the Baalbek temples was the apogee of Roman engineering. In Joe Tainter's office, I'd been intrigued by David Roberts's famous lithograph of the gate of the Temple of Bacchus. Later, when I learned that the complex included the largest stones ever moved by human beings, I knew I had to see the temples for myself. Luckily, being Canadian, I have less trouble traveling in the Bekaa than an American might.

As my flight descended into Beirut, I realized that the trip was more than another chance to marvel at the Roman empire's prowess with rock. The contrast between ancient Rome and modern Lebanon could tell a larger story—a story about, once again, the links between energy, social complexity, and political stability.

Just days before, Beirut had been in turmoil. The highway from the city's center to the airport had been cut when riots spread across the city. The trigger was a sharp rise in fuel prices, as oil reached a record of nearly US$42 a barrel. A confederation of trade unions had called a strike over high prices and government mismanagement of the economy, demanding a 40 percent cut in the price of gasoline. Soon protesters were burning tires in the streets, stoning soldiers, and setting government buildings ablaze. In the southern Shiite district of Hay al-Silom, the army opened fire on rioters, killing five and sending dozens to hospital. It was the worst violence that the deeply divided city had seen in a decade.[2]

Lebanon and the city of Beirut itself are microcosms of our cockeyed world—an incredibly complex mosaic of religions, ethnicities, and identities that's fractured by culture, political interest, and economic disparity. Shiite and Sunni Muslim, Druze, and Maronite Christian—seventeen different sects in all—are packed together in a land that has few natural resources and that's rife with ancient rivalries and enmity. Lebanon sits at the intersection of East and West, Islam and Christianity, and the rich and poor worlds. It's also squeezed between dangerous neighbors: to the north and east lies Syria, an avaricious meddler, and to the south is Israel, an insecure regional superpower. In 1975, the combination of internal and external stresses proved too much, and the country spiraled into a fifteen-year civil war that cost 150,000 lives and reduced much of Beirut to rubble.[3] The day I arrived in the city, everyone was talking about whether the country would slide back into that horror.

The next morning, the sky was crystal blue. After an early walk along one of the city's glorious oceanfront promenades and on through the downtown—still marred by bullet-scarred buildings—I hired a car to take me to Baalbek. We drove south and then turned sharply east to head inland through Beirut's predominantly Shiite districts. Skirting the Sabra and Shatila refugee camps—site of a massacre of Palestinians in September 1982—we continued through the suburb of Ghobeiri and up the twisting highway into the Lebanese Mountains above the city. We crossed the height of land, at 1,600 meters, and wound our way down into the broad Bekaa Valley. The road narrowed to two lanes and made a long, slow curve to the north, toward Baalbek.

Kilometer by kilometer, the landscape and character of the towns and villages changed dramatically. As we traveled up the valley, we passed through Lebanon's wine-growing region, but soon the scattered vineyards petered out. Trees disappeared from the hills on each side of the valley, the soil along the valley bottom became rockier, and the countryside started to look desiccated. Villages and houses were noticeably poorer, cars were dilapidated, and the road was full of potholes. This part of the valley had been a major source of hashish and opium for Europe and North America until the end of the civil war. Then, with U.S. encouragement and funding, the Syrians and the Lebanese government crushed the business, sending the local economy reeling.

We were now in territory controlled by the Hezbollah—a Shiite Islamist group that to much of the Muslim world is a resistance movement and to the United States and many other Western countries is a terrorist organization.* English and French and any non-Arabic script

---

*As the hardcover edition of this book went to press in late July 2006, Israeli forces were bombing Hezbollah targets in Beirut, southern Lebanon, and the Bekaa Valley—including Baalbek—in retaliation for Hezbollah rocket attacks on northern Israel.

vanished from all road and store signs. The urban buzz of central Beirut—a city that's palpably Western in architecture, dress, and style—seemed light-years away.

I thought about my journey from Rome and California to southern France and Lebanon—and about the implications of what I'd learned along the way. It seemed almost impossible to make sense of it all, and I wasn't sure I'd put the story's pieces together in the right way. Still, every shred of my intuition told me that humankind is approaching a monumental shift in its trajectory and prospects and also in the mental tools through which it sees, understands, and copes with its problems. A similar shift has happened before. Between 900 and 200 BCE, a revolutionary transformation of ideas occurred simultaneously in civilizations across Eurasia, from China to Greece. This "Axial Age," as the German existential philosopher Karl Jaspers famously labeled it, produced new categories and cosmologies to guide people's thought, self-understanding, and religious and ethical views. During these centuries, people came to understand that they could use reason and reflection to see beyond their immediate reality, that societies exist in an extended flow of time, that we all have some capacity to exercise our agency to change our circumstances and fate, and that the mundane and spiritual worlds are fundamentally separate.[4] In Jaspers's opinion, those of us alive today are the direct heirs of this transformation; we still think using fundamental categories and assumptions that emerged over two millennia ago.[5]

Could we be, I wondered, on the cusp of a new Axial Age—a transformation, simultaneously around the world, of the deepest principles guiding humankind's diverse civilizations? And if it occurs, what might it look like?

More important, what would I want to come from this transformation? In planning for renewal after breakdown—for catagenesis—we need to have some idea of where we want to go in the future, even if it's just an ideal goal that we know we can only partly achieve. And to figure out what our goal is, we need to be clearheaded about our values.

In Western liberal societies, public discussion of values is dreadfully impoverished. As long as one person's values don't interfere with other people's interests, they're usually considered a private matter. To the extent that we discuss values at all, we do so only when provoked by hot-button issues like abortion, drugs, sexuality, or political corruption. And we often assume that most things around us in our societies don't involve value judgments of any kind. They're just value-neutral facts of life. But whether or not we recognize the fact, people's values—often those of the most powerful among us—shape everything from the income gaps in our economies to the intimate details of how we lead our lives, such as where we live, work, and send our children to school.

In trying to make sense of values, I've long found it helpful to distinguish between three kinds. *Utilitarian values* are simple likes and dislikes—whether one likes chocolate ice cream more than vanilla ice cream, for instance. These are the values most familiar to economists, and they dominate our consumerist culture. *Moral values* concern fairness and justice, especially regarding things like the distribution of wealth, power, and opportunity among people across space and time. Last but definitely not least, *existential values* apply to things that give our lives significance and meaning. Some people might call them spiritual values. They help us understand how we fit in the larger scheme of the universe, and they provide answers to questions like Why are we here? and What is the purpose of my life?

We all ask these questions when we're young, till the obvious discomfort of adults around us makes us stop. Usually the adults tell us that we're naive or that we should take our annoying questions to a religious institution. When we get in the door of our nearest church, mosque, or synagogue, we find there's no real opportunity for discussion. Instead, we're handed a creed of some kind. We're told *what* to think about values, not *how* to think about them.

Because we're reluctant or unable to talk about moral and existential values—and these values remain largely unexplored—utilitarian values fill the void. This is one reason why consumerism has developed such a firm grip on the psyches of so many of us in the West. Without a coherent notion of what will give our lives meaning, we try to satisfy our need for meaning by buying ever more stuff. In the process, the mental muscle that allows us to think and talk about values in complex and sophisticated ways atrophies. Reduced to walking appetites, we lose resilience. We risk becoming hollow people with no character, substance, or core—like eggshells that can be shattered or crushed with one sharp shock.

We finally arrived at Baalbek. The main street was lined with electricity poles decorated with Hezbollah flags, behind which stood dreary low-rise concrete buildings. After passing a traffic island occupied by a large model of Jerusalem's Al Aqsa Mosque, we pulled up in front of the Palmyra Hotel. I stepped out of the car into a bizarre jumble of times and cultures.

The dusty square in front of the hotel was lined with small shops, including an Internet café and a butcher that sold meat pizzas, a local favorite. People and cars milled about. Two teenage boys drove by in an old Toyota, and I noticed that the passenger was fidgeting with a pistol in his lap. Then a well-dressed thirtysomething fellow pulled up in a late-model Mercedes and asked if I wanted to buy some hash. Beyond the square I could just glimpse the ruins, dominated by the last standing pillars of the Temple of Jupiter.

I checked into the Palmyra Hotel. It must have been magnificent in its heyday. Built by a Greek from Istanbul in the 1870s to cater to wealthy Europeans who came by stagecoach from Damascus to visit the ruins, it was Lebanon's first Western-style hotel. Now it looked like a dilapidated time capsule, empty of life (it seemed as though I was the sole guest), and home mainly to the ghosts of more glamorous times. The manager sat in a dimly lit office with two 1940s-era Bakelite rotary phones on his desk. The hallways and rooms featured weary Ottoman-era antiques and dusty display cabinets full of bowls, figurines, and glassware. On the crumbling plaster walls were faded photos of the dig-nitaries, artists, and actors who had stayed there over the decades. There was something indescribably sad about the place.

But any sadness lifted when I saw the view from my sitting room. There before me was one of the world's greatest archaeological mar-vels—the temples of Baalbek. An hour later, I was making my way up the remains of the monumental staircase that led to the forecourt of the Temple of Jupiter. And this time, despite the fact that I'd now seen many examples of extraordinary Roman engineering, I found my imagination and comprehension pushed to their limits, as I tried to conceive how mere muscles could have built stone buildings of such size, complexity, and elegance.

Likely completed one or two decades before the Colosseum in Rome, the Temple of Jupiter must have seemed incredible to a visitor in the late first century.[6] Although similar in design to the Parthenon in Athens, it was twice its size in area and four times its total volume. Fifty-four beautifully sculpted Corinthian columns, each twenty meters high and 135 tons in weight, stood around the temple's perimeter. Across the top of these columns was a delicately carved entablature nearly six meters high. Each block of this entablature, spanning the distance between two columns, weighed almost sixty tonnes. And the corners of the temple's pediment—the triangular gablelike structure that supported the temple's roof at the front—weighed an astonishing seventy-five tonnes.[7] Only six of the columns still stand—those that I'd spied from my hotel—while many bits and pieces of entablature and pediment litter the ground. As I wandered around the site, I kept looking at these enormous pieces of rock and then at the tops of the standing columns. How, I asked myself, could the Romans have possibly hoisted such behemoths . . . up there?

That day and the next I spent hours clambering around the ruins, and then more hours sitting in various spots that offered a good vantage, trying to imagine what the buildings must have looked like when they were intact. I also investigated the adjacent Temple of Bacchus, the front

The last standing columns of the Temple of Jupiter, Baalbek, Lebanon

gate of which—with its keystone hanging precariously in midair—was depicted in the nineteenth-century lithograph I'd seen in Tainter's office. In 1901, German archaeologists lifted the keystone back into place—probably a good thing, from the point of view of the gate's long-term resilience. The temple as a whole is a gem, with splendidly intricate carving over many of its inside and outside surfaces.

The thing that astonished me most about the ruins was hidden from view. I had to leave the temple complex and walk along its north side through a small forest and a field overgrown with brush. When I came to the west side, I found three rocks that have dumbfounded people for centuries. Each is nearly twenty meters long, four meters high, and four meters deep, which means each weighs about eight hundred tonnes, or roughly the same as two fully loaded 747s. Called the "trilithon," the three rocks are laid end to end to form part of a western extension of the platform that supports the temple. They're positioned with the same kind of precision that I'd seen in the Colosseum; the seams between them are so tight that I had to look closely to see them.

The feat of engineering required to cut, move, and hoist into place these rocks is—on the face of it—so implausible that the trilithon has become something of a cult topic among people inclined to believe in the supernatural or ancient visits by aliens. According to such a view, the Romans simply appropriated a platform of rocks that had been constructed long before, perhaps by a race of giants or by aliens as a launching pad for rocket ships.[8] But this is silliness: serious archaeologists don't doubt the Romans did the deed, and they're pretty sure they know how.[9] After all, by the first century CE, the Romans were transporting huge carved stones—including mammoth columns and obelisks—all over the Mediterranean basin.

I was willing to accept the archaeologists' claim that the Romans had put the rocks here, but as I stood looking up at them in awe, I did wonder about something else—about the values that would drive people to invest energy in such a task. What would possess the Romans to build the largest temple complex in the empire at a remote junction along a trading route in a distant part of Syria? What kind of existential values—what kind of understanding of one's purpose in life and spiritual place in the cosmos—would motivate cutting and moving these rocks? And the phenomenal sense of purpose endured: the entire temple complex was built over a period of two centuries. Thousands of tonnes of granite were brought from as far away as the Aswan quarries in Egypt, and the buildings were finished with huge quantities of marble, semiprecious stone, bronze, and gold. The project must have sucked wealth—and, in the final analysis, energy—from across the empire.

A civilization's values powerfully influence what form the civilization takes as well as what kind of evidence of its existence it leaves behind. For the Romans, this evidence is mostly in rock. Barring a staggering earthquake, the rocks of the trilithon will likely stay exactly where they are for millions of years. Who or what would bother, or have the energy, to move them?

I asked myself what kind of long-lasting evidence today's planetary civilization would leave behind—with its overriding values of material growth, production, and consumption. My immediate answer wasn't a happy one. Because little of our modern construction will have the durability of ancient rock buildings, the main evidence of our existence on the planet will probably be damage to Earth's life and environment. A hothouse climate and a 25 to 50 percent decline in the number of species on the planet could last for millions of years, just like the trilithon.

I left the hidden west wall of the temple complex and walked back through the forest. There was one more thing I wanted to see. It was just down Baalbek's main street, at the quarry that produced some of these rocks. As I picked my way along the street's broken sidewalk and past a new branch of a Lebanese bank with a modern ATM behind the front door, I reflected on alternative value systems that could help us achieve different futures. One thing is clear to me now, I thought: our values must be compatible with the exigencies of the natural world we live in and depend on.[10] They must implicitly recognize the laws of thermodynamics, energy's role in our survival, the dangers of certain kinds of connectivity, and the nonlinear behavior of natural systems like the climate. The endless material growth of our economies is fundamentally inconsistent with these physical facts of life.[11] Period. End of story. And a value system that makes endless growth the primary source of our social stability and spiritual well-being will destroy us.[12]

Our current values serve the interests of today's political and economic elites, and so are aggressively defended by these elites. Growth, even in already obscenely rich societies, is sacrosanct. This central value won't really change until it's discredited by some kind of major shock, which probably means some kind of system breakdown. Then, alternative values that are centered on the idea of resilience might flower, not just at the fringes of our societies but also at their core. They might, for instance, promote the merit of smaller populations that tread lightly on nature, of decentralized communities that can take better care of their own needs, and of lifestyles that are far less complex and fast paced.[13]

Alternative values might also promote broader, fairer, and more vigorous democracy, maybe using some kind of open-source approach. New forms of democracy are essential, because we need as many heads as possible working together to solve our common problems, and because the larger the number of people involved in making crucial decisions that affect everyone, the less likely that narrow elite interests will dominate.[14] And only through much broader and deeper democratic practice will humankind likely develop the expansive "moral commonwealth" essential to our collective survival.[15] Only when we all grasp that we're in one boat together—that together we're one *we* with an indivisible fate—will we be serious about making the concessions to each other that are essential if we're to address global challenges like climate change and energy scarcity.[16] And any kind of new democracy must encompass not only communities, towns, cities, and societies, but humankind as a whole. In fact, it's hard to imagine how we'll prosper together on this tiny planet if we don't eventually have some kind of democratic world government.[17] Of course, many hard-nosed realists would say that this is an implausible and even scary idea. Maybe that's only because alternatives to our current trajectory remain so difficult to imagine.

I turned left off the main street and, after walking up a cul-de-sac, came to the quarry. No one was around. Before me was one of the most amazing things I've ever seen. In a large depression in the ground, rearing out of the dirt like a doomed ship about to slide beneath the waves for the last time, was the renowned Hajar el Hibla, or "Stone of the pregnant woman."[18] Nearly twenty-two meters long and wider than the trilithon stones on each side, this monster weighs one thousand metric tons. It appeared to have been cut, dressed, and prepared for transport to join its comrades in the extension of the temple platform, but for some mysterious reason it had been left in the quarry, its lower end still attached to the underlying bedrock.

No one knows why the Romans wanted to extend the temple platform. In addition to their work along its west side, they placed a string of

The Hajar el Hibla rests in a quarry near the Temple of Jupiter. (The flag at the right end of the rock is one meter high.)

huge rocks, each weighing hundreds of tonnes, along the platform's base to the north. Today, these rocks look as if they were meant to be the foundation of a northward extension, yet nothing was ever put on top of them.

An explanation of what had happened came to me, based on what I'd learned from my travels. In Roman times the Bekaa valley's fertile soils made it one of the richest parts of the empire. Yet over the centuries, in their quest for energy, the valley's inhabitants likely felled the forests on the valley's slopes and overused the valley's soils. So the land became poorer and drier. Almost certainly, just as had happened elsewhere in the empire when Romans overtaxed the land, the valley's output of grain fell, and the energy available from wood probably became scarce too. These changes would have made it far harder to support the temple complex, let alone continue to expand it by, for instance, extending its foundations with rocks weighing hundreds of tonnes.

If my supposition was right, the rock before me—the last rock—was a powerful symbol of the exhaustion of an enormous social and political enterprise. It was enduring evidence of overreach. The Romans had sought to cut and move a stone unlike anything they'd ever moved before. And they couldn't do it. In the end, Rome's existential values—values that said, among other things, that life's meaning could be found partly in monumental efforts of engineering—led the empire into a dead end from which it couldn't escape.

As I stood looking at the Hajar el Hibla, I wondered, Will our civilization have its last rock too—some mighty and misguided project left unfinished because we couldn't muster the energy, will, or competence to complete it? Or maybe, just maybe, before we've exhausted nature and ourselves in a futile effort to produce meaning from material things, we'll reconsider our values and recognize that we can choose another path into the future.

Our first step down this path must be to acknowledge that our global situation is urgent—that we're on the cusp of a planetary emergency—and then to begin a wide-ranging and vigorous conversation now about what we can and should do.

We have an advantage over the Romans that gives us a head start: we understand much better how the complex systems around us behave. Rome failed in the end partly because it didn't—and couldn't—understand that it had doomed itself to fail. It couldn't see clearly the multiple stresses converging on it; that it was bounded by the exigencies of its natural world; and that, as complexities and entrenched power accumulated, it was inexorably becoming a static, brittle system. We don't have the same excuse. We know that static, brittle systems don't survive. We also understand that in any complex adaptive system, breakdown, if limited, can be a key part of that system's long-term resilience and renewal.

This knowledge—if we adopt the prospective mind that allows us to use it effectively—enables us to act with vastly greater wisdom than did the Romans. It also tells us that if we want to thrive, we need to move from a growth imperative to a resilience imperative. Some form of economic growth is absolutely essential for billions of people, but for the world as a whole, and even for individual societies, it must not be at the expense of the overarching principle of resilience, so needed for any coming transformation of human civilization.

We still can't see beyond the white wall of fog in front of us, but now we have knowledge we can use like a compass—to help us, together, choose our path through a future full of surprises, danger, and opportunity.

NOTES

PROLOGUE

1. The use of field artillery to destroy buildings along Van Ness Avenue is not widely noted in histories of the great San Francisco fire, even in the U.S. Army's documentation of its role in the disaster. In the fire's aftermath, great controversy surrounded the decision to destroy buildings to create firebreaks. For this reason, the use of artillery was likely downplayed. The army's report on the dynamiting of buildings was prepared by Captain Le Vert Coleman, and is available from the San Francisco Museum at http://www.sfmuseum.org/1906/coleman.html. The account of the great San Francisco fire in these paragraphs is drawn largely from reports in *The New York Times* on April 19, 20, and 21, 1906, especially "Bombardment a Mile Long Fails to Save San Francisco: Mansions Wrecked by Cannon in Last Stand on Nob Hill," *New York Times*, April 20, 1906, 1. For a fascinating recent treatment of the quake and fire, see Simon Winchester, *A Crack in the Edge of the World: America and the Great California Earthquake of 1906* (New York: HarperCollins, 2005), especially page 290, which explains how the quake caused the fire.

2. Rome's ruins have inspired literature, art, and scholarly investigation for millennia. Perhaps most famously, Edward Gibbon said he was prompted to write his monumental history on October 15, 1764, as he "sat musing amidst the ruins of the Capitol, while the barefooted fryars were singing Vespers in the temple of Jupiter." Unfortunately, modern scholars aren't sure exactly where Gibbon was sitting at the time; there were no visible ruins of the Temple of Jupiter's superstructure in the eighteenth century. Gibbon's remark from his autobiography is quoted in David Womersley's introduction to Edward Gibbon, *The History of the Decline and Fall of the Roman Empire*, edited and abridged by David Womersley (London: Penguin, 2000 [1776]), xvi.

3. See Joseph Tainter, "Post-Collapse Societies," *Companion Encyclopedia of Archaeology*, Graeme Barker, ed. (London: Routledge, 1999), 988-1039. As we will see later in this book, there's scholarly controversy about the extent to which one can say the Roman empire "fell," "declined," or "collapsed," and also about whether people's average standards of living fell.

CHAPTER ONE

1. James Burke writes marvelously about urban dwellers' dependence on technologies that they don't comprehend, their vulnerability to failure of these technologies, and the implications of a mass exodus from cities in the event of such failure. See Burke, *Connections* (Boston: Little, Brown, 1978), 4–7.

2. A survey article that provides a somewhat similar breakdown of stresses is Robert Kates and Thomas Parris, "Long-term Trends and a Sustainability Transition," *Proceedings of the National Academy of Sciences* 100, no. 14 (July 8, 2003): 8062–67.

3. Harvard University's John Holdren, a physicist and environmental scientist, notes, "Civilization remains dependent on nature, for most of the cycling of nutrients on which food production depends, for most of the regulation of crop pests and agents and vectors of human disease, for most of the detoxification and disposal of wastes, and for the maintenance of climate conditions within limits conducive to all these other environmental services and to the human enterprise more generally." Holdren, "Environmental Change and the Human Condition," *Bulletin of the American Academy of Arts and Sciences* 57, no. 1 (Fall 2003): 25.

4. Two good examples of such arguments are Jack Hollander, *The Real Environmental Crisis: Why Poverty, Not Affluence, Is the Environment's Number One Enemy* (Berkeley: University of California Press, 2003); and Peter Huber, *Hard Green: Saving the Environment from Environmentalists, a Conservative Manifesto* (New York: Basic, 1999).

5. For a discussion of these constraints, see Thomas Homer-Dixon, *The Ingenuity Gap: Facing the Economic, Environment, and Other Challenges of an Increasingly Complex and Unforgettable World* (New York: Vintage, 2002).

6. Jared Diamond, "The Last Americans," *Harper's Magazine* (June 2003): 45.

7. Eric Hobsbawm, *Age of Extremes: The Short Twentieth Century, 1914–1991* (London: Abacus, 1994), 15.

8. Christopher Chase-Dunn, Yukio Kawano, and Benjamin Brewer, "Trade Globalization Since 1795: Waves of Integration in the World-System," *American Sociological Review* 65 (February 2000): 77–95.

9. In his autobiography, the Manhattan Project physicist Luis Alvarez writes, "With modern weapons-grade uranium, the background neutron rate is so low that terrorists, if they had such material, would have a good chance of setting off a high-yield explosion simply by dropping one half of the material onto the other half. . . . Even a high school kid could make a bomb in short order." Alvarez, *Alvarez: Adventures of a Physicist* (New York: Basic Books, 1987), 125.

10. In the 1970s, the Italian electrical engineer and futurist Roberto Vacca presented an argument about breakdown in complex systems that's similar in some respects to the one offered in these pages, especially in its focus on interactions between stresses and the possibility of "coincident breakdown." According to Vacca, systems break down when their complexity and congestion exceed managers' control. Although his argument doesn't reflect recent research on self-organizing complex systems, it still bears close attention. Vacca has developed considerable renown for his ability to predict system failures, including the collapse of the

Soviet Union. See Vacca, *The Coming Dark Age: What Will Happen When Modern Technology Breaks Down*, trans. J. S. Whale (Garden City, NY: Anchor, 1974).

11. James Howard Kunstler makes an argument along these lines in *The Long Emergency: Surviving the Converging Catastrophes of the Twenty-First Century* (New York: Atlantic Monthly Press, 2005).

12. Jack Goldstone, *Revolution and Rebellion in the Early Modern World* (Berkeley: University of California Press, 1991).

13. This is especially true because such events might not be independent of each other: in some circumstances, the occurrence of one kind of shock could boost the likelihood of others.

14. For instance, the American writer Gregg Easterbrook contends that concerns about the rising likelihood of social breakdown are simply a product of "collapse anxiety"—a generalized fear in rich countries that high standards of living can't be sustained. He thus manages to disparage such concerns by labeling them a psychopathology, without really explaining their source. See Easterbrook, *The Progress Paradox: How Life Gets Better While People Feel Worse* (New York: Random House, 2003).

15. Some thoughtful people have reached similar conclusions about our future. See, for example, Martin Rees, *Our Final Hour: A Scientist's Warning: How Terror, Error, and Environmental Disaster Threaten Humankind's Future in This Century—On Earth and Beyond* (New York: Basic, 2003); Robert Harvey, *Global Disorder: How to Avoid a Fourth World War* (New York: Carroll & Graf , 2003); Jared Diamond, *Collapse: How Societies Choose to Fail or Succeed* (New York: Viking, 2005); and Didier Sornette,"2050: The End of the Growth Era," chapter 10 in *Why Stock Markets Crash: Critical Events in Complex Financial Systems* (Princeton: Princeton University Press, 2003).

16. In statistical terms, catastrophic events lie in the tail of a "power-law frequency distribution." This means they're very rare but not impossible. As we'll see in chapter 5, the same is true of highly connected hubs in scale-free networks. Richard Posner uses cost-benefit analysis (technically, expected value calculations) to argue that it makes economic sense to invest in preventing rare, large-scale catastrophes. See Posner, *Catastrophe: Risk and Response* (New York: Oxford, 2005).

17. Amory Lovins and Hunter Lovins, *Brittle Power: Energy Strategy for National Security* (Andover, MA: Brick House, 1982), 1.

18. Amory Lovins, correspondence with the author, July 23, 2002. Permission granted for quotation.

19. The word *catagenesis* is also used in petroleum geology: categenesis happens when organic compounds are "cracked" or broken down into oil under conditions of high pressure and temperature deep underground.

20. For a thorough survey of the history of systems thinking, see Charles François, "Systemics and Cybernetics in a Historical Perspective," *Systems Research and Behavioral Science, Syst. Res.* 16 (1999): 203–19.

21. We'll learn in chapter 2 that complex adaptive systems are orderly, thermodynamically open, and far from thermodynamic equilibrium, that their parts are diverse and specialized, and that they exhibit self-organization. In chapter 5, we'll learn that the parts of complex systems are often connected together in dense,

scale-free networks that produce feedbacks and synergies. A serviceable indicator of a system's complexity is its "algorithmic complexity," which is the length of a computer program, or algorithm, that can reproduce the system's behavior (the longer the algorithm, the more complex the system). On measures of complexity, see Homer-Dixon, *The Ingenuity Gap*, 115–16.

22. A good overview of this research is C. S. Holling, "Understanding the Complexity of Economic, Ecological, and Social Systems," *Ecosystems* 4 (2001): 390–405. See also Lance Gunderson and C. S. Holling, *Panarchy: Understanding Transformations in Human and Natural Systems* (Washington, DC: Island Press, 2002).

23. The notion of constrained breakdown may seem odd because most of us assume that breakdown has to be—almost by definition—sudden, thoroughgoing, and catastrophic. But in reality there are lots of gradations along a continuum between catastrophic collapse at one extreme and straight-line stability at the other.

24. When small changes produce very large effects, specialists say the system is "sensitive to initial conditions." Such sensitivity is a key feature of systems that exhibit chaotic behavior.

25. Figures from the Internet Systems Consortium, available at http://www.isc.org/index.pl?/ops/ds/.

26. Henry Mintzberg surveys the dismal record of our best forecasters in "The Performance of Forecasting," in *The Rise and Fall of Strategic Planning: Reconceiving Roles for Planning, Plans, and Planners* (New York: Free Press, 1994), 228–30.

27. An informative attempt at long-range forecasting is the *State of the Future* project of the Millennium Project of the American Council for the United Nations University. This project produces annual reports that are among the best efforts at synthesizing large amounts of data and expert opinion on humankind's future. Once again, though, forecasts tend to be straight-line extrapolations of current trends. Further information is available at http://www.acunu.org.

28. James William Sullivan, "The Future Is a Fancyland Palace," in Dave Walter, ed., *Today Then: America's Best Minds Look 100 Years into the Future on the Occasion of the 1893 World's Columbian Exposition* (Helena, MT: American & World Geographic Publishing, 1992), 27.

CHAPTER TWO

1. The architectural historian Frank Sear writes, "[The voussoir arch] was not a Roman invention. Probably of eastern origin, it was making a tentative appearance in Hellenistic and Etruscan architecture by the fourth century." According to Jean-Pierre Adam, "It can be established that the technique of the true arch arrived in the Italian peninsula gradually and that the Greeks and the Etruscans, more advanced in the art of stone-work, worked out the first models known to the Romans." See Sear, *Roman Architecture*, revised edition (London: B. T. Batsford, 1989), 17–18; and Adam, *Roman Building: Materials and Techniques*, trans. Anthony Mathews (London: Routledge, 2001), 158–63.

2. Travertine has a density of about 2.7 grams per cubic centimeter, giving a mass of 2.7 metric tons per cubic meter. I estimated that the keystone in question was about 2.1 cubic meters in volume, giving a total mass of 5.7 metric tons.

3. Rabun Taylor, *Roman Builders: A Study in Architectural Process* (Cambridge: Cambridge University Press, 2003), 135.

4. Ibid., 8, 134.

5. Stone clamps were made of forged iron; after their insertion, molten lead was poured in the void around them to fix them in place and deter rusting.

6. The data and calculations for stone, concrete, and brick can be found at www.theupsideofdown.com/rome/colosseum. The figure for the number of tons of marble was taken from John Pearson, *Arena: The Story of the Colosseum* (London: Thames and Hudson, 1973), 84, and that for the quantity of metal from Sear, *Roman Architecture*, 138.

7. Joseph Tainter, subsection II, "Energy Basis of Ancient Societies," in "Sociopolitical Collapse, Energy and," *Encyclopedia of Energy*, Cutler Cleveland, ed. (San Diego: Academic Press/Elsevier Science, 2004), 529–43.

8. "The history of human culture can be viewed as the progressive development of new energy sources and their associated conversion technologies." Charles Hall et al., "Hydrocarbons and the Evolution of Human Culture," *Nature* 426, no. 6964 (November 20, 2003): 318–22. See also Alfred Crosby, *Children of the Sun: A History of Humanity's Unappeasable Appetite for Energy* (New York: Norton, 2006).

9. M. S. Spurr, *Arable Cultivation in Roman Italy: c. 200 B.C.–c. A.D. 100* (London: Society for the Promotion of Roman Studies, 1986), 1–16.

10. Ian Graham and Joseph Tainter generously contributed to the ideas and text in the following paragraphs. The Swedish human ecologist Alf Hornborg develops an argument similar to the argument in this subsection in *The Power of the Machine: Global Inequalities of Economy, Technology, and Environment* (Walnut Creek, CA: AltaMira Press, 2001).

11. An early statement of this argument can be found in Leslie White, *The Science of Culture: A Study of Man and Civilization* (New York: Farrar, Straus and Giroux, 1949 [1969]), especially chapter 13, "Energy and the Evolution of Culture," 363–93, in particular 367.

12. There is a vast amount of heat energy in the ground—heat that has been absorbed from the sun above or that has percolated from Earth's core below—but we can't use it to power our cars or light our streets. We can use it to heat our buildings, but we first need a high-quality form of energy, like electricity, to drive a heat pump (basically a refrigerator operating in reverse) to concentrate the ground's diffuse heat into useful building heat.

13. There are at least three independent concepts of energy quality. First, and perhaps most fundamental, is energy quality in terms of capacity to do work—or what physicists call "exergy"—as measured by some thermodynamic indicator like Gibbs free energy. Energy quality, by this conception, depends in part on physical context, especially the boundary conditions between two systems. For instance, the amount of work that can be done by energy flowing from a heat source to a heat sink is proportional to the temperature difference between the two systems. A second concept is energy quality in terms of energy density, as

defined by a measure like calories/unit of mass or calories/unit of volume. And third is energy quality in terms of usability of the energy resource, which is a function of human technology. Oil wasn't a high-quality energy resource until human beings developed the technologies to exploit it. This tripartite distinction means that we can have a degradation of energy quality in a system in thermodynamic terms, while energy quality in parts of the system nevertheless increases in terms of density and usability. For discussions of energy quality, see Charles Hall, Cutler Cleveland, and Robert Kauffman, *Energy and Resource Quality: The Ecology of the Economic Process* (Niwot, Colorado: University Press of Colorado, 1992); Howard Odum, *Environment, Power, and Society* (New York: Wiley-Interscience, 1971); and Joseph Tainter et al., "Resource Transitions and Energy Gain: Contexts of Organization," *Conservation Ecology* 7, no. 3 (2003), available at www.ecologyandsociety.org/vol7/iss3/art4/print.pdf.

14. This first law of thermodynamics is a special case of the more general principle of the interchangeability of matter and energy stated by Albert Einstein in his special theory of relativity, $E = mc^2$.

15. I'm assuming here that the physical processes in question are not "reversible," which is almost always true. In special circumstances, however, some processes are reversible: for instance, chemical reactions, if run slowly enough, can be reversible, and some coherent quantum-mechanical processes like tunneling and superconductivity are reversible. In these cases, entropy stays constant. See Seth Lloyd, "Going into Reverse," *Nature* 430, no. 7003 (August 26, 2004): 971.

16. Bruce Frier, "Demography," chapter 27 in Alan Bowman, Peter Garnsey, and Dominic Rathbone, *The Cambridge Ancient History, Second Edition, Vol. XI, The High Empire, A.D. 70–192* (Cambridge: Cambridge University Press, 2000), 787–816, especially 793.

17. A classic discussion is Ludwig von Bertalanffy, "An Outline of General System Theory," *British Journal for the Philosophy of Science* 1, no. 2 (August 1950): 134–165, especially 162. Bertalanffy distinguishes between "catamorphosis" (the inevitable degradation of inorganic matter) and "anamorphosis" (the spontaneous creation of complexity and diversity in living nature).

18. The relatively new branch of physics of non-equilibrium thermodynamics studies how orderly, complex, and self-organizing systems are possible. Classical thermodynamics assumes that things happen slowly and that the flow of energy in and out of a system is very small relative to the energy inside the system itself. The system is in equilibrium with its surrounding environment, which means, in simplest terms, that it has the same temperature as that environment. But non-equilibrium systems like steel mills, societies, or ecosystems take in and expel vast amounts of energy and have a much higher temperature than their surroundings. Physicists call such systems "dissipative structures" because the energy that sustains them is dissipated to waste heat. The great theoretical physicist Erwin Schrödinger pioneered non-equilibrium thermodynamics. For a lay account, see his book, *What Is Life?* (Cambridge: Cambridge University Press, 1944), especially chapter 6, "Order, Disorder and Entropy," 72–80. For more technical details, see Grégoire and Ilya Prigogine, *Self-Organization in Nonequilibrium Systems* (New York: John Wiley and Sons,

1977). The field of non-equilibrium thermodynamics remains contentious, and some scientists argue it explains little. For a skeptical review, see Philip Anderson and Daniel Stein, "Broken Symmetry, Emergent Properties, Dissipative Structures, Life: Are They Related?" in F. Eugene Yates, ed., *Self-Organizing Systems: The Emergence of Order* (New York: Plenum Press, 1987), 445–57. In contrast, Eric Schneider and James Kay argue that self-organization functions to increase the rate of energy degradation, which in turn helps move the overall system back toward thermodynamic equilibrium. "No longer is the emergence of coherent self-organizing structures a surprise, but rather it is an expected response of the system as it attempts to resist and dissipate externally applied [energy] gradients which would move the system away from equilibrium." See their article, "Complexity and Thermodynamics: Towards a New Ecology," *Futures* 26, no. 6 (1994): 626–47.

19. For a fascinating account of the emergence of an early form of the corporation in the form of *societas publicanorum* or "society of publicans," see Ulrike Malmendier, "Roman Shares," in William Goetzmann and K. Geert Rouwenhorst, eds., *The Origins of Value: The Financial Innovations That Created Modern Capital Markets* (Oxford: Oxford University Press, 2005), 31–42.

20. "Many once-proud ancient cultures have collapsed, in part, because of their inability to maintain energy resources and societal complexity." Hall et al., "Hydrocarbons."

21. Further details on these calculations can be found at www.theupsideofdown.com/rome/colosseum.

22. Similarly, the lime used in concrete was slaked in wood-fired kilns.

23. Jean-Pierre Adam provides a wonderfully detailed and illustrated account of Roman techniques for cutting rock in "Materials," chapter 2 of *Roman Building*, 22–40.

24. Sear, *Roman Architecture*, 139. It has also been estimated that 292,000 cartloads were required to transport the materials to build the Colosseum. Over five years, assuming 220 working days a year, this translates into about 265 cartloads for each working day. See William MacDonald, *The Architecture of the Roman Empire: I. An Introductory Study* (London: Yale University Press, 1965), 148; and Pearson, *Arena*, 84.

25. Janet DeLaine points out that Roman "concrete" was really mortared rubble construction. The Romans didn't combine mortar with aggregate and then pour the mixture into forms, as do modern builders. Instead they built walls, for instance, by first using mortar and something like brick to make the two faces of the wall; after the faces dried to create a permanent form, they then laid between them "alternate layers of rubble and a stiff mortar" to form the wall's core. See DeLaine, "Bricks and Mortar: Exploring the Economics of Building Techniques at Rome and Ostia," chapter 11 in David Mattingly and John Salmon, eds., *Economies beyond Agriculture in the Classical World* (London: Routledge, 2001), 230–68.

26. Adam, *Roman Building*, 174–77.

27. On Roman cranes in general, see Adam, *Roman Builders*, 43–48.

28. Ibid., 43.

29. Taylor, *Roman Builders*, 170–72.

30. Karen broke down the building's three-dimensional volume into smaller geometric forms, like cubes, cylinders, sections of cones, and three-dimensional ellipses. The pillars of the arch containing my keystone, for example, were two

rectangular cubes standing on end, while the arch itself was another rectangular cube laid across the top of the pillars with a cross-sectional slice of a cylinder subtracted to account for the arch's open space. Then, by calculating the volume of each of these geometric objects and multiplying the volume by travertine's mass per cubic meter, she estimated the arch's total mass. Finally, by adding up the mass of all the building's components made of the same material, she produced an estimate of the Colosseum's total mass of travertine.

31. To keep a steady stream of materials coming into the Colosseum, we assumed that two oxen were on the road to the Colosseum for every ox making the return journey to the quarry.

32. Pearson, *Arena*, 85. On the guilds involved in Roman construction in general, see MacDonald, *The Architecture of the Roman Empire*, 144.

33. We based these calculations on detailed estimates of the labor requirements in Roman construction developed by the Oxford scholar of Roman engineering Janet DeLaine. See DeLaine, *The Baths of Caracalla: A Study in the Design, Construction, and Economics of Large-scale Building Projects in Imperial Rome* (Portsmouth, RI: Journal of Roman Archaeology, Supplementary Series no. 25, 1997).

34. On the number of fountains, see Taylor, *Roman Building*, 143–44.

35. The standard unit of human energy consumption is the kilocalorie, which is sometimes called the Calorie (in upper case) by human nutrition specialists. The kilocalorie corresponds to 1,000 calories (in lower case), where a calorie is the amount of heat required to raise one gram of water one degree Celsius. Dieters commonly talk about calories, and so do just about all media commentators on dieting and human health, but they're really referring to kilocalories (or Calories).

36. A number of important assumptions are implicit in these figures. For instance, we assumed that the resting caloric consumption, which specialists call "the basal metabolic rate," is 1,694 kilocalories a day for human beings and 6,261 kilocalories a day for oxen, while the heavy-work caloric consumption is 3,015 kilocalories for human beings and 11,144 kilocalories for oxen. We also assumed frictional coefficients of 0.1 for cart transport offsite and 0.3 for sliding of materials on site; an efficiency of conversion of calories to work of 40 percent for both humans and oxen; and a general inefficiency factor of 50 percent for all hoisting and sliding by laborers. To ascertain the robustness of our conclusions in the event of changes in some of these assumptions, we conducted a sensitivity analysis that varied both the inefficiency factor and frictional coefficients. For further details, see www.theupsideofdown.com/rome/colosseum. Calculations of basal metabolic rate and adjustments for activity level are based on Vaclav Smil, *Feeding the World: A Challenge for the Twenty-First Century* (Cambridge, MA: MIT Press, 2000), 149, 215–23.

37. Scholars generally believe that the complex of chambers and passageways under the arena's floor was added after the building's initial construction, probably by Emperor Domitian, Vespasian's younger son.

38. "Wheat was the staple diet of the vast majority of the people, and far and away the largest item in their food bill." A. H. M. Jones, *The Roman Economy: Studies in Ancient Economic and Administrative History*, P. A. Brunt ed. (Totawa, NJ: Rowman and Littlefield, 1974), 192.

39. See M. P. Cato, chapter 54 of *Cato, The Censor, On Farming [De agricultura]*, trans. Ernest Brehaut (New York: Octagon Books, 1966), 77–78.

40. The value of alfalfa as fodder in Roman times is discussed in Michael Russelle, "Alfalfa," *American Scientist* 89, no. 3 (May–June 2001): 252, available at http://www.americanscientist.org/template/AssetDetail/assetid/14349/page/1;jsessionid=aaa94U-2fac3-m.

41. See M. P. Cato, chapter 54, *On Farming [De agricultura]*, in *Roman Farm Management: The Treatises of Cato and Varro done into English, with Notes of Modern Instances by a Virginia Farmer*, trans. and ed. Fairfax Harrison (New York: MacMillan, 1918), 45.

42. Geoffrey Rickman, chapter 5, "The Corn Lands," in *The Corn Supply of Ancient Rome* (Oxford: Clarendon Press, 1980), 94–119; and Greg Aldrete and David J. Mattingly, "Feeding the City: The Organization, Operation, and Scale of the Supply System for Rome," in D. S. Potter and D. J. Mattingly, eds., *Life, Death, and Entertainment in the Roman Empire* (Ann Arbor: University of Michigan Press, 1999), 171–204. See also Emin Tengström, *Bread for the People: Studies of the Corn-Supply of Rome during the Late Empire* (Stockholm: Paul Åströms Förlag, 1974).

43. Despite Rome's outstanding road system, overland transport of grain was prohibitively expensive. The Roman historian A. H. M. Jones writes, "Wheat seems in fact rarely or never to have been transported any distance by land, except by the imperial government, which did not have to count the cost." According to Michael Fulford, however, archaeological evidence indicates there were substantial shipments of grain by sea around the Mediterranean basin, and not just to the city of Rome. See Fulford, "Economic Interdependence among Urban Communities of the Roman Mediterranean," *World Archaeology* 19 (1987): 58–75; and Jones, *The Roman Economy*, 37. On the role of Etruria, Campania, and Latium in supplying food to Rome, see Rickman, *The Corn Supply of Ancient Rome*, 14.

44. A. H. M. Jones, *The Later Roman Empire, 284–602: A Social, Economic, and Administrative Survey*, Vol. 1 (Baltimore: Johns Hopkins University Press, 1964), 698. Aldrete and Mattingly, "Feeding the City," 179–84.

45. Jérôme Carcopino, *Daily Life in Ancient Rome: The People and the City at the Height of the Empire*, edited with bibliography and notes by Henry Rowell and translated by E. O. Lorimer (New Haven: Yale University Press), 18. On the logistics of grain storage, see Rickman, chapter 6, "Transport, Storage, and Prices," *The Corn Supply of Ancient Rome*, 120–55.

46. On wheat yields, see M. T. Varro, Book I, Chapter 64 of *On Farming (Rerum rusticarum)*, trans. Lloyd Storr-Best (London: G. Bell and Sons, Ltd., 1912), 92 and the calculations at www.theupsideofdown.com/rome/colosseum. Our estimate of wheat yields, based on Varro, is high compared with some other authors and therefore leads to a conservative estimate of the land needed to build the Colosseum. For instance, Smil provides a figure of 400 kilograms of wheat per hectare, and Jones interprets tax and rent records in Roman Egypt to suggest yields around 1,000 kilograms. See Table A3.9, "Labor Requirements and Energy Costs of European Wheat Harvests, 200–1800," in Vaclav Smil, *Energy in World History* (Boulder: Westview, 1994), 89; and Jones, *The Roman Economy*, 83. On alfalfa yields, see Purdue University, Centre for New Crops and Plants Products,

"Medicago sativa L.," sections on "Energy" and "Yields and Economics," based on James A. Duke, "Handbook of Energy Crops," unpublished, 1983, available from http://www.hort.purdue.edu/newcrop/duke_energy/Medicago_sativa.html.

47. Our calculations assumed that 10 percent of the grain produced in any year was held back for seed for the following year's planting, and 30 percent of the grain produced was lost to spoilage and pests. The estimate of holdback for seed is likely conservative; for instance, during the Middle Ages farmers in Europe usually put aside at least a third of their crop for this purpose. Our spoilage estimate is drawn from Vaclav Smil's discussion. In modern-day agriculture, Smil writes, cumulative losses—from harvesting and threshing through storage, transport, and milling—range "from well below 10 percent to as much as 40 percent (and even higher figures [have been] reported for some African crops)." See Smil, *Feeding the World*, 185–86.

48. We estimated that a kilogram of wheat contains 3,420 kilocalories, while a kilogram of dry alfalfa contains 2,557 kilocalories. For the caloric content of wheat, see USDA Agricultural Research Service, Nutrient Data Laboratory. *USDA National Nutrient Database for Standard Reference*. Release 18, 2006, "Wheat, hard white," NDB no. 20074, available from http://www.ars.usda.gov/main/site_main.htm?modecode=12354500. For caloric content of alfalfa hay, see Douglas M. Considine, ed., "Feedstuffs," *Foods and Food Production Encyclopedia* (New York: Van Nostrand Reinhold, 1982), 616–63, especially 621.

49. The EROI concept was first introduced in Cutler Cleveland, Robert Costanza, Charles Hall, and Robert Kaufmann, "Energy and the U. S. Economy: A Biophysical Perspective," *Science* 255 (1984): 890–97. See also Hall, Cleveland, and Kauffman, *Energy and Resource Quality*, 27–29. Calculating an EROI raises some difficult issues of aggregation across different qualities of energy. For a discussion, see Cutler Cleveland, "Net Energy from the Extraction of Oil and Gas in the United States," *Energy* 30, no. 5 (April 2005): 769–82.

50. Hall, "Hydrocarbons," 320.

51. Chris Wickham, "The Other Transition: From the Ancient World to Feudalism," *Past and Present* 103, (May 1984): 3–36, especially 6.

52. M. S. Spurr, "Arable Cultivation in Roman Italy, c.200 B.C.–c. A.D. 100," *Journal of Roman Studies*, Monograph no. 3.(London: Society for the Promotion of Roman Studies, 1986): 138–39.

53. The difference in EROIs for wheat and alfalfa partly explains why Romans preferred to use, wherever possible, draft animals like oxen rather than slaves or laborers to work. The EROI figures provided here for wheat are considerably lower than those suggested by other scholars, such as Smil. The latter's calculations, however, apparently do not include losses due to spoilage and vermin or the need to put aside a portion of the harvest for the subsequent year's seeding; nor do they take into account the energy cost of keeping laborers alive on off days. See Table A3.9, "Labor Requirements and Energy Costs of European Wheat Harvests, 200–1800," in Smil, *Energy in World History*, 89.

54. This estimate is conservative: the actual amount the Romans needed was almost certainly larger. As mentioned earlier in this chapter, a number of things were left out of the calculation of the energy needed to build the Colosseum. Also, the calculation of the amount of grain needed to supply this energy excludes the energy needed

to move that grain itself. In a sense, the calculation implicitly assumes that the food grew right next to the construction site. But, in reality, transporting it from Africa, Egypt, Sicily, and Spain—or even from Etruria—would have required a great deal of work. This is, once again, an open system problem: not only did the grain have to be shipped but the shippers had to be fed, as did the people who fed the shippers, and the people who constructed the shippers' boats and carts, and so on.

55. The details of these calculations can be found at www.theupsideofdown.com/rome/colosseum. The figures in this paragraph incorporate the assumption that at any one time 50 percent of the land had to be left fallow to maintain its fertility. To calculate the farm area needed to carve, move, and place the keystone, we assumed that the food energy for these tasks was generated over one growing season only.

56. Rome was the "most urbanized state of the Western world before modern times." Roger S. Bagnall and Bruce W. Frier, *The Demography of Roman Egypt* (Cambridge: Cambridge University Press, 1994), 56.

57. See Lewis Mumford's marvelous discussion of Rome in "The Natural History of Urbanization," in William L. Thomas Jr., ed., *Man's Role in Changing the Face of the Earth* (Chicago and London: University of Chicago Press, 1956), available at http://habitat.aq.upm.es/boletin/n21/almum.en.html. See also Sander van der Leeuw and Bert de Vries, "Empire: The Romans in the Mediterranean," chapter 7 in Bert de Vries and Johan Goudsblom, eds., *Mappae Mundi: Humans and Their Habitats in a Long-Term Socio-Ecological Perspective: Myths, Maps and Models* (Amsterdam: RIVM, Amsterdam University Press, 2002), 209–56.

58. The maximum size of the city's population is open to dispute. Estimates range from a low of 250,000 to a high of 1.6 million. After a balanced assessment, Carcopino arrives at an estimate in the middle of the range. "The available data," he writes, "combine to force us to conclude that the inhabitants of Rome must have reached nearly a million." See Carcopino, *Daily Life in Ancient Rome*, 10–21, especially 18. Walter Scheidel identifies the consensus estimate: "[A] million is conventionally taken to represent a credible peak value. . . ." See Scheidel, "Progress and Problems in Roman Demography," in *Debating Roman Demography*, ed. Walter Scheidel (Leiden: Brill, 2001), 51.

59. For an excellent account of these difficulties, see Pierre Salmon, *Population et depopulation dans l'Émpire romain* (Brussels: Latomus, Revue d'études Latines, 1974); Salmon lists the most important scholarly estimates of the city of Rome's population (up to the date of his writing) on pp. 11–12.

60. Bruce Frier, "Demography," in *The Cambridge Ancient History*, 813–14. See also Bruce Frier, "Roman Demography," in Potter and Mattingly, eds., *Life, Death, and Entertainment in the Roman Empire*, 101.

61. Bagnall and Frier argue that Alexandria (with a population of around half a million, one of the largest cities in the empire after Rome) and the other cities in Egypt together made up an urban population of 1.75 million out of a total provincial population in the neighborhood of 4.75 million, for a rate of urbanization that was approximately 37 percent. Although they acknowledge that the figure seems high, they conclude that a third of Egypt's population is likely to have lived in cities. Egypt exhibited, they write, "a degree of urbanization . . . that was high

even by the standards of the Roman empire. . . ." Russell convincingly argues that the population of the Italian peninsula in the time of Augustus was around 7 million. If Rome's population was around 1 million, and that of each of Italy's other urban areas totaled around half a million, the region's level of urbanization would have been over 20 percent. See Bagnall and Frier, *The Demography of Roman Egypt*, 56; and J. C. Russell, "Late Ancient and Medieval Population," *Transactions of the American Philosophical Society*, New Ser., 48, no. 3. (1958): 1–152, specifically 72–73.

62. Walter Scheidel, "Progress and Problems in Roman Demography" in *Debating Roman Demography*, ed. Walter Scheidel (Leiden: Brill, 2001), 51. Historical demographers Richard Lawton and Robert Lee write, "By 1750 nearly nine million people lived in the 261 European cities of over 10,000 inhabitants. By then one-third was in central European towns, over one-quarter (28.7 per cent) in northern and western Europe and only 35.8 per cent in the Mediterranean: moreover, the percentage of the total population in such towns in these three regions was 7.5, 13.6, and 11.8, respectively, a substantial growth in the urbanized population of north-west and central Europe, but a relative fall in the Mediterranean." See Lawton and Lee, chapter 1, "Introduction: The Framework of Comparative Urban Population Studies in Western Europe, c. 1750–1920," in Richard Lawton and Robert Lee, eds. *Urban Population Development in Western Europe* (Liverpool: Liverpool University Press, 1989), 1.

63. E. A. Wrigley, "Brake or Accelerator? Urban Growth and Population Growth before the Industrial Revolution," chapter 7 in Ad van der Woude, Akira Hayami, and Jan de Vries, *Urbanization in History: A Process of Dynamic Interactions* (Oxford: Clarendon Press, 1990), 101–12.

64. Some of the most impressive construction took place in the vicinity of the port of Ostia, at the mouth of the Tiber. Greg Aldrete and David Mattingly write, "About four kilometers north of Ostia, Claudius excavated out of the coastline a gigantic basin over 1,000 meters wide. He also cut canals connecting the new harbor with the Tiber and had two moles [earthworks] built up to shelter the harbor from the sea. Even in this new harbor . . . ships were still not immune from storms. . . . The problem of storms was finally solved by Trajan, who excavated a hexagonal inner harbor basin 700 meters in diameter." Aldrete and Mattingly, "Feeding the City," 179.

65. "The wheat supply of Rome would necessitate a minimum of 948 shiploads of wheat each year." The shipping was done by the private sector, though the government offered incentives: "The state did not seek initially to develop its own merchant fleet, relying instead on private shipping agencies to transport the goods to Rome. It is clear, however, that a number of inducements were offered to encourage this trade. . . ." Aldrete and Mattingly, "Feeding the City," 177, 193. See also Rickman, *The Corn Supply of Ancient Rome*, 17.

66. Jones, *The Roman Economy*, 228–56 and 405–406. See also van der Leeuw and de Vries, "Empire: The Romans in the Mediterranean," 234–36.

67. Jones, *The Roman Economy*, 83. For a discussion of the importance of the agricultural tax to the empire's finances, see Jones, *The Later Roman Empire*, Vol. 1, 464–65.

CHAPTER THREE

1. The emperor Antoninus Pius even minted coins embellished with the words *felicitas temporum*, literally "the happiness of the times."

2. Edward Luttwak, *The Grand Strategy of the Roman Empire: From the First Century A.D. to the Third* (Baltimore: Johns Hopkins University Press, 1976), 128–29. See also A. H. M. Jones, *The Later Roman Empire, 284–602: A Social, Economic, and Administrative Survey*, Vol. 1 (Baltimore: Johns Hopkins University Press, 1964), 14–36.

3. David Whitehouse, "Archaeology and the Pirenne Thesis," in Charles Redman, ed., *Medieval Archaeology: Papers of the Seventeenth Annual Conference of the Center for Medieval and Early Renaissance Studies* (Binghamton: State University of New York at Binghamton, 1989), 6–7.

4. Roberto Luciani, *Roma Sotterranae* (Rome: Fratelli Palombi, 1984), 9; Jérôme Carcopino, *Daily Life in Ancient Rome: The People and the City at the Height of the Empire*, ed. Henry Rowell, trans. E. O. Lorimer (New Haven: Yale University Press, 1940), 16–21; Whitehouse, "Archaeology and the Pirenne Thesis," 4–21; and Richard Hodges and David Whitehouse, *Mohammed, Charlemagne & the Origins of Europe: Archaeology and the Pirenne Thesis* (London: Duckworth, 1983), 51. According to Carcopino, the 1939 census estimated that Rome's population was almost 1.3 million people, which suggests that Luciani's estimate of 1 million inhabitants in 1950 is too low.

5. Luciani, *Roma Sotterranae*, 9–15; and Jones, *The Later Roman Empire*, 1043.

6. Jones, *The Later Roman Empire*, 1044.

7. Bruce Frier, "Demography," chapter 27 in Alan Bowman, Peter Garnsey, and Dominic Rathbone, *The Cambridge Ancient History, Second Edition, Vol. XI, The High Empire, A.D. 70–192* (Cambridge: Cambridge University Press, 2000), 787–816, especially 813–15.

8. Jones, *The Later Roman Empire*, 1040–44.

9. Chris Wickham, *Early Medieval Italy: Central Power and Local Society, 400–1000* (Ann Arbor: University of Michigan Press, 1989), 27, 40–41; and Joseph Tainter, "Post-Collapse Societies," *Companion Encyclopedia of Archaeology*, Graeme Barker, ed. (London: Routledge, 1999), 988–1039.

10. Frier, "Demography," 815. For estimates of world population through history, see the summary provided by the U.S. Census Bureau available at http://www.census.gov/ipc/www/worldhis.html.

11. The statistics in this and the following paragraphs were extracted from the United Nations Department of Economic and Social Affairs, Population Division, "World Population Prospects: The 2004 Revision, Highlights" and "Data Online" (New York: United Nations, Department of Economic and Social Affairs, 2005), available at http://www.un.org/esa/population/unpop.htm. See also "Demographic Prospects 2000–2050 According to the 2002 Revision of the United Nations Population Projections," *Population and Development Review* 29, no. 1 (March 2003): 139–45.

12. This estimate assumes a world population in 2055 (based on the 2004 UN medium variant) of 9.2 billion.

13. For example, see Ben Wattenberg, *Fewer: How the New Demography of Depopulation*

Will Shape Our Future (Chicago: Ivan R. Dee, 2004); and Nicholas Eberstadt, "The Population Implosion," *Foreign Policy* (March/April 2001): 42–53.

14. Branko Milanovic, lead economist at the World Bank, uses Portuguese per capita income—calculated at "purchasing power parity"—to define the threshold at which countries become "rich." See Milanovic, *Worlds Apart: Measuring International and Global Inequality* (Princeton: Princeton University Press, 2005), 130–31.

15. On the challenges of low fertility rates, see Peter Peterson, *Gray Dawn: How the Coming Age Wave Will Transform America—and the World* (New York: Times Books, 1999). Demographers generally assume that a rate of 2.1 births a female ensures the replacement of the mother and her reproductive partner, with allowance (0.1) for the children's possible mortality before they reach their reproductive years. Recently, however, they have recognized that the true replacement rate is higher in some societies.

16. John Bongaarts, "Demographic Consequences of Declining Fertility," *Science* 282, no. 5388 (October 16, 1998): 419–20. In India and China, various forms of sex selection prior to birth, often using ultrasound technologies, have significantly decreased the ratio of girls to boys, reducing the force of the population momentum described here.

17. Paul Demeny, "Population Policy Dilemmas in Europe at the Dawn of the Twenty-First Century," *Population and Development Review* 29, no. 1 (March 2003): 4.

18. These countries are, Demeny says, the European Union's "southern hinterland—a kind of near-abroad to the continent's western half." He continues: "The obviously Eurocentric label is justified by the chosen topic of the present discussion. Seen from a different vantage point, the European Union could be described with equal accuracy as the hinterland of North Africa and West Asia." Ibid., 11. Demeny somewhat arbitrarily excludes from his list Muslim black Africa, also a large source of migration to Europe.

19. The UN's 2050 population estimate for Europe incorporates a number of key assumptions that have the effect of boosting Europe's predicted population, so Demeny's projection of the population imbalance between Europe and its neighbors is highly conservative. Specifically, the UN assumes that European birthrates will increase from the current 1.4 children a woman to 1.82 by 2050, that the average European life expectancy will increase to eighty-three years, and that there will be an influx of 25 million immigrants into Europe in the first half of this century (a number equivalent to more than 40 percent of France's current population).

20. Demeny, "Population Policy Dilemmas in Europe," 14.

21. Many analysts use a tripartite distinction between low-, middle-, and high-income countries; such a distinction, however, would not materially affect my argument. See Geoffrey Garrett, "Globalization's Missing Middle," *Foreign Affairs* 83, no. 6 (November–December 2004): 84–97. On the disparity in population growth between rich and poor countries, see Nathan Keyfitz's review of Alfred Sauvy, *L'Europe submergée: Sud ‡ Nord dans 30 ans* (Paris: Dunod, 1987), in *Population and Development Review* 15, no. 2 (1989): 359–62.

22. Norimitsu Onishi, "Out of Africa or Bust, with a Desert to Cross," *New York Times*, January 4, 2001, national edition, A1 and A21.

23. Reuters, "241 Illegal Migrants Reach Italian Island in a Fishing Vessel," *New York Times*, August 30, 2004, national edition, A7; and Ian Fisher and Richard Bernstein, "On Italian Isle, Migrant Debate Sharpens Focus," *New York Times*, October 5, 2004, national edition, A1.

24. Frank Bruni, "Off Sicily, Tide of Bodies Roils the Debate over Immigrants," *New York Times*, September 23, 2002, national edition, A1; Al Baker, "Body Falls As Jet Nears Kennedy," *New York Times*, August 9, 2001, national edition, A18.

25. Suzanne Daley, "African Migrants Risk All on Passage to Spain," *New York Times*, July 10, 2001, national edition, A1 and A6.

26. In a controversial 2004 article in *Foreign Policy*, the Harvard political scientist Samuel Huntington talked about the eventual result of this flow, with a revealing hint of fear. "In California—as in Hawaii, New Mexico, and the District of Columbia—non-Hispanic whites are now a minority," he wrote. "Demographers predict that, by 2040, non-Hispanic whites could be a minority of all Americans." Huntington, ""The Hispanic Challenge," *Foreign Policy* (March/April 2004): 41.

27. Bruni, "Off Sicily."

28. Jagdish Bhagwati, "Borders Beyond Control," *Foreign Affairs* 82, no. 1 (January/ February 2003): 98–104.

29. The statistic on U.S. border apprehensions was obtained from the Web site of the U.S. Customs and Border Protection, at http://www.cbp.gov/xp/cgov/home.xml.

30. Rachel Swarns, "Tight Immigration Policy Hits Roadblock of Reality," *New York Times*, January 20, 2006, national edition, A12.

31. Ginger Thompson and Sandra Ochoa, "By a Back Door to the U.S.: A Migrant's Grim Sea Voyage," *New York Times*, June 13, 2004, national edition, 1.

32. Charlie LeDuff, "Holidays Inspire a Rush to the Border," *New York Times*, December 23, 2004, national edition, A12.

33. Doug Saunders, "European Dream Relies on Immigrant Workers' Nightmares," *Globe and Mail* (Toronto), September 4, 2004, F3.

34. On the dangers in Europe, see Timothy Savage, "Europe and Islam: Crescent Waxing, Cultures Clashing," *Washington Quarterly* 27, no. 3 (Summer 2004): 25–50. In the United States, Huntington suggests, the country's rapid demographic change could eventually cause "the rise of an anti-Hispanic, anti-black, and anti-immigrant movement composed largely of white, working- and middle-class males, protesting their job losses to immigrants and foreign countries, the perversion of their culture, and the displacement of their language." Huntington, "The Hispanic Challenge," 41.

35. See, for instance, Ester Boserup, *The Conditions of Agricultural Growth: The Economics of Agrarian Change under Population Pressure* (Chicago: Aldine, 1965); and Julian Simon, *The Ultimate Resource* 2 (Princeton: Princeton University Press, 1996). In poor countries, larger populations can provide labor to improve and protect cropland, for instance by building terraces and retaining walls to reduce soil erosion.

36. For a survey of population growth's manifold social and economic effects, see Dennis Ahlberg, Alan Kelley, and Karen Oppenheim Mason, eds., *The Impact of Population Growth on Well-being in Developing Countries* (Berlin: Springer-Verlag, 1996).

37. Theodore Panayotou, "An Inquiry into Population, Resources and Environment," in Ahlberg, Kelley, Mason, eds., *The Impact of Population Growth*, 259–98. An

excellent study of the complex relationship between population growth and local natural resources in poor countries is Scott Templeton and Sara Scherr, "Effects of Demographic and Related Microeconomic Change on Land Quality in Hills and Mountains of Developing Countries," *World Development* 27, no. 6 (1999): 903–18.

38. Dennis Ahlburg reviews recent research and thinking about the relationship between population growth and economic performance in "Does Population Matter? A Review Essay," *Population and Development Review* 28, no. 2 (June 2002): 329–50.

39. An early statement of this argument is found in Ansley Coale and Edgar Hoover, *Population Growth and Economic Development in Low-income Countries: A Case Study of India's Prospects* (Princeton: Princeton University Press, 1959). More recently, some analysts have concluded that lower fertility rates enabled East Asian countries to significantly boost domestic savings, capital investment and, consequently, economic growth from the 1970s into the 1990s. See Matthew Higgins and Jeffrey Williamson, "Age Structure Dynamics in Asia and Dependence on Foreign Capital," *Population and Development Review* 23 (1997): 261–93.

40. On rural-urban migration, see Richard Bilsborrow, "Migration, Population Change, and the Rural Environment," in *Environmental Change and Security Program Report 8* (Washington, DC: Woodrow Wilson International Center for Scholars, Environmental Change and Security Program, 2002).

41. United Nations Human Settlements Program, *The Challenge of Slums: Global Report on Human Settlements 2003* (London: UN-Habitat and Earthscan Publications, 2003), xxv.

42. The best current survey of world urbanization is United Nations Human Settlements Programme, *The State of the World's Cities 2004/2005: Globalization and Urban Culture* (London: Earthscan, 2004), available at http://www.unhabitat.org/mediacentre/sowckit.asp. On urban population growth projections, see Martin Brockerhoff, "Urban Growth in Developing Countries: A Review of Projections and Predictions," *Working Paper: Policy Research Division*, no. 131 (New York: Population Council, 1999).

43. United Nations, Department of Economic and Social Affairs, Population Division, Table 1 of *World Urbanization Prospects: The 2003 Revision, Data Table and Highlights*, available at http://www.un.org/esa/population/publications/wup2003/2003WUPHighlights.pdf, 4; and United Nations Human Settlements Program, *The Challenge of Slums*, xxv.

44. United Nations, Department of Economic and Social Affairs, Population Division, Table 7 of *World Urbanization Prospects*, 7.

45. Ibid., Table 8, 8.

46. Martin Brockerhoff and Ellen Brennan, "The Poverty of Cities in Developing Regions," *Population and Development Review* 24, no. 1 (1998): 75–114.

47. The resourcefulness and cooperative spirit of slum communities in the face of such obstacles is invariably remarkable. For a detailed account, see Robert Neuwirth, *Shadow Cities: A Billion Squatters, a New Urban World* (New York: Routledge, 2005).

48. Ellen Brennan, "Population, Urbanization, Environment, and Security: A Summary of the Issues," *Comparative Urban Studies: Occasional Paper Series*, no. 22 (Washington, DC: Woodrow Wilson International Center for Scholars, 1999), 12.

49. Seth Mydans, "Eking Out a Living, of Sorts, From a Mountain of Muck," *New York Times*, 23 May 2006, national edition, p. A4.

50. This problem is particularly acute between Central America and the United States. See Ginger Thompson, "Gangs without Borders, Fierce and Resilient, Confound the Law," *New York Times*, September 26, 2004, national edition, A1.

51. Larry Rohter, "Ipanema Under Siege: Rio's Gangs Flex Harder," *New York Times*, Sunday Week in Review, October 20, 2002, national edition, 4.

52. As quoted by Rohter, in "Ipanema Under Siege."

53. On the correlation between youth bulges and political violence, see Henrik Urdal, "A Clash of Generations? Youth Bulges and Political Violence" (Oslo: Centre for Study of Civil War, International Peace Research Institute, Oslo, 2005).

54. Richard Cincotta and Robert Engelman, "Conflict Thrives Where Young Men Are Many," *International Herald Tribune*, March 2, 2004. For a more developed treatment of the relationship between youth bulges and civil violence, see Richard Cincotta, Robert Engelman, and Daniele Anastasion, "Appendix 4: Country Data Table," in *The Security Demographic: Population and Civil Conflict after the Cold War* (Washington, DC: Population Action International, 2003).

55. Cincotta, Engelman, and Anastasion, "Appendix 4: Country Data Table," in *The Security Demographic*, 96–101.

56. The relationship between urban growth and violence is discussed in Thomas Homer-Dixon, *Environment, Scarcity, and Violence* (Princeton: Princeton University Press, 1999), 155–66.

57. Ibid., 162–63.

58. Between 1976 and 1992, over 140 separate incidents of strikes, riots, and demonstrations took place, mainly in Latin America. See John Walton and David Seddon, *Free Markets and Food Riots: The Politics of Global Adjustment* (Cambridge: Blackwell Publishers, 1994), 39–40. See also John Walton and Charles Ragin, "Global and National Sources of Political Protest: Third World Responses to the Debt Crisis," *American Sociological Review* 55, n. 6 (1990): 876–90; and John Walton, "Debt, Protest and the State in Latin America," in *Power and Popular Protest: Latin American Social Movements*, ed. Susan Eckstein (Berkeley: University of California Press, 1989), 299–328.

59. Lewis Mumford discusses cities' extraction of resources from their hinterlands in "The Natural History of Urbanization," in William L. Thomas Jr., ed., *Man's Role in Changing the Face of the Earth* (Chicago and London: University of Chicago Press, 1956), available at http //habitat.aq.upm.es/boletin/n21/almum.en.html.

60. Greg Aldrete and David J. Mattingly, "Feeding the City: The Organization, Operation, and Scale of the Supply System for Rome," in D. S. Potter and D. J. Mattingly, eds., *Life, Death, and Entertainment in the Roman Empire* (Ann Arbor: University of Michigan Press, 1999), 174. See also Geoffrey Rickman, *The Corn Supply of Ancient Rome* (Oxford: Clarendon Press, 1980), 14–17.

61. Emin Tengström, *Bread for the People: Studies of the Corn-Supply of Rome during the Late Empire* (Stockholm: Paul Åströms Förlag, 1974), 47–48.

62. Quoted in Aldrete and. Mattingly, "Feeding the City," 176–77.

CHAPTER FOUR

1. On the basis of figures provided by the Harvard energy expert John Holdren, assuming a "business-as-usual economic and energy scenario," global energy use between 2000 and 2050 will rise by a factor of 2.46. See Holdren, "Environmental Change and the Human Condition," *Bulletin of the American Academy of Arts and Sciences* 57, no. 1 (Fall 2003): 27.

2. On India's energy needs, see Somini Sengupta, "Hunger for Energy Transforms How India Operates," *New York Times*, June 5, 2005, national edition, 2.

3. Jim Yardley, "China's Economic Engine Needs Power (Lots of It)," *New York Times*, Week in Review, March 14, 2004, national edition, 3.

4. China's desperation is reflected in its leaders' words. Prime Minister Wen Jiabo declared in early 2004, "We must speed up the development of large coal mines, important power generating facilities and power grids, [and] the exploration and exploitation of petroleum and other important resources." Quoted in Yardley, "China's Economic Engine." See also Keith Bradsher, "China Struggles to Cut Reliance on Mideast Oil," *New York Times*, September 3, 2002, national edition, A1; James Kynge, "China Continues Its Quest for Secure Energy Supplies with Variety of Sources As the Aim," *Financial Times*, May 25, 2004, 6; Paul Roberts, "The Undeclared Oil War," *Washington Post*, June 28, 2004, online edition; Simon Romero, "Canada's Oil: China in Line As U.S. Rival," *New York Times*, December 23, 2004, national edition, A1; and Chris Buckley, "Venezuela Agrees to Export Oil and Gas to China," *New York Times*, December 28, 2004, national edition, W1.

5. In the past few years, as the price of oil has shot up, a flurry of articles and books has appeared on the issue of future oil availability. They include James Howard Kunstler, *The Long Emergency: Surviving the Converging Catastrophes of the Twenty-First Century* (New York: Atlantic Monthly Press, 2005); Kenneth Deffeyes, *Beyond Oil: The View from Hubbert's Peak* (New York: Hill and Wang, 2005); David Goodstein, *Out of Gas: The End of the Age of Oil* (Norton, 2004); Paul Roberts, *The End of Oil: On the Edge of a Perilous New World* (Boston: Houghton Mifflin, 2004); Kenneth Deffeyes, *Hubbert's Peak: The Impending World Oil Shortage* (Princeton: Princeton University Press, 2003); and Richard Heinberg, *The Party's Over: Oil, War and the Fate of Industrial Societies* (Gabriola Island, British Columbia: New Society Publishers, 2003).

6. William Nordhaus, "Do Real-Output and Real-Wage Measures Capture Reality? The History of Lighting Suggests Not," in *The Economics of New Goods*, Timothy Bresnahan and Robert Gordon, eds. (Chicago: University of Chicago Press, 1997), 29–66.

7. For a fascinating account of the beginnings of the oil age, see Daniel Yergin, *The Prize: The Epic Quest for Oil, Money & Power* (New York: Free Press, 1992).

8. Specialists use the term "proximate solar energy" for forms of energy derived more or less directly from the sun. The food we eat is proximate solar energy. For an accessible discussion of the transitions described in this paragraph, see Alfred Crosby, *Children of the Sun: A History of Humanity's Unappeasable Appetite for Energy* (New York: Norton, 2006).

9. Ian Graham suggested this metaphor of Earth as a fossil-fuel battery charger.

10. The calorific value of crude oil is roughly 12,000 watt hours per kilogram. A fit human male can generate about 500 watt hours of work in an eight-hour day of manual labor. So a kilogram of oil contains heat energy equivalent to about twenty-four days of human manual labor, and about forty grams of oil (roughly equivalent to three tablespoons) contain the equivalent of a day's labor. Gasoline has about the same calorific value as crude oil. A car's average forty-litre tank of gasoline weighs about thirty kilograms, and is therefore equivalent to about 720 days of labor.

11. Vaclav Smil, *Energy in World History* (Boulder, CO: Westview, 1994), 190–91.

12. Transcript of a Global Vision interview with Colin Campbell, available at http://www.global-vision.org/wssd/campbell.html.

13. David Rosenbaum, "As 2 Sides Push, Arctic Oil Plan Seems Doomed," *New York Times*, April 18, 2002, national edition, A19

14. Morris Adelman, *The Economics of Petroleum Supply* (Cambridge, MA: MIT Press, 1993), xi. Emphasis in original text.

15. On U.S. oil imports, see United States Energy Information Administration, "Table 1.7: Overview of U.S. Petroleum Trade," available at http://www.eia.doe.gov/emeu/mer/pdf/pages/sec1_15.pdf.

16. OPEC currently has eleven member countries: Algeria, Libya, and Nigeria in Africa; Iran, Iraq, Kuwait, Qatar, Saudi Arabia, and the United Arab Emirates in the Middle East; Venezuela in South America; and Indonesia in Southeast Asia. (Indonesia's membership may soon end because the country no longer exports oil.) At the time of the 1979–81 oil shock, Gabon and Ecuador were also members.

17. Colin Campbell and Jean Laherrère note that "after the price of crude hit all-time highs in the early 1980s, explorers developed new technology for finding and recovering oil, and they scoured the world for new fields. They found few." Campbell and Laherrère, "The End of Cheap Oil," *Scientific American* 278, no. 3 (March 1998): 81.

18. According to the International Energy Agency, as soon as 2015, Russia, the Persian Gulf, and West Africa will provide 80 percent of the world's traded oil, and at least a fifth of that oil will have to come from Saudi Arabia alone. See Lord John Browne, "Beyond Kyoto," a speech to the Council on Foreign Relations, New York, June 24, 2004, available at http://www.cfr.org/pub7148/john_browne/beyond_kyoto.php.

19. R. W. Bentley, "Global Oil & Gas Depletion: An Overview," *Energy Policy* 30 (2002): 204. On Shell's censorship of Hubbert's address, see Jeremy Leggett, *Half Gone: Oil, Gas, Hot Air and the Global Energy Crisis* (London: Portobello Books, 2006), excerpted in "What They Don't Want You to Know about the Coming Oil Crisis," *Independent* (London), January 20, 2006.

20. Robert Kaufman points out that this prediction was "part genius and part luck." The luck came in the form of the Texas Railroad Commission, an organization that controlled the rate of extraction from Texas wells from the 1930s till the 1970s. If the Commission's decisions had been different, U.S. production could have peaked earlier or later. Still, Kauffmann concludes, the underlying logic of Hubbert's approach is correct, and its predictions on the timing of peak output are remarkably robust, even when the total oil in a basin turns out to be substantially larger

than originally estimated. See Kaufmann, "Planning for the Peak in World Oil Production," *World Watch* 19, no. 1 (January/February 2006): 19–21.

21. Between 1970 and 2005, total U.S. production of crude oil and natural gas liquids declined from 11.3 to 6.8 million barrels a day. See "Table 5.1, Petroleum Overview, Selected Years, 1949–2004, available at http://www.eia.doe.gov/emeu/ aer/pdf/pages/sec5_5.pdf; and "Table 3.1a, Petroleum Overview: Supply," available at http://www.eia.doe.gov/emeu/mer/pdf/pages/sec3_2.pdf.

22. When estimating the amount of oil in a region, petroleum geologists and energy specialists distinguish between at least five different quantities: 1. original oil-in-place (the total amount of oil in a basin or region, regardless of its recoverability); 2. cumulative production (the total of all oil extracted so far from a given basin); 3. proved reserves (oil already found in that basin that can be recovered with today's technologies at a profit); 4. undiscovered reserves (recoverable oil in basins or fields yet to be discovered and explored); and 5. reserve growth (the future increase in the amount of recoverable oil from known basins because of higher prices and new extraction technologies). While estimates of cumulative production for a given basin or region are often fairly solid, those for the other four categories often vary widely. The ultimately recoverable resource, or URR, is the sum of estimates for categories 2 through 5, while the oil recovery factor is this URR divided by the original oil-in-place (i.e., category 1). Estimates of oil recovery currently range from 30 to 70 percent.

23. Estimates of a region's URR are generally revised as continuing exploration better reveals the full volume of the region's individual oilfields.

24. Detailed treatments of the mathematical techniques Hubbert used to arrive at his estimates are available in "Hubbert Revisited," chapter 7 in Deffeyes, *Hubbert's Peak*, 133–49; and "The Hubbert Method," chapter 3 in Deffeyes, *Beyond Oil*, 35–51.

25. Transcript of a Global Vision interview with Colin Campbell.

26. In the United Kingdom's North Sea fields, oil discovery peaked in the early 1970s, while production peaked in 1999. Norway reached its discovery peak in 1979 and is passing its production peak right about now. Indonesia's discovery peaked in 1955 and output peaked in 1977. And in the Persian Gulf, Oman's oil discovery peaked in 1962, while its output peaked in 2001. Oman's largest field is now declining at an astonishing 12 percent a year. This country's experience is a cautionary tale for many in the oil industry because the rapid decline of the Yibal field has occurred despite, and perhaps even because of, aggressive use of new extraction technologies.

Russian discovery peaked in 1960, while its production peaked in 1987. Many analysts, such as Daniel Yergin, head of Cambridge Energy Research Associates (CERA), believe that Russia has huge undiscovered and untapped reserves of oil and will have the capacity to meet a large portion of increased world oil demand in coming decades. Other analysts believe that Russian's reserves are much more limited and that the country has already passed peak output. For example, R. W. Bentley writes, "There is recognition within the industry that [Russia] is now past its physical resource peak."

All figures in this note were obtained from the newsletters of the Association for the Study of Peak Oil (available at www.asponews.org) and cross-checked against data from the International Energy Agency's Oil Market Report (available at http://omrpublic.iea.org/supplysearch.asp). On Oman, see Jeff Gerth and Stephen Labaton, "Oman's Oil Yield Long in Declining, Shell Data Show," *New York Times*, April 8, 2004, online edition. See also Bentley, "Global Oil & Gas Depletion," 191.

27. Conventional oil consists of light, short-chained molecules and is usually found on land or in shallow waters offshore; non-conventional oil is found in deep water or consists of heavier, longer-chained molecules.

28. Campbell's estimates as of April 2006, as presented by the Association for the Study of Peak Oil and Gas, can be found in the summary chart on the second page of the Association's newsletter, available at http://www.peakoil.ie/downloads/newsletters/newsletter64_200604.pdf.

29. USGS World Energy Assessment Team, *U.S. Geological Survey World Petroleum Assessment 2000* available at http://greenwood.cr.usgs.gov/energy/WorldEnergy/DDS-60/. See also Thomas Ahlbrandt et al., "Future Oil and Gas Resources of the World," *Geotimes* (June 2000): 24–25.

30. John Wood and Gary Long, *Long Term World Oil Supply: A Resource Base/Production Path Analysis* (Washington, DC: Energy Information Administration, 2000), available at http://www.eia.doe.gov/pub/oil_gas/petroleum/presentations/2000/long_term_supply/index.htm. On the Saudi use of the USGS estimate, see Jeff Gerth, "Doubts Raised on Saudi Vow for More Oil," *New York Times*, October 27, 2005, national edition, A1.

31. Some critics contend that the USGS study doesn't properly account for declining discovery trends in many oil basins and that it uses an unsound method to forecast how much geologists' estimates of current reserves will grow as they better understand these reserves' true extent. See Bentley, "Global Oil & Gas Depletion," 200–201. The USGS study also seems to overstate the likelihood of major new discoveries. So, for example, it includes oil production from a possible basin in East Greenland. On the assumption that there's a 95 percent chance of finding at least 1 barrel and a 5 percent chance of finding 112 billion barrels, the USGS calculates that the basin will produce 47 billion barrels— a figure that's then folded into the forecast of the world's total oil endowment. Yet this Greenland basin hasn't been explored or even properly tested for its oil content. Colin Campbell comments, "Since the [USGS] numbers were quoted to three decimal places, the reader could be forgiven for assuming them to be accurate. But a moment's reflection would question the very concept of a sub-jective 5 percent probability. In plain language, it was a guess that could as well be the half or the double, yet it entered the calculations distorting the critical mean value." Campbell, "Forecasting Global Oil Supply 2000–2050," *Hubbert Center Newsletter* 3 (2002): 2.

32. "USGS Study Revisited," *ASPO Newsletter* 63 (March 2006), 10, available at http://www.peakoil.ie/downloads/newsletters/newsletter63_200603.pdf. For an assessment of the USGS's study's predictions by the study's authors themselves, see T. R. Klett, Donald Gautier, and Thomas Ahlbrandt, "An Evaluation of the U.S. Geological Survey World Petroleum Assessment 2000," *American Association of Petroleum Geologists (AAPG) Bulletin* 89, no. 8 (August 2005): 1033–42.

33. Some critics note that several of Hubbert's less well known predictions were seriously off the mark. Others argue that there's no theoretical reason to assume that oil production will follow a bell curve, so Hubbert's success in predicting U.S. peak output was just dumb luck. For instance, Michael Lynch writes, "[Hubbert's] forecast of U.S. gas production in 2000 was 65 percent too low and his world oil production forecast for 2000 was 50 percent too low. Even production in Texas is now about twice the amount he forecast." To be fair, however, Hubbert admitted that he had far less knowledge of the URR for the world than he did for the lower forty-eight states, and he was therefore reluctant to predict a global peak. See Lynch, "Forecasting Oil Supply: Theory and Practice," *The Quarterly Review of Economics and Finance* 42 (2002): 373–89, especially 377. On the lack of a sound theoretical basis for the Hubbert model, see Robert Kauffman and Cutler Cleveland, "Oil Production in the Lower 48 States: Economic, Geological, and Institutional Determinants, *The Energy Journal* 22, no. 1 (2001): 27–49.

34. For example, see Vaclav Smil, *Energy at the Crossroads: Global Perspectives and Uncertainties* (Cambridge, MA: MIT Press, 2003), 195–201.

35. Lynch, "Forecasting Oil Supply," 377–78.

36. Lynch, for example, roundly criticizes all estimates of oil output based on Hubbert's methodology, declaring that "no major region has yet shown signs of such behavior, outside of the U.S." Yet in the very next sentence he acknowledges that Egypt, Argentina, and the North Sea "appear to be reaching their peaks," and one page later he further acknowledges that "Hubbert-style production profiles" have proved "very accurate" for Southeast Asia. Furthermore, the data series that he presents to refute Campbell's projection for the United Kingdom omits the sharp downturn in UK production since 1999. See Lynch, "Forecasting Oil Supply," 376–77, 381. Kauffman and Cleveland assert that three "stochastic trends" that critically affected lower forty-eight output "are not present in the deterministic [Hubbert] Bell-shaped curve." But their incorporation of these trends in their model simply allows them to reproduce the variance in lower forty-eight production around an overall Hubbert bell-shaped trend line. Moreover, one of the key pieces of evidence these authors use to refute a Hubbert-style model of U.S. oil output is that "production in the lower 48 States stabilizes in the late 1970s and early 1980s, which contradicts the steady decline forecast by the Hubbert model." But the Hubbert model doesn't preclude considerable variation around a long-term secular trend of declining output; also, the particular episode of stabilized production that the authors highlight lasted only five years, whereas the trend of declining production in the United States has been otherwise uninterrupted for thirty-five years. See Kauffman and Cleveland, "Oil Production," 46–47, especially Figure 3.

37. In the past, tough U.S. federal regulations encouraged major oil-producing companies to underreport their oil reserves. Underreporting didn't cause problems when newly discovered oil fields tended to be large and their oil easy to extract. Once the oil started flowing, the producers could then revise upward their fields' reserves, which kept shareholders, bankers, and other investors happy. Unfortunately, though, these steady upward revisions encouraged investors and commentators to become excessively optimistic about oil's long-term availability

and about the ability of new technologies to boost the fraction of oil that could be recovered from existing fields.

As large oil fields became depleted and supplies tightened, the scope for underreporting narrowed. Major oil companies thus had less slack in their reserve estimates to counterbalance, in the eyes of investors, any of the bad news from failed exploration that's inevitably part of the oil business. And as new discoveries became smaller, with harder-to-extract oil, companies came under pressure to exaggerate the size of these discoveries to attract investor interest at all. As a result, some companies appear to have moved from systematic underreporting of their reserves, at least within the U.S., to the exact opposite practice—systematic overreporting—especially of overseas reserves. In early 2004, for instance, Royal Dutch/Shell, the world's third-largest oil company, had to revise its figures for its oil and gas reserves downward by a stunning 22 percent. Overall, too, the long-standing link between large corporate reserves and high production has weakened: while the stated reserves of major oil-producing companies have risen in recent years, these companies' oil output has generally declined, which suggests reserve figures have been inflated. On this last point, see Alex Berenson, "An Oil Enigma: Production Falls Even as Reserves Rise," *New York Times*, June 12, 2004, online edition.

38. Quoted in ibid.

39. See, for example, John F. Bookout, "Two Centuries of Fossil Fuel Energy," *Episodes* 12 4 (1989): 257–62.

40. Harry Longwell backdated upward revisions (to a given oil field's total size) to the year that the field was discovered. This procedure requires some explanation. When a field is discovered, geologists estimate its size in terms of its total content of oil (usually assuming a certain oil price, which determines how hard producers would be willing to work to extract the oil). Over time, as field-imaging technology improves, and as production wells are drilled into the field, geologists update the estimated total amount in the field, which in the past has usually meant an upward revision (sometimes substantial). Longwell has attributed these revisions to the year that the original field was discovered rather than to the year that a given revision was made. This allows the graph of global oil discovery shown here to represent the fact that most—if not all—of the really big fields have already been discovered. In recent years, geologists' techniques for predicting the ultimate size of a field have greatly improved, which means that their estimates of the size of any newly discovered fields are less likely to be revised upward substantially in the future (to the extent that revisions occur, they are just as often downward now). So we can be reasonably confident that the declining tail of the world discovery trend in the graph represents a genuine long-term diminishment of oil discovery.

41. This point is well established in the energy literature. See, for instance, John Hallock Jr. et al., "Forecasting the Limits to the Availability and Diversity of Global Conventional Oil Supply," *Energy* 29, no. 11 (September 2004): 1673–96, especially 1681; and Robert Hirsch, Roger Bezdek, and Robert Wendling, *Peaking of World Oil Production: Impacts, Mitigation, & Risk Management* (San Diego: Science Applications International Corporation, 2005), 17.

42. Harry Longwell, "The Future of the Oil and Gas Industry: Past Approaches, New Challenges," *World Energy* 5, n. 3 (2002): 100–104.

43. Charles Hall et al., "Hydrocarbons and the Evolution of Human Culture," *Nature* 426, no. 6964 (November 20, 2003): 320; and Cutler Cleveland, "Net Energy from the Extraction of Oil and Gas in the United States," Working Paper 0101, Center for Energy and Environmental Studies and Department of Geography, Boston University. Available at http://www.bu.edu/cees/research/workingp/ pdfs/Net_Energy_w=figures.doc.pdf. The final version of this paper was published as Cutler Cleveland, "Net Energy from the Extraction of Oil and Gas in the United States," *Energy* 30, no. 5 (April 2005): 769–82. On the worldwide trend toward higher costs of exploration, see Jad Mouawad, "Oil Explores Searching Ever More Remote Areas," *New York Times*, September 9, 2004, national edition, C1.

44. Data on the average cost (in inflation-adjusted dollars) of producing oil in the U.S. are available from Kauffman and Cleveland, "Oil Production," Figure 2, 43.

45. Matthew Simmons, *The World's Giant Oilfields* (Houston: Simmons & Company International, 2001).

46. Campbell and Laherrère, "The End of Cheap Oil," 82.

47. In 2005, ExxonMobil issued a report, *The Outlook for Energy: A 2030 View*, indicating that non-OPEC conventional oil production will peak in 2010. Afterward, the report indicates, rising demand will have to be met by non-conventional oil sources, especially oil sands and natural gas liquids, and OPEC conventional production, mainly from the Persian Gulf. On the important role of the Middle East in meeting future demand, see also International Energy Agency, *World Energy Outlook 2005: Middle East and North Africa* (Paris: IEA, 2005). The report states, "The oil and gas resources of the Middle East and North Africa (MENA) will be critical to meeting the world's growing appetite for energy. The greater part of the world's remaining reserves lie in that region. They are relatively under-exploited and are sufficient to meet rising global demand for the next quarter century and beyond."

48. For example, the U.S. Department of Energy (DOE) estimated in 2005 that Saudi Arabia will be able to boost its output from the current 9.5 million to 14 million barrels a day as soon as 2010 and to more than 15 million barrels a day by 2020. This 2005 DOE estimate of Saudi 2020 production is substantially lower than a 2002 DOE estimate of 20 million barrels a day that was widely thought unrealistic by specialists. The 2005 estimates are available at http://www.eia.doe.gov/oiaf/ieo/pdf/ ieooiltab_1.pdf. On the 2002 estimates, see Jeff Gerth, "Growing U.S. Need for Oil from the Mideast Is Forecast," *New York Times*, December 26, 2002, national edition, A16. More recently, the International Energy Agency in its *World Energy Outlook 2005* has estimated Saudi output at over 18 million barrels a day in 2030.

49. The energy investment banker Matthew Simmons writes, "The accounts of Saudi exploration activities as related in technical papers from Aramco [the Saudi national oil company] confirm that there has . . . been intensive exploration in Saudi Arabia for the past thirty years, and that the effort has brought only marginal success." Simmons, *Twilight in the Desert: The Coming Saudi Oil Shock and the World Economy* (Hoboken, NJ: Wiley, 2005), 241.

50. The Association for the Study of Peak Oil, "Saudi Arabia," *ASPO Newsletter* 21 (September 2002), 3–6, available at http://www.peakoil.ie/downloads/newsletters/ newsletter21_200209.pdf.

51. The best publicly accessible account of the state of the Ghawar field is provided by Simmons in "Ghawar, the King of Oilfields," chapter 7 in *Twilight in the Desert*, 151–79. Simmons provides production estimates on 152–54.

52. Ibid., 161–65.

53. The Association for the Study of Peak Oil, "Saudi Arabia."

54. The country's actual reserves are unlikely to be more than 200 and could even be less than 90 billion barrels. See ibid; and also the Association for the Study of Peak Oil and Gas, "Saudi Reserves," *ASPO Newsletter* 40 (April 2004): 3, available at http://www.peakoil.ie/downloads/newsletters/newsletter40_200404.pdf.

55. Jeff Gerth, "Forecast of Rising Oil Demand Challenges Tired Saudi Fields," *New York Times*, February 24, 2004, national edition, A1.

56. In 2004, the U.S. administration received a key intelligence report questioning Saudi Arabia's long-term capacity to meet global demand for conventional oil. See Gerth, "Doubts Raised on Saudi Vow for More Oil." See also Peter Mass, "The Breaking Point," *New York Times Magazine* (August 21, 2005): 30.

57. Jim Giles, "Every Last Drop," *Nature* 429, no. 6993 (June 17, 2004): 694–95.

58. According to Matthew Simmons, who has been a financial adviser to oil-field service companies for three decades, oil-field managers have learned that "these advances combined to extract the easily recoverable oil from giant fields even faster and led to decline curves, once reservoir pressures depleted, steeper than the industry had ever experienced before." Elsewhere he writes, "The industry is beginning to appreciate that advanced technologies . . . are essentially turbo-charged super-straws designed to suck out the recoverable oil faster—not miracle drugs that prolong field life and recover far higher percentages of the original oil-in-place." Simmons, *Twilight in the Desert*, 337, 279. See also Bentley, "Global Oil & Gas Depletion," 195.

59. Oil production (encompassing conventional and unconventional oil, including natural gas liquids) currently averages about 80 million barrels a day. The U.S. Energy Information Agency estimates that this production will need to grow to 120 million barrels a day by 2025 to meet the growing demand especially of China and India. See Energy Information Administration, "International Energy Outlook 2005," available at http://www.eia.doe.gov/oiaf/ieo/oil.html.

60. See, for instance, Cambridge Energy Research Associates (CERA), "Oil & Liquids Capacity to Outstrip Demand Until at Least 2010: New CERA Report," press release, June 21, 2005, available at http://www.cera.com/news/details/1,2318,7453,00.html.

61. Simon Romero, "Mr. Sandman, Bring Me Some Oil," *New York Times*, August 31, 2004, national edition, C1.

62. The calorific value of thirty cubic meters of natural gas is about one-fifth of that of a barrel of crude oil. Any complete EROI estimate would also have to include the energy involved in mining and transporting the raw tar sands, as well as energy to pump water and deal with wastes.

63. Technologies are now being developed for "in situ" production of tar sands oil using energy in the tar sands themselves, but they are still experimental.

64. With somewhat tongue-in-cheek immodesty, the Princeton University petroleum engineer Kenneth Deffeyes has used a Hubbert-type analysis to specify November 24, 2005 (American Thanksgiving Day), as the likely date of peak output. Because

it takes a while for a trend to appear in oil production data, we won't know a couple of years whether this prediction is correct. Deffeyes, *Beyond Oil*, 3.

65. See CERA, "Oil & Liquids Capacity."

66. "Relatively large uncertainties about recoverable oil supply have relatively little effect on the timing of the peak. . . . Optimistic estimates for the amount of oil that remains only postpone the peak slightly." See Robert Kaufmann, "Planning for the Peak in World Oil Production," *World Watch* 19, no. 1 (January/February 2006): 19. The energy analyst John Hallock Jr. and his colleagues have predicted the time frame within which peak output is likely to occur. They write, "Global production of conventional oil will almost certainly begin an irreversible decline somewhere between 2004 and 2037." Hallock Jr. et al., "Forecasting," 1673.

67. Bentley, "Global Oil & Gas Depletion," 202. The post-peak depletion rate could be much higher: executives of oil-field service companies have noted that the average post-peak depletion rate for older reservoirs is around 8 percent.

68. Daniel Yergin and his researchers at Cambridge Energy Research Associates (CERA) have proposed that global oil output will not peak but reach an "undulating plateau" in the third or fourth decade of this century, and that this plateau will continue for several decades. See CERA, "Oil & Liquids Capacity."

69. As quoted in Dow Jones Newswires and reproduced at the Culture Change website: http://www.culturechange.org/fall_of_petroleum_DowJones.html.

70. Some of the material in the following paragraphs was written in collaboration with S. Julio Friedmann of Lawrence Livermore Laboratory. For a survey of advantages and disadvantages of the energy sources discussed here, see chapters 4 through 9 of Deffeyes, *Beyond Oil*.

71. For data on the number of natural gas wells drilled in the U.S. and on natural gas production and price, see U.S. Energy Information Administration, "Table 5.2: Crude Oil and Natural Gas Wells Drilled," available at http://www.eia.doe.gov/emeu/mer/pdf/pages/sec5_4.pdf; "Table 4.2: Natural Gas Production," available at http://www.eia.doe.gov/emeu/mer/pdf/pages/sec4_4.pdf; and "Table 9.11: Natural Gas Prices," available at http://www.eia.doe.gov/emeu/mer/pdf/pages/sec9_17.pdf.

72. Although the natural gas industry has downplayed the danger of LNG explosions, Jerry Havens, a professor of chemical engineering at the University of Arkansas and an adviser to the U.S. Coast Guard, argues that there is a genuine danger of "pool fires" caused by LNG that accumulates on the surface of the ground or the sea after catastrophic release from a tanker: "Most predictions suggest that even the largest LNG tankers (typically more than 900 feet in length) might be completely enveloped in a pool fire following a complete spill of a single 6.5 million gallon tank. . . . A typical LNG tanker contains as many as five tanks with a combined capacity of 33 million gallons. [Such fires] would be expected to burn more rapidly and with greater intensity than crude oil or even gasoline fires." See Jerry Havens, "Terrorism: Ready to Blow?" *Bulletin of the Atomic Scientists* 59, no. 4 (July/August 2003): 16–18.

73. This technology is described in S. Julio Friedmann and Thomas Homer-Dixon, "Out of the Energy Box," *Foreign Affairs* 83, no. 6 (November/December 2004): 72–83.

74. Tad Patzek, "Thermodynamics of the Corn-Ethanol Biofuel Cycle," *Critical Reviews in Plant Sciences* 23, no. 6 (2004): 519–67, available at http://petroleum.berkeley.edu/papers/patzek/CRPS416-Patzek-Web.pdf.

75. Arthur Ragauskas, "The Path Forward for Biofuels and Biomaterials," *Science* 311, no. 5760 (January 27, 2006): 484–89.

76. An excellent discussion of the power-density issue is available in Smil, *Energy at the Crossroads*, 240–44.

77. Declan Butler, "Nuclear Power's New Dawn," *Nature* 429, no. 6989 (May 20, 2004): 238–40.

78. Paul Grant, "Hydrogen Lifts Off—With a Heavy Load," *Nature* 424, no. 424 (July 10, 2003): 129–30.

79. Holdren, "Environmental Change and the Human Condition," 30. Ernst von Weizsäcker, Amory Lovins, and Hunter Lovins argue that resource and energy productivity can increase 4 to 5 percent a year. See Weizsäcker, Lovins, and Lovins, *Factor Four: Doubling Wealth, Halving Resource Use, The New Report to the Club of Rome* (London: Earthscan, 1997), 142.

80. American Iron and Steel Institute, "U.S. Steel Industry: World Leaders in Energy Efficiency," available at http://www.steel.org/AM/Template.cfm?Section=Home&TEMPLATE=/CM/ContentDisplay.cfm&CONTENTID=13399.

81. Statistics available from the U.S. Energy Information Administration, "Table 1.5: Energy Consumption, Expenditures, and Emissions Indicators, 1949–2004," available at http://www.eia.doe.gov/emeu/aer/txt/ptb0105.html.

82. Campbell, "Forecasting."

CHAPTER FIVE

1. M. J. R. Wortel and W. Spakman, "Subduction and Slab Detachment in the Mediterranean-Carpathian Region," *Science* 290, no. 5498 (December 8, 2000): 1910–17.

2. Renato Funiciello et al., "Seismic Damage and Geological Heterogeneity in Rome's Colosseum Area: Are They Related?" *Annali di Geofisica* 38, no. 5–6 (November–December 1995): 927–37.

3. V. I. Keilis-Borok, "The Concept of Chaos in the Problem of Earthquake Prediction," in *The Impact of Chaos on Science and Society*, ed. Celso Grebogi and James Yorke (Tokyo: United Nations University Press, 1997), 243–54, especially 245.

4. Ross Stein, "Earthquake Conversations," *Scientific American* 288, no. 1 (January 2003): 72–79. On the consequences of the great December 2004 Indonesian earthquake, see Kenneth Chang, "Post-Tsunami Earthquakes Rumbled around the Globe," *New York Times*, May 24, 2005, national edition, D3.

5. Matthew Gerstenberger et al., "Real-Time Forecasts of Tomorrow's Earthquakes in California," *Nature* 435 (May 19, 2005): 328–31. The current debate about the prospects for earthquake prediction is summarized in David Cyranoski, "A Seismic Shift in Thinking," *Nature* 431, no. 7012 (October 28, 2004): 1032–34.

Some of today's most advanced methods of prediction rely on analysis of earthquake precursors—patterns of seismicity—to identify regions with a high likelihood of large earthquakes. See, in particular, Vladimir Keilis-Borok, "Reverse Tracing of Short-term Earthquake Precursors," *Physics of the Earth and Planetary Interiors* 145, no. 1–4 (July 30, 2004): 75–85, and Vladimir Keilis-Borok, "Intermediate-term Earthquake Prediction," *Proceedings of the National Academy of Sciences, USA* 93 (April 1996): 3748–55.

6. Susan Hough, *Earthshaking Science: What We Know (and Don't Know) about Earthquakes* (Princeton: Princeton University Press, 2002), 111.

7. In seismically active zones, small tremors occur all the time, yet an earthquake follows only about 5 percent of them.

8. "Some experts have speculated that once all the damages from the California Wildfires of 2003 are tabulated (provisionally estimated at about $1.7–$3.5 billion), [the total] may rival the costliest fire incident in California's history, which occurred following the 1906 San Francisco Earthquake, which caused $5.7 billion in damages, in inflation-adjusted dollars." Federal Emergency Management Agency, *The California Fires Coordination Group: A Report to the Secretary of Homeland Security* (Washington, DC: U.S. Department of Homeland Security, 2004), 9; available at http://permanent.access.gpo.gov/websites/www.fema.gov/pdf/library/draft_cfcg_report_0204.pdf.

9. Although social science research often fails to recognize the danger of negative synergy, an exception is research on the causes of urban decay in the United States. See, for instance, R. Wallace and D. Wallace, "Resilience and Persistence of the Synergism of Plagues: Stochastic Resonance and the Ecology of Disease, Disorder, and Disinvestment in US Urban Neighborhoods," *Environment and Planning A* 29 (1997): 789–804. The risk posed by negative synergy among environmental stresses is also discussed in Will Steffen et al., "Abrupt Changes: The Achilles' Heels of the Earth System," *Environment* 46, no. 3 (April 2004): 8–20. The authors write, "How many [environmental] stresses, occurring when and where would it take for the global economic system to begin a downward, self-reinforcing spiral that would lead to a rapid collapse?"

10. Federal Emergency Management Agency, *The California Fires Coordination Group*, 11.

11. Says Craig Allen, a research ecologist with the United States Geological Survey in New Mexico, "As the climate is changing, these ecosystems are rearranging themselves. Massive forested die-back is one way these systems will reassemble." Quoted in Jim Robbins, "Beetles Take a Devastating Toll on Western Forests," *New York Times*, July 13, 2004, national edition, D4.

12. As Mike Davis, professor of history at the University of California in Irvine and author of well-known books about urbanization in Southern California, says, "These dead forests represent an almost apocalyptic hazard to more than 100,000 mountain and foothill residents, many of whom depend on a single, narrow road for their fire escape." Davis, "The Perfect Fire," History News Network, Center for History and New Media, George Mason University, October 27, 2003, available at http://hnn.us/articles/1761.html.

13. This definition of breakdown actually subsumes the word's conventional meaning: a system's simplification will likely disrupt its regular functions, while a breakdown

of a system's regular functions will, at least over time, lead to its simplification. My definitions of breakdown and collapse in these paragraphs are adapted from Joseph Tainter's "Comments on the Symposium 'I Fall to Pieces: Global Perspectives on the Collapse of Complex Systems,'" presented at the 65th Annual Meeting of the Society for American Archaeology, Philadelphia, April 8, 2000. See also Tainter, *Collapse*, 4, 31. Jared Diamond offers a similar definition in *Collapse: How Societies Choose to Fail or Succeed* (New York: Viking, 2005), 3.

14. Note that simpler doesn't mean easier. In fact, in most cases everyday tasks became much harder. We usually introduce complexity into our lives in order to make things easier for us, so when this complexity fails, life often becomes harder.

15. For overviews of common explanations of societal collapse, see "The Study of Collapse," chapter 3 in Joseph Tainter, *The Collapse of Complex Societies* (Cambridge: Cambridge University Press, 1988), 39–90; Tainter, "Theories of the Collapse of States," *The Oxford Companion to Archeology*, ed. Brian Fagan et al. (Oxford: Oxford University Press, 1996), 688–90; and Norman Yoffee, "Orienting Collapse," chapter 1 in Norman Yoffee and George Cowgill, eds., *The Collapse of Ancient States and Civilizations* (Tucson: University of Arizona Press, 1988), 1–19. Bert Useem reviews recent sociological theories of civil violence such as riots and rebellions in "Breakdown Theories of Collective Action," *Annual Review of Sociology* 24 (1998): 215–38.

16. Overload-breakdown theories of this form have an honorable pedigree derived from at least three distinct lines of research and thought: the functionalist sociology of Émile Durkheim and Talcott Parsons; the systems theory and cybernetics pioneered by Ludwig von Bertalanffy and Norbert Wiener; and the information-processing and computational theories of cognitive scientists and organizational theorists like Herbert Simon. Durkheim's most relevant work is his study of suicide; ideas of overload and breakdown are implicit in his theories of both "egoistic" and "anomic" suicide. Parsons, in his discussion of social change, argues that "strain" can upset the "balance between forces tending toward reequilibration of the previous structure and toward transition to a new structure." In political science, Karl Deutsch, Samuel Huntington, and Alexander Motyl, among others, have proposed overload-breakdown theories of social and political change. Huntington, for instance, argues that societies are vulnerable to instability when their level of political participation exceeds their level of political institutionalization. See Émile Durkeim, *Suicide: A Study in Sociology*, trans. John A. Spaulding and George Simpson (Glencoe, IL: Free Press, [1897] 1951); Talcott Parsons, chapter 11, "The Processes of Change of Social Systems," in *The Social System* (Glencoe, IL: Free Press, 1951), especially 493; Ludwig von Bertalanffy, "An Outline of General System Theory," *British Journal for the Philosophy of Science* 1, no. 2 (August 1950): 134–65; Norbert Wiener, *Cybernetics: Or Control and Communication in the Animal and the Machine* (Cambridge, MA: MIT Press, 1961); Herbert Simon, *Reason in Human Affairs* (Stanford: Stanford University Press, 1983), especially chapter 3, "Rational Processes in Social Affairs," 75–107; Karl Deutsch, "Cracks in the Monolith: Possibilities and Patterns of Disintegration in Totalitarian Systems," in Carl J. Friedrich, ed., *Totalitarianism: Proceedings of a Conference Held at the American Academy of Arts and Sciences* (Cambridge, MA: Harvard University Press, 1954), 308–33; Samuel Huntington,

Political Order in Changing Societies (New Haven: Yale University Press, 1968), especially 79 and 274–78; and Alexander Motyl, *Imperial Ends: The Decay, Collapse, and Revival of Empires* (New York: Columbia University Press, 2001), 50–53.

17. For a discussion of the sources of this extraordinary adaptability, see Thomas Homer-Dixon, *The Ingenuity Gap: Facing the Economic, Environmental, and Other Challenges of an Increasingly Complex and Unpredictable Future* (New York: Vintage, 2002), 306–307.

18. The political scientist Robert Jackson calls these countries "quasi-states." See Jackson, "Quasi-states, Dual Regimes, and Neoclassical Theory: International Jurisprudence and the Third World," *International Organization* 41, no. 4 (Autumn 1987): 519–49.

19. "Speed," as I use the term here, is a composite variable incorporating both the velocity of movement of material, energy, or information along a link between two nodes and the average density of each "package" of material, energy, or information that moves along the link. It is, therefore, essentially the link's transmission capacity.

20. Also, the nodes themselves tend to become more complex, as the people who create and operate them try to make them perform better. For example, a manufacturing company might improve the efficiency of its production processes by adopting more sophisticated methods for inventory control. W. Brian Arthur shows that competition among entities in a co-evolutionary environment (for instance, among corporations in a market or among organisms in an ecosystem) boosts the complexity of the entities as they try to survive by improving their performance, a process he calls "structural deepening." See Arthur, "On the Evolution of Complexity," in *Complexity: Metaphors, Models, and Reality*, ed. G. Cowan, D. Pines, and D. Meltzer, Santa Fe Institute Studies in the Sciences of Complexity, Proceedings, Vol. 19 (Reading, MA, 1994), 65–78.

21. About a century ago, the great French sociologist Émile Durkheim, who dominated French social thought during the late nineteenth and early twentieth centuries, referred to the general phenomenon of growing connectivity and speed as the rising "dynamic density" of human societies. He argued that greater dynamic density resulted from the growth in human population, its increasing concentration in cities, and the rapid development of communication and transportation technologies; dynamic density in turn stimulated the economic division of labor. See Durkheim, *The Division of Labor in Society*, trans. George Simpson (New York: Free Press, 1968 [1933]), 257.

22. The classic discussion of tight coupling is Charles Perrow, *Normal Accidents: Living with High-Risk Technologies* (New York: Basic, 1984).

23. For a discussion of key factors, including human demographics and travel, promoting the emergence of infectious disease, see chapter 3, "Factors in Emergence," in Mark Smolinski, Margaret Hamburg, and Joshua Lederberg, eds., *Microbial Threats to Health: Emergence, Detection, and Response* (Washington, DC: Institute of Medicine of the National Academies, National Academy Press, 2003), 53–148.

24. For an explanation of stock market booms and crashes that highlights self-reinforcing feedback loops in complex systems, see Didier Sornette, *Why Stock Markets Crash: Critical Events in Complex Financial Systems* (Princeton: Princeton University Press, 2003).

25. On failures arising from new links between previously isolated systems, see Dietrich Dörner, *The Logic of Failure: Recognizing and Avoiding Error in Complex Situations*, trans. Rita and Robert Kimber (Cambridge, MA: Perseus, 1996). On failures arising from new links *inside* systems (that is, new links between previously separated system components), which the Yale sociologist Charles Perrow famously calls "normal accidents," see Perrow, *Normal Accidents*, especially chapter 3, "Complexity, Coupling, and Catastrophe," 62–100. Scott Sagan reviews the influence of Perrow's book in Sagan, "Learning from *Normal Accidents*," *Organization & Environment* 17, no. 1 (March 2004): 15–19.

26. Researchers have found that human, organizational, and sociocultural factors are often deep causes of this kind of breakdown. Organizations, including NASA before the Columbia disaster, can have structures or cultures that get in the way of communication between groups responsible for designing, maintaining, and running a system or that encourage people to neglect proper testing, training, and safety procedures. See William Evan and Mark Manion, *Minding the Machines: Preventing Technological Disasters* (Upper Saddle River, NJ: Prentice Hall PTR, 2002); and James Reason, *Human Error* (Cambridge: Cambridge University Press, 1990), 173. The evolution of thinking about organizational errors is summarized in Karlene H. Roberts, "Organizational Errors: Catastrophic," in Neil J. Smelser et al., eds., *International Encyclopedia of the Social and Behavioral Sciences*, Vol. 16 (Amsterdam: Elsevier, 2001), 10942–45. For a remarkable account of these organizational causes in the case of the Columbia disaster, see the Columbia Accident Investigation Board, Report Volume I (Government Printing Office, Washington, DC: NASA, August 2003).

27. Eric Lerner, "What's Wrong with the Electric Grid?" *The Industrial Physicist* (American Institute of Physics, October–November 2003): 8–13; and James Glanz and Andrew Revkin, "Set of Rules Too Complex to Be Followed Properly, or Not Complex Enough," *New York Times*, August 19, 2003, national edition, A20.

28. Andrew Revkin, "Experts Point to Strains on Electric Grid's Specialists," *New York Times*, September 2, 2003, national edition, A12. For a complete assessment of the causes of the 2003 blackout, see U.S.–Canada Power System Outage Task Force, *Final Report on the August 14, 2003, Blackout in the United States and Canada: Causes and Recommendations* (Washington, DC, and Ottawa: U.S. Department of Energy and Ministry of Natural Resources Canada, April 2004).

29. Accessible summaries of this research are Albert-László Barabási, *Linked: The New Science of Networks* (Cambridge, MA: Perseus, 2002); and Duncan Watts, *Six Degrees: The Science of a Connected Age* (New York: Norton, 2003). See also Albert-László Barabási and Eric Bonabeau, "Scale-Free Networks," *Scientific American* 288, no. 5 (May 2003): 60–69.

30. The airline financial crisis following the 9/11 attacks encouraged many U.S. airlines to move away from a hub-and-spoke routing arrangement.

31. In the language of statistics, the frequency distribution of nodes in a random network (when graphed according to the number of links per node) forms a bell-shaped curve. The distribution of a scale-free network, in contrast, forms a declining curve that drops quickly at first and then has a long, slowly diminishing tail, as illustrated ahead. (In technical terms, this is a "power-law" distribution.) The network's hubs

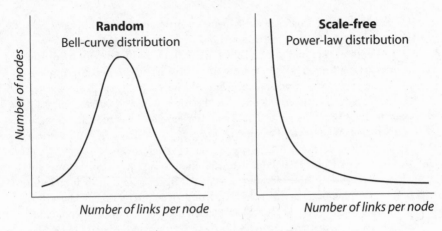

Frequency distributions of random and scale-free networks

are at the end of the tail. Such networks are called scale-free because, in contrast to random networks, there's no typical or "average" number of links between nodes.

32. There is some dispute among specialists about whether the North American electricity grid is scale-free. L. A. N. Amaral et al. argue that it is not in "Classes of Small-World Networks," in *Proceedings of the National Academy of Sciences of the United States of America* 97, no. 21 (October 10, 2000): 11149–52, while Albert-László Barabási and Réka Albert present evidence indicating that it is in "Emergence of Scaling in Random Networks," *Science* 286 (October 15, 1999): 509–12. A close look at the respective authors' interpretation of the data supports, I believe, the assessment of Barabási and Albert, although all these authors undertake only a static analysis of network architecture that neglects the variable carrying capacities of different links.

33. Ricard Solé and José Montoya, "Complexity and Fragility in Ecological Networks," *Proceedings of the Royal Society of London* 268 (2001): 2039–45.

34. On the role of urbanization in the emergence and spread of infectious disease, see Smolinski, Hamburg, and Lederberg, eds., *Microbial Threats to Health*, 81–85.

35. Some of the material in the following paragraphs appeared in Thomas Homer-Dixon, "The Rise of Complex Terrorism," *Foreign Policy*, no. 128 (January/February 2002): 52–62. For similar arguments, see the comments by John Robb, a security analyst with a background in counterterrorism and complex systems analysis, on his blog at http://globalguerrillas.typepad.com/globalguerrillas/.

36. In 1997, a special investigative commission set up by President Bill Clinton reported that "growing complexity and interdependence, especially in the energy and communications infrastructures, create an increased possibility that a rather minor and routine disturbance can cascade into a regional outage." Technical complexity, the Commission continued, echoing Charles Perrow's concept of normal accidents, "may also permit interdependencies and vulnerabilities to go unrecognized until a major failure occurs." The Commission concluded, "We are convinced that our vulnerabilities are increasing steadily, that the means to exploit those weaknesses are readily available and that the costs associated with an effective attack continue to

drop." Report of the President's Commission on Critical Infrastructure Protection, *Critical Foundations: Protecting America's Infrastructures* (Washington, DC: 1997): x. See also Massoud Amin, "National Infrastructures as Complex Interactive Networks," in Tariq Samad and John Weyrauch, eds, *Automation, Control, and Complexity: New Developments and Directions* (John Widely and Sons, 1999).

37. I describe a scenario for a terrorist attack against the U.S. electricity grid in the opening paragraphs of Homer-Dixon, "The Rise of Complex Terrorism."

38. Adilson Motter and Ying-Cheng Lai, "Cascade-Based Attacks on Complex Networks," *Physical Review E* 66, 065102 (Rapid Communication) (2002). See also Steven Rinaldi, James Peerenboom, and Terrence Kelly, "Identifying, Understanding, and Analyzing Critical Infrastructure Interdependencies," *IEEE Control Systems Magazine* (December, 2001): 11–25.

39. Langdon Winner, "Complexity and the Limits of Human Understanding," in *Organized Social Complexity: Challenge to Politics and Policy*, ed. Todd La Porte (Princeton: Princeton University Press, 1975): 69–70.

40. McKinsey & Company, Banking & Securities Practice, "Impact of Attack on New York Financial Services" (November, 2001), available at http://www.mckinsey.com/ideas/articles/ImpactofAttack.asp.

41. In its December 2001 statement on prospects for the world economy, the International Monetary Fund estimated the global GDP would be 1 percent lower in 2002 than it would have been in the absence of the attacks. This works out to a loss, in one year, of around $300–400 billion. Such lost productivity produces a stream of future losses (because of forgone investment, among other things). In 2002, the discounted present value of that year's loss plus future losses would have easily exceeded $1 trillion. See International Monetary Fund, *World Economic Outlook: The Global Economy after September 11* (Washington, DC: IMF, December 2001), 4, available at http://www.imf.org/external/pubs/ft/weo/2001/03/index.htm.

42. Jack Goldstone, *Revolution and Rebellion in the Early Modern World* (Berkeley: University of California Press, 1991), 469. (Emphasis in original.) Goldstone explicitly adopts an earthquake analogy for revolution on page 35. See also Jack Goldstone and Bert Useem, "Prison Riots as Microrevolutions: An Extension of State-Centered Theories of Revolution," *American Journal of Sociology* 104, no. 4 (January 1999): 985–1029; and Jack Goldstone, "Toward a Fourth Generation of Revolutionary Theory," in Nelson Polsby, ed., *Annual Review of Political Science*, Vol. 4, 2001 (Palo Alto, CA: Annual Reviews, 2001), 139–87.

43. Goldstone, *Revolution and Rebellion*, 36.

44. Marten Scheffer et al., "Catastrophic Shifts in Ecosystems," *Nature* 413 (October 11, 2001): 591–96.

45. The practice of "sustainable yield" resource management, as often applied to fisheries and other renewable-resource systems, frequently sets an allowable rate of extraction above the level the system can sustain given its natural fluctuations in productivity. See Donald Ludwig, Ray Hilborn, and Carl Walters, "Uncertainty, Resource Exploitation, and Conservation: Lessons from History," *Science* 260, no. 5104 (April 2, 1993): 17, 36. On sensitivity of cod stocks to fluctuations in temperature and salinity, see Alida Bundy and L. Paul Fanning, "Can Atlantic Cod (*Gadus morhua*) Recover? Exploring Trophic Explanations for the

Non-Recovery of the Cod Stock on the Eastern Scotian Shelf, Canada" *Canadian Journal of Fisheries and Aquatic Sciences* 62, no. 7 (July, 2005): 1474–90; and E. Meltzer, *Overview of the East Coast Marine Environment* (Ottawa: Canadian Arctic Resources Committee/Canadian Nature Federation, 1995).

46. For further details, see David Webster, *The Fall of the Ancient Maya: Solving the Mystery of the Maya Collapse* (London: Thames and Hudson, 2002); and Richardson Gill, *The Great Maya Droughts: Water, Life, and Death* (Mexico City: University of New Mexico Press, 2000). Webster emphasizes the role of population growth and Gill the role of drought in the Mayan collapse. Gill also uses an energy-based, complex-systems model similar to that adopted in this book. Jared Diamond summarizes the findings of these and many other works and their implications for modern industrial society in "The Last Americans: Environmental Collapse and the End of Civilization," *Harper's Magazine* (June 2003): 43–51. See also chapter 5, "The Maya Collapses," in Diamond, *Collapse*, 157–77. For an overview of the relationship between climate change and societal collapse, see Harvey Weiss and Raymond Bradley, "What Drives Societal Collapse?" *Science* 291, no. 5504 (January 26, 2001): 609–10.

47. The great Belgian historian Henri Pirenne argued in the 1930s that the Germanic invasions of the West in the fifth century did not destroy the "essential features" of Roman society, economy, and culture. It was only the advance of Islam in the seventh and early eighth centuries that isolated Western Europe from the Mediterranean and the eastern empire and brought the final demise of classical civilization in the West. Archaeological research, however, has shown this thesis to be largely incorrect. David Whitehouse writes, "Mediterranean civilization experienced a long process of change, in the course of which the old political and economic unity disintegrated. This process . . . was well advanced long before the seventh century." Whitehouse, "Archaeology and the Pirenne Thesis," in Charles Redman, ed., *Medieval Archaeology: Papers of the Seventeenth Annual Conference of the Center for Medieval and Early Renaissance Studies* (Binghamton, NY: State University of New York at Binghamton, 1989), 10. See also Richard Hodges and David Whitehouse, *Mohammed, Charlemagne & the Origins of Europe: Archaeology and the Pirenne Thesis* (London: Duckworth, 1983).

48. A leading advocate of these views is the historian Glen Bowersock of the Princeton Institute for Advanced Study. "Social, political, and intellectual reorganization was accomplished entirely within the framework of what had been there before," he writes. "As far as the internal functioning of the Roman empire in the sixth century is concerned," he goes on, "there is no clear indication of any substantial instability or depression in the social, economic, and political life of the time." G. W. Bowersock, "The Dissolution of the Roman Empire," in Yoffee and Cowgill, eds., *The Collapse of Ancient States and Civilizations*, 170–72.

49. The historian Chris Wickham writes, "Early medieval Italy was a very much simpler world than late Roman Italy." Wickham, *Early Medieval Italy: Central Power and Local Society, 400–1000* (Ann Arbor : University of Michigan Press, 1989), 27, 40–41. Says the archaeologist David Whitehouse, "the political fragmentation of the Roman empire was accompanied by a long process of economic decline and urban decay." Whitehouse, "Archaeology and the Pirenne Thesis," 6–7. On

archaeological evidence, see also Hodges and Whitehouse, *Mohammed, Charlemagne & the Origins of Europe.*

50. On Roman and later lighting technology, see William Nordhaus, "Do Real-Output and Real-Wage Measures Capture Reality? The History of Lighting Suggests Not," in *The Economics of New Goods*, Timothy Bresnahan and Robert Gordon, eds. (Chicago: University of Chicago Press, 1997), 29–66, especially 33.

51. Rein Taagepera has conducted the most important research on changes in the geographical area of historical empires. See Taagepera, "Growth Curves of Empires," *General Systems* 13 (1968): 171–75; Taagepera, "Size and Duration of Empires: Systematics of Size," *Social Science Research* 7 (1978): 108–27; Taagepera, "Size and Duration of Empires: Growth-Decline Curves, 3000 to 600 B.C.," *Social Science Research* 7 (1978): 180–96; Taagepera, "Size and Duration of Empires: Growth-Decline Curves, 600 B.C. to 600 A.D.," *Social Science History* 3, no. 3/4 (1979): 115–38; and Taagepera, "Expansion and Contraction Patterns of Large Polities: Context for Russia," *International Studies Quarterly* 41 (1997): 475–504.

52. An exception is the historian Aurelio Bernardi, who writes, "Thus the fall of an immense State that had lasted a thousand years was completed in the course of not much more than half a century." Bernardi, "The Economic Problems of the Roman Empire at the Time of Its Decline," in Carlo Cipolla, ed., *The Economic Decline of Empires* (London: Methuen, 1970), 25.

53. This conclusion can be drawn from examining the graphs in Rein Taagepera, "Expansion and Contraction Patterns of Large Polities," 482–84, and in Alexander Moytl, *Imperial Ends: The Decay, Collapse, and Revival of Empires* (New York: Columbia University Press, 2001), 41–45.

54. Some historians, however, have lately betrayed a distinct nostalgia for empires. For example, see Niall Ferguson, *Colossus: The Rights of America's Empire* (New York: Penguin, 2004), especially chapter 5, "The Case for Liberal Empire," 169–99.

55. General Accounting Office, *Emerging Infectious Diseases: Asian SARS Outbreak Challenged International and National Responses* (Washington, DC: GAO, April 2004), 4.

56. Arthur Koestler, *The Sleepwalkers: A History of Man's Changing Vision of the Universe* (London: Hutchison & Co., 1959), 48.

CHAPTER SIX

1. Ian Stirling, Nicholas Lunn, and John Iacozza, "Long-Term Trends in the Population Ecology of Polar Bears in Western Hudson Bay in Relation to Climatic Change," *Arctic* 52, no. 3 (September 1999): 294–306. Percentage changes in bear condition were calculated using figure 6 on page 302.

2. While most polar bears hunt on the ice after freeze-up in the fall, pregnant females remain on shore to give birth to their cubs, usually around December. To ensure the health and survival of their cubs, the females must begin the long winter with as much body fat as possible.

3. Peter Clarkson and Doug Irish, "Den Collapse Kills Female Polar Bear and Two Newborn Cubs," *Arctic* 44, no. 1 (March 1991): 83–84; and Ian Stirling and

Andrew Derocher, "Possible Impacts of Climatic Warming on Polar Bears," *Arctic* 46, no. 3 (September 1993): 240–45, especially 244.

4. Natalie Angier, "Built for the Arctic: A Species' Splendid Adaptations," *New York Times*, January 27, 2004, national edition, D1. While there is a general consensus among scientists and wildlife experts that global warming is disrupting polar bear ecology, some skeptics have challenged this consensus. For a summary of the debate, see Clifford Krauss, "Debate on Global Warming Has Polar Bear Hunting in Its Sights," *New York Times*, May 27, 2002, national edition, A1.

5. Camille Parmesan and Gary Yohe, "A Globally Coherent Fingerprint of Climate Change Impacts across Natural Systems," *Nature* 421, no. 6918 (January 2, 2003): 37–42; and Terry Root et al., "Fingerprints of Global Warming on Wild Animals and Plants," *Nature* 421, no. 6918 (January 2, 2003): 57–60.

6. On sardine catches in Africa, see Dirk Verschuren, "The Heat on Lake Tanganyika," *Nature* 424, no. 6950 (August 14, 2003): 731–32.

7. Andrew Blaustein and Pieter Johnson, "Explaining Frog Deformities," *Scientific American* 288, no. 2 (February 2003): 60–65; and Stephen Buchmann and Gary Nabhan, *The Forgotten Pollinators* (Washington, DC: Island Press, 1996).

8. The Harvard ecologist E. O. Wilson writes, "We evolved here, one among many species, across millions of years, and exist as one organic miracle linked to others. The natural environment we treat with such unnecessary ignorance and recklessness was our cradle and nursery, our school, and remains our one and only home. To its special conditions we are intimately adapted in every one of the bodily fibers and biochemical transactions that give us life." Edward O. Wilson, "The Bottleneck," *Scientific American* 286, no. 2 (February 2002): 91.

9. Bjørn Lomborg, *The Skeptical Environmentalist: Measuring the Real State of the World* (Cambridge: Cambridge University Press, 2001).

10. Lomborg writes, "We will not lose our forests; we will not run out of energy, raw materials or water. We have reduced atmospheric pollution in the cities of the developed world and have good reason to believe that this will also be achieved in the developing world. Our oceans have not been defiled, our rivers have become cleaner and support more life. . . . Acid rain did not kill off our forests, our species are not dying out as many have claimed. . . . The problem of the ozone layer has been more or less solved. The current outlook on the development of global warming does not indicate a catastrophe." Ibid., 329.

11. Lomborg simplistically extrapolates past trends into the future. He looks mainly at global averages, which often obscure key developments at the regional level. He frequently uses a resource's price as an objective indicator of its scarcity, when in fact price often reflects a multitude of political, economic, and social factors that have little to do with underlying scarcity or abundance. And he underplays the possibility of nonlinear shifts in ecosystems, like the collapse of fisheries or a sudden climate flip. Lomborg is also breathtaking in his hypocrisy: he too manipulates statistics, uses evidence selectively, and employs straw-man argumentation—just like the worst environmental ideologue. "Every class of mistake of which he accuses environmentalists and environmental scientists," writes John Holdren of Harvard University, "is in fact committed prolifically and indiscriminately in *The Skeptical Environmentalist*." John Holdren, "A Response to Bjørn

Lomborg's Response to My Critique of His Energy Chapter," *Scientific American.com*, April 15, 2002, 5. Available at http://www.scientificamerican.com/print_version.cfm?articleID=000DC658-9373-1CDA-B4A8809EC588EEDF.

12. See, for instance, Stuart Pimm and Jeff Harvey, "No Need to Worry about the Future," review in *Nature* 414, no. 6860 (November 8, 2001): 149–50; Michael Grubb, "Relying on Manna from Heaven," review in *Science* 294, no. 5545 (November 9, 2001): 1285–87; Douglas Kysar, "Some Realism about Environmental Skepticism: The Implications of Bjørn Lomborg's *The Skeptical Environmentalist* for Environmental Law and Policy," *Ecology Law Quarterly* 30 (2003): 223–80; and "Misleading Math about the Earth," a compilation of critiques of Lomborg's arguments by Stephen Schneider (on global warming), John Holdren (on energy), John Bongaarts (on population), and Thomas Lovejoy (on biodiversity), in *Scientific American* 286, no. 1 (January 2002): 61–71.

13. The charge that Lomborg sometimes engages in outright deceit is justified. To take one of many examples in *The Skeptical Environmentalist*, he claims on page 113 that forest loss in the tropics is not nearly as severe as often claimed: the rate of loss is only 0.46 percent a year, he says, not 0.7 to 0.8 percent as widely reported. To support this claim, he refers to the Summary Report of *The Global Forest Resources Assessment 2000* produced by the UN's Food and Agriculture Organization (FAO) in 2001. But a close look at this document shows that he cites only the *Assessment's* data from a satellite survey of tropical forests. He completely ignores the *Assessment's* main conclusions about tropical forest loss, generated by combining the results of the satellite survey with a painstaking country-by-country on-the-ground inventory of forests. Based on these two methods, the FAO concluded that the rate of tropical forest loss during the 1990s was 0.73 percent a year, not significantly different from the previous decade's rate. This conclusion appears in the Summary Report's paragraphs immediately preceding the satellite data—so it's impossible to miss. Moreover, the Summary Report's abstract and the FAO's annual survey, *State of the World's Forests 2001*, cite only the combined results of the two methods, not the satellite data by themselves. So the only reasonable interpretation of Lomborg's omission of the *Assessment's* main findings is that he deliberately intended to mislead his readers. See Committee on Forestry, *The Global Forest Resources Assessment 2000, Summary Report* (Rome: FAO, 2001) available at ftp://ftp.fao.org/unfao/bodies/cofo/cofo15/X9835e.pdf; and Food and Agriculture Organization, *State of the World's Forests 2001* (Rome: FAO, 2001), available at ftp://ftp.fao.org/docrep/fao/003/y0900e/y0900e00.pdf.

14. A classic discussion of our chronic denial of environmental problems is David Orr and David Ehrenfeld, "None So Blind: The Problem of Ecological Denial," *Conservation Biology* 9, no. 5 (October 1995): 985–87.

15. I'm indebted to John Holdren for pointing out these stages to me, although the labels are mine.

16. "One of the great success stories of the recent half-century is . . . the remarkable progress the industrial societies have made, during a period of robust economic growth, in reversing the negative environmental impacts of industrialization." Jack Hollander, *The Real Environmental Crisis: Why Poverty, Not Affluence, Is the Environment's Number One Enemy* (Berkeley: University of California Press, 2003), 3.

17. "A Project to Grow Fish in Once-Polluted Boston Harbor Waters," *New York Times*, December 28, 1997, national edition, 22.

18. Felicity Barringer, "California Air Is Cleaner, but Troubles Remain," *New York Times*, August 3, 2005, national edition, A1.

19. This is often called the Environmental Kuznets Curve (EKC) hypothesis. Simon Kuznets, one of the twentieth century's great economists, proposed that a country's income inequality rises and subsequently declines as its average income rises. The EKC hypothesis, although not proposed by Kuznets himself, postulates that pollution and other forms of environmental damage will also rise and then decline as average income rises. Although the hypothesis has become a staple of conservative commentary on environmental issues, researchers have shown that it's invalid in important respects. For a discussion and critique, see Cutler Cleveland and Matthias Ruth, "Indicators of Dematerialization and the Materials Intensity of Use: A Critical Review with Suggestions for Future Research," *Journal of Industrial Ecology* 2, no. 3 (Summer 1998): 15–50. See also Dale Rothman and Sander de Bruyn, eds., "The Environmental Kuznets Curve," special issue of the journal *Ecological Economics* 25 (1998). Evidence in favor of the EKC is presented in Gene Grossman and Alan Krueger, "Economic Growth and the Environment," *NBER Working Paper* #4634, National Bureau of Economic Research (February 1994). The quotation from Wilfred Beckerman can be found in Beckerman, "Economic Growth and the Environment: Whose Growth? Whose Environment?" *World Development* 20, no. 4, Special Issue (April 1992): 481–96.

20. Although in recent decades most companies have sharply reduced resource inputs to production, they've invariably done so to reduce costs and not to reduce their environmental impact.

21. National Association of Home Builders, "New Home Characteristics," *Housing 2004: Facts, Figures & Trends* (Washington, DC: NAHB, 2004), 11; Joy Nielsen and Barry Popkin, "Patterns and Trends in Food Portion Sizes, 1977–1998," *Journal of the American Medical Association* 289, no. 4 (January 22, 2003): 450–53.

22. Matthew Wald, "Oil Crises: Which One Is Worse," *New York Times*, Week in Review, April 21, 2002, national edition, 4. According to the U.S. Bureau of Transportation Statistics, in 2001 there were about 230 million registered cars, trucks, and motorcycles in United States (see the data table at http://www.bts.gov/publications/national_transportation_statistics/2002/html/table_automobile_profile.html).

23. For a detailed treatment of the factors that influence the environmental impact of such migrations, see Richard Bilsborrow, "Migration, Population Change, and the Rural Environment," in *Environmental Change and Security Program, Report 8* (Washington, DC: Woodrow Wilson International Center for Scholars, Environmental Change and Security Program, 2002), available at http://www.wilsoncenter.org/topics/pubs/Report_8_BIlsborrow_article.pdf.

24. The most comprehensive assessment of humankind's impact on the global environment is the Millennium Ecosystem Assessment, an international work program sponsored and coordinated by the United Nations and "designed to meet the needs of decision makers and the public for scientific information concerning

the consequences of ecosystem change for human well-being and options for responding to those changes." Almost two thousand authors from nearly one hundred countries have been involved in preparing this assessment, which has been summarized in fifteen reports. Further information is available at http://www.maweb.org//en/index.aspx.

25. William Ruddiman, however, puts the date much earlier, about eight thousand years ago, reckoning that large-scale deforestation for agriculture in Eurasia caused, around that time, a fundamental shift in Earth's carbon and methane cycles. See W. F. Ruddiman, "The Anthropogenic Greenhouse Era Began Thousands of Years Ago," *Climatic Change* 61 (2003): 261–93; Betsy Mason, "The Hot Hand of History," *Nature* 427, no. 6975 (February 12, 2004): 582–83; and Paul Crutzen, "Geology of Mankind," *Nature* 415, no. 6867 (January 3, 2002): 23.

26. Robert Berner, "The Long-term Carbon Cycle, Fossil Fuels and Atmospheric Composition," *Nature* 426, no. 6964 (November 20, 2003): 323–26.

27. David Schimel and David Baker, "The Wildfire Factor," *Nature* 420, no. 6911 (November 7, 2002): 29–30; and Susan Page et al., "The Amount of Carbon Released from Peat and Forest Fires in Indonesia during 1997," *Nature* 420, no. 6911 (November 7, 2002): 61–65.

28. Andrew Revkin, "Sunken Fires Menace Land and Climate," *New York Times*, January 15, 2002, national edition, D1. Estimates of the quantity of coal burned by these fires are prone to large errors because they require multiple assumptions about such things as the average thickness of the coal seams, the rate of combustion, and the combustion temperature.

29. See chapter 6 in Vaclav Smil, *Cycles of Life: Civilization and the Biosphere* (New York: Scientific American Library, 1997), 141–69.

30. Reactive or *fixed* nitrogen (as opposed the form of nitrogen that's abundant in air) allows plants to build proteins and so is essential to all higher life. There are large uncertainties in estimates of total natural nitrogen fixation: although the range is commonly put at 90 to 100 million tons, the figure could range as high as 250 million tons. Smil, *The Earth's Biosphere*, 248–51; and Smil, personal correspondence with the author, March 28, 2004.

31. Robert May, "Melding Heart and Head," *Our Planet* (2000), available at http://www.ourplanet.com/imgversn/111/may.html; and Vaclav Smil, "Global Population in the Nitrogen Cycle," *Scientific American* 277, no. 1 (July 1997): 76–81.

32. Nicola Nosengo, "Fertilized to Death," *Nature* 425, no. 6961 (October 30, 2003): 894–95.

33. United Nations Environment Programme, *Global Environment Outlook 2003* (Nairobi: UNEP, 2003). See also Emily Matthews and Allen Hammond, *Critical Consumption Trends and Implications: Degrading Earth's Ecosystems* (Washington, DC: World Resources Institute, 1999), 11–30; and David Malakoff, "Death by Suffocation in the Gulf of Mexico," *Science* 281, no. 5374 (July 10, 1998): 190–92.

34. "Humans are now an order of magnitude more important at moving sediment than the sum of all other natural processes operating on the surface of the planet." Bruce Wilkinson, "Humans as Geologic Agents: A Deep-Time Perspective," *Geology* 33, no. 3 (March 2005): 161–64. See also B. L. Turner et al., eds., *The Earth As Transformed by Human Action: Global and Regional Changes in the Biosphere*

over the Past 300 Years (Cambridge: Cambridge University Press with Clark University, 1990), 13.

35. Vitousek and his colleagues estimate that humans have transformed about a third to a half of Earth's total land surface, while Smil estimates that we have "strongly or partially imprinted" some 55 percent of non-glaciated land. See Peter Vitousek, Harold Mooney, Jane Lubchenco, and Jerry Melillo, "Human Domination of Earth's Ecosystems," *Science* 277, no. 5325 (July 25, 1997): 494–9; and Vaclav Smil, *The Earth's Biosphere: Evolution, Dynamics, and Change* (Cambridge, MA: MIT, 2002), 239–40.

36. Vitousek et al., "Human Domination," 498.

37. James Gustave Speth, "A New Green Regime," *Environment* (Spring 2002): 18.

38. Peter Vitousek, Paul Ehrlich, Anne Ehrlich, and Pamela Matson, "Human Appropriation of the Products of Photosynthesis," *BioScience* 36, no. 6 (June 1986): 368–73. The authors examine "human impact on the biosphere by calculating the fraction of net primary production (NPP) that humans have appropriated. NPP is the amount of energy left after subtracting the respiration of primary producers (mostly plants) from the total amount of energy (mostly solar) that is fixed biologically." For a more recent analysis that uses an alternative methodology but arrives at similar conclusions, see Marc Imhoff et al., "Global Patterns in Human Consumption of Net Primary Production," *Nature* 429, no. 6994 (June 24, 2004): 870–73. See also, Helmut Habert, "Human Appropriation of Net Primary Production as an Environmental Indicator: Implications for Sustainable Development," *Ambio* 26, no. 3 (May 1997): 143–46.

39. Vitousek et al., "Human Appropriation," 372. The date of presumed plant diversification was derived from the discussion in Paul Kenrick and Peter R. Crane, chapter 7, "Early Evolution of Land Plants," *The Origin and Early Diversification of Land Plants: A Cladistic Study* (Washington, DC: Smithsonian Institution Press, 1997), 226–310.

40. Food and Agriculture Organization, "Wood Energy: Promoting Sustainable Wood Energy Systems (SWES)," report available at http://www.fao.org/forestry/foris/webview/energy/index.jsp?siteId=3281&langId=1.

41. In 2001, the United Nations Food and Agriculture Organization (FAO) released the results of the Global Forest Resources Assessment 2000, which provided a comprehensive account of global forest loss based on a country-by-country inventory and a satellite survey. The results can be found in Committee on Forestry, *The Global Forest Resources Assessment 2000, Summary Report* (Rome: FAO, 2001) available at ftp://ftp.fao.org/unfao/bodies/cofo/cofo15/X9835e.pdf; and at FAO, *State of the World's Forests 2001* (Rome: FAO, 2001), available at ftp://ftp.fao.org/docrep/fao/003/y0900e/y0900e00.pdf.

42. Currently, forest loss is concentrated in certain countries. In Asia, these include Indonesia, Malaysia, Thailand, Myanmar, and the Philippines; in Africa, forests are disappearing at a high rate in Zambia, Malawi, and Zimbabwe and across a swath of West African countries from Nigeria through Guinea; and in Latin America, deforestation is severe in Brazil, Argentina, and Mexico. On virgin forests, see James Gustave Speth, "Recycling Envronmentalism," *Foreign Policy* (July/August 2002): 74–75; and on mangroves, see FAO, "Part 1: The Situation and

Developments in the Forest Sector," *State of World Forests 2003* (Rome: FAO, 2003), available at http://www.fao.org/DOCREP/005/Y7581E/y7581e04.htm#Po_4.

43. "Making Mincemeat out of the Rainforest," *Environment* 46, no. 5 (June 2004): 5. Larry Rohter, "Loggers, Scorning the Law, Ravage the Amazon," *New York Times*, October 16, 2005, national edition, 1; Rohter, "Deep in Amazon, Vast Questions about Climate," *New York Times*, November 4, 2003, national edition, D1; and Rohter, "Amazon Forest Is Still Burning, Despite Pledges," *New York Times*, August 23, 2002, national edition, A1.

44. Raymond Bonner, "Indonesia's Forests Go Under Ax for Flooring," *New York Times*, September 13, 2002, national edition, A3; and Jane Perlez, "Forests in Southeast Asia Fall to Prosperity's Ax," *New York Times*, April 29, 2006, national edition, A1.

45. Vitousek et al., "Human Domination," 496–97; and Smil, *The Earth's Biosphere*, 246.

46. United Nations Educational, Scientific, and Cultural Organization, *The UN World Water Development Report Water for People, Water for Life* (Paris: UNESCO, 2003), 10; available at http://www.unesco.org/water/wwap/wwdr/table_contents.shtml.

47. Tom Gardner-Outlaw and Robert Engleman, *Sustaining Water, Easing Scarcity: A Second Update* (Washington, DC: Population Action International, 1997).

48. "Outside China, the world's population has been increasing more quickly than the total food fish supply from production, resulting in a decreased global per capita fish supply from 14.6 kg in 1987 to 13.1 kg in 2000." Food and Agriculture Organization, *The State of World Fisheries and Aquaculture*), "Part 1: World Review of Fisheries and Aquaculture" (Rome: FAO, 2002), available at http://www.fao.org/docrep/005/y7300e/y7300e04.htm#P5_111. See also Figure 2 in Reg Watson, "The Sea Around Us Project Runs a Successful Marine Symposium at AAAS," *The Sea Around Us Project Newsletter* 11 (May/June 2002): 4, available at http://saup.fisheries.ubc.ca/Newsletters/Issue11.pdf.

49. Ecologists speak of the upper "trophic levels" of the fisheries ecosystem. Trophic levels are "ranked according to how many steps they are removed from the primary producers at the base of the web, which generally consists of phytoplanktonic algae." Daniel Pauly and Reg Watson, "Counting the Last Fish," *Scientific American* 289, no. 1 (July 2003): 43–47, especially 45; and Daniel Pauly et al., "Fishing Down Marine Food Webs," *Science* 279, no. 5352 (February 6, 1998): 860–63.

50. Ransom Myers and Boris Worm, "Rapid Worldwide Depletion of Predatory Fish Communities," *Nature* 423, no. 6937 (May 15, 2003): 280–83. For a critical response, see John Hampton et al., "Fisheries: Decline of Pacific Tuna Populations Exaggerated?" *Nature* 434, no. 7037 (April 28, 2005): E1–E2, and the response by Myers and Worm in the same edition. Also see Andrew Revkin, "Atlantic Sharks Found in Rapid Decline," *New York Times*, January 17, 2003, national edition, A16; and Andrew Revkin, "Commercial Fleets Slashed Stocks of Big Fish by 90%, Study Says," *New York Times*, May 15, 2003, national edition, A1.

51. Jeffrey Hutchings, "The cod that got away," *Nature* 428, no. 6986 (April 29, 2004): 899–900.

52. Villy Christensen et al., "Hundred-year Decline of North Atlantic Predatory Fishes," *Fish and Fisheries* 4, no. 1 (March 2003): 1. See also Craig Smith, "North Sea Cod Crisis Brings Call for Nations to Act," *New York Times*, November 7, 2002, national edition, A3.

53. "Trawlers trailing dredges the size of football fields have literally scraped the bottom clean," write Daniel Pauly and Reg Watson of the University of British Columbia. These practices harvest "an entire ecosystem—including supporting substrates such as sponges—along with the catch of the day. Farther up the water column, long lines and drift nets are snagging the last sharks, swordfish and tuna. The hauls of these commercially desirable species are dwindling, and the sizes of individual fish being taken are getting smaller; a large number are even captured before they have time to mature." Pauly and Watson, "Counting the Last Fish," 43–47.

54. These practices have caused "the wholesale destruction of many deep-water environments," says Callum Roberts, a marine biologist at England's University of York. He points to the example of orange roughy, an exotic deep-water fish that was once abundant in the waters off Australia and New Zealand. In just a few years in the 1970s and 1980s, trawlers—some of which could land sixty metric tons of fish in as little as twenty minutes—depleted stocks by 80 percent. Because individuals in this species grow slowly and can be seventy to one hundred years old, the devastated stocks won't recover for decades, if ever. And the assault has affected much more than orange roughy: "In the sea mounts where the orange roughy is hunted, there were once sea fans, black corals, hydroids, invertebrates. Yet these centers of life have frequently been stripped down to the rock. . . . On land, if we thought we would destroy an entire forest just to catch a few deer, there'd be an outcry. Yet we are doing something like that in the deep sea." Claudia Dreifus, "A Biologist Decries the 'Strip Mining' of the Deep Sea," *New York Times*, March 5, 2002, national edition, D4. See also Jennifer Devine, Krista Baker, and Richard Haedrich, "Deep-Sea Fishes Qualify as Endangered," *Nature* 439, no. 7072 (January 5, 2006): 29.

55. Rosamond Naylor et al., "Effect of Aquaculture on World Fish Supplies," *Nature* 405, no. 6790 (June 29, 2000): 1017–24; and Kendall Powell, "Eat Your Veg," *Nature* 426, no. 6965 (November 27, 2003): 378–79.

56. "[Human] demand may well have exceeded the biosphere's regenerative capacity since the 1980s. According to this preliminary and exploratory assessment, humanity's load corresponded to 70 percent of the capacity of the global biosphere in 1961, and grew to 120 percent in 1999." Mathis Wackernagel et al., "Tracking the Ecological Overshoot of the Human Economy," *Proceedings of the National Academy of Sciences of the United States of America* 99, no. 14 (July 9, 2002): 9266–71.

57. Tim Wiener, "In Mexico, Greed Kills Fish by the Seafull," *New York Times*, April 10, 2002, national edition, A1.

58. Jessica Tuchman Mathews, "Redefining Security," *Foreign Affairs* 68, no. 2 (1989): 168.

59. Tim Wiener, "Life Is Hard and Short in Bleak Villages of Haiti," *New York Times*, March 14, 2004, national edition, 1.

60. Asian Development Bank, *Asian Environment Outlook 2001* (Manila: ADB, 2001), xiii.

61. For an analysis of the links between environmental stress and violent conflict, including details on many of the cases mentioned in this paragraph, see Thomas Homer-Dixon, *Environment, Scarcity, and Violence* (Princeton: Princeton University Press, 1999). See also Colin Kahl, *States, Scarcity, and Civil Strife in the Developing World* (Princeton: Princeton University Press, 2006); and Richard Cincotta, Robert Engelman, and Daniele Anastasion, *The Security Demographic:*

Population and Civil Conflict after the Cold War (Washington, DC: Population Action International, 2003).

62. UN Integrated Regional Information Networks (IRIN), November 20, 2001.

63. Tim Weiner, "87 Orphans Will Be Told of the Killers Next Door," *New York Times*, June 4, 2002, national edition, A4.

64. Howard French, "Riots in Shanghai Suburb as Pollution Protest Heats Up," *New York Times*, July 19, 2005, national edition, A5.

65. Philip Howard, *Environmental Scarcities and Conflict in Haiti: Ecology and Grievances in Haiti's Troubled Past and Uncertain Future* (Ottawa: Canadian International Development Agency, 1998); and Ginger Thompson, "A New Scourge Afflicts Haiti: Kidnappings," *New York Times*, June 6, 2005, national edition, A1.

66. Kahl, "Green Crisis, Red Rebels: Communist Insurgency in the Philippines," in *States, Scarcity, and Civil Strife*, 65–116; and Seth Mydans, "Communist Revolt Is Alive, and Active, in the Philippines," *New York Times*, March 26, 2003, national edition, A3.

67. Jean Bigagaza, Carolyne Abong, and Cecile Mukarubuga, "Land Scarcity, Distribution and Conflict in Rwanda," chapter 2 in Lind and Sturman, eds., *Scarcity and Surfeit: The Ecology of Africa's Conflict* (Pretoria: Institute of Security Studies, 2002), 51–84; and James K. Gasana, "Natural Resource Scarcity and Violence in Rwanda," in Richard Matthew, Mark Halle, and Jason Switzer, eds., *Conserving the Peace: Resources, Livelihoods and Scarcity* (Winnipeg: International Institute of Sustainable Development, 2002), 199–246.

68. Marc Lacey, "In Sudan, Militiamen on Horses Uproot a Million," *New York Times*, May 4, 2004, national edition, A1.

CHAPTER SEVEN

1. Lonnie Thompson et al., "Kilimanjaro Ice Core Records: Evidence of Holocene Climate Change in Tropical Africa," *Science* 298, no. 5593 (October 18, 2002): 589–93; and Andrew Revkin, "Climate Debate Gets Its Icon: Mt. Kilimanjaro," *New York Times*, March 23, 2004, national edition, D1.

2. Richard Bernstein, "Melting Mountain Majesties: Warming in Austrian Alps," *New York Times*, August 8, 2005, national edition, A4.

3. T. P. Barnett, J. C. Adam, and D. P. Lettenmaier, "Potential Impacts of a Warming Climate on Water Availability in Snow-Dominated Regions," *Nature* 438, no. 7066 (November 17, 2005): 303–9.

4. Juan Forero, "As Andean Glaciers Shrink, Water Worries Grow," *New York Times*, November 24, 2002, national edition, A3.

5. Howard French, "A Melting Glacier in Tibet Serves as an Example and a Warning," *New York Times*, November 9, 2004, national edition, D1.

6. Eugene Domack et al., "Stability of the Larsen B Ice Shelf on the Antarctic Peninsula during the Holocene Epoch," *Nature* 436, no. 7051 (August 4, 2005): 681–85; and Larry Rohter, "Antarctic, Warming, Looks More Vulnerable," *New York Times*, January 25, 2005, national edition, D1.

7. Matthew Sturm, Donald Perovich, and Mark Serreze, "Meltdown in the North," *Scientific American* 289, no. 4 (October 2003): 63; and John Whitfield, "Too Hot to Handle," *Nature* 425, no. 6956 (September 25, 2003): 338–39. A comprehensive assessment of the effects of climate change in the Arctic can be found in Arctic Climate Impact Assessment, *Impacts of a Warming Arctic* (Cambridge: Cambridge University Press, 2004), available at http://amap.no/acia/.

8. Timothy Egan, "Now, in Alaska, Even the Permafrost Is Melting," *New York Times,* June 16, 2002, national edition, 1.

9. Timothy Egan, "On Hot Trail of Tiny Killer In Alaska," *New York Times,* June 25, 2002, national edition, D1.

10. Eric Rignot and Pannir Kanagaratnam, "Changes in the Velocity Structure of the Greenland Ice Sheet," *Science* 311, no. 5763 (February 17, 2006): 986–90; Julian Dowdeswell, "The Greenland Ice Sheet and Global Sea-Level Rise," *Science* 311, no. 5763 (February 17, 2006): 963–64; and Michael Lemonick, "Has the Meltdown Begun," *Time,* February 27, 2006, Canadian edition, 39.

11. On surface ice temperatures in summer, see Josefino Comiso, "A Rapidly Declining Perennial Sea Ice Cover in the Arctic" *Geophysical Research Letters* 29 20 1956 (October 18, 2002): 3. On the twenty-year trend, see "Josefino Comiso, "Warming Trends in the Arctic from Clear Sky Satellite Observations," *Journal of Climate* 16, no. 21 (2003): 3498–3510.

12. National Snow and Ice Data Center, "Winter Sea Ice Fails to Recover, Down to Record Low," press release, April 5, 2006, available at http://www.nsidc.org/news/press/20060404_winterrecovery.html. The extent of end-of-summer sea ice—measured as the area of ocean covered by at least 15 percent ice on September 21—is now declining 8 percent a decade. Since 1978, the Arctic has lost an area of ice twice the size of Texas. See National Snow and Ice Data Center, "Sea Ice Decline Intensifies," press release, September 28, 2005, available at http://www.nsidc.org/news/press/20050928_trendscontinue.html. See also Comiso, "A Rapidly Declining Perennial Sea Ice Cover in the Arctic"; and "No Sno-Cones for Santa," *Environment* 45, no. 2 (March 2003): 7.

13. J. T. Overpeck et al., "Arctic System on Trajectory to New, Seasonally Ice-Free State," *Eos* 86, no. 34 (August 23, 2005): 309–16.

14. The 2005 scientific findings are summarized in Kelly Levin and Jonathan Pershing, "Climate Science 2005: Major New Discoveries," *WRI Issue Brief* (Washington, DC: World Resources Institute, March 2006), available at http://pdf.wri.org/climatescience_2005.pdf.

15. This difference is roughly equivalent to the energy released by a small Christmas tree lightbulb. So, in terms of Earth's energy balance, our perturbation of the atmosphere is having the same effect as a network of tiny lightbulbs, one per square meter, strung across Earth's lands and seas. James Hansen, "Diffusing the Global Warming Time Bomb," *Scientific American* 290, no. 3 (March 2004): 72.

16. "Except for a possible brief period following the next large volcanic eruption, the Earth's positive energy imbalance is now continuous, relentless, and still growing." James Hansen, "A Slippery Slope: How Much Global Warming Constitutes 'Dangerous Anthropogenic Interference'?" editorial essay, *Climate*

Change 68, no. 3 (February 2005): 269–79. See also James Hansen et al., "Earth's Energy Imbalance: Confirmation and Implications," *Science* 308, no. 5727 (June 3, 2005): 1431–35.

17. See, for example, Ronald Bailey, ed., *Global Warming and Other Eco Myths: How the Environmental Movement Uses False Science to Scare Us to Death* (Roseville, California: Prima Lifestyles, 2002); and "Global Warming," chapter 24 in Bjørn Lomborg, *The Skeptical Environmentalist: Measuring the Real State of the World* (Cambridge: Cambridge University Press, 2001), 258–324.

18. Quoted in BBC News, U.K. edition, "2003 Climate Havoc 'Cost $60 Billion,'" December 11, 2003, report on Milan conference on Kyoto Protocol, available at http://news.bbc.co.uk/1/hi/world/americas/3308959.stm.

19. The best accounts of this effort are Ross Gelbspan, *The Heat Is On: The High Stakes Battle over Earth's Threatened Climate* (Reading, MA: Addison-Wesley, 1997); and Gelbspan, *Boiling Point: How Politicians, Big Oil and Coal, Journalists, and Activists are Fueling the Climate Crisis—And What We Can Do to Avert Disaster* (New York: Basic, 2004).

20. Maxwell Boykoff and Jules Boykoff, "Balance as Bias: Global Warming and the U.S. Prestige Press," *Global Environmental Change* 14 (2004): 125–36.

21. Michael Crichton develops his argument in his novel *State of Fear* (New York: HarperCollins, 2004). For a critical review, see Michael Crowley, "Michael Crichton's Scariest Creation: Jurassic President," *The New Republic*, March 20, 2006.

22. On how corporations "manufacture uncertainty" to delay action on complex public policy issues involving scientific judgment, see David Michaels, "Doubt Is Their Product," *Scientific American* 262, no. 6 (June 2005): 96–101.

23. Urs Siegenthaler et al., "Stable Carbon Cycle-Climate Relationship during the Late Pleistocene," *Science* 310, no. 5752 (November 25, 2005): 1313–17. Scientists are able to determine the atmosphere's levels of carbon dioxide far into the past by extracting long cores from the ice sheets of Antarctica and Greenland. The sheets have been created by snow falling over the aeons that has been compressed into ice, and the air bubbles trapped in each layer of ice provide scientists with samples of Earth's atmosphere from roughly the time the snow fell.

24. Intergovernmental Panel on Climate Change, "Summary for Policymakers: A Report of Working Group 1 of the Intergovernmental Panel on Climate Change," *IPCC Third Assessment Report: Climate Change 2001* (Geneva: 2001), available at http://www.ipcc.ch/pub/spm22-01.pdf; and J. Hansen, "Diffusing the Global Warming Time Bomb," 75.

25. See the data on the carbon dioxide readings at the Mauna Loa observatory in Hawaii, available from the U.S. National Oceanic & Atmospheric Administration, at http://www.cmdl.noaa.gov/ccgg/trends/. Paul Brown, "Climate Fear as Carbon Levels Soar," *Guardian* (U.K.), October 11, 2004.

26. The original argument that satellite data show no warming of the lower atmosphere appeared in Roy W. Spencer and John R. Christy, "Precise Monitoring of Global Temperature Trends from Satellites," *Science* 247, no. 4950 (1990): 1558–62. For recent and decisive evidence to the contrary, see B. D. Santer et al., "Influence of Satellite Data Uncertainties on the Detection of Externally Forced Climate Change," *Science* 300, no. 5623 (May 23, 2003): 1280–84; and Carl Mears

and Frank Wentz, "The Effect of Diurnal Correction on Satellite-Derived Lower Tropospheric Temperature," *Science* 309, no. 5740 (September 2, 2005): 1548–51. On errors in interpreting weather balloon data, see Steven Sherwood, John Lazante, and Cathryn Meyer, "Radiosonde Daytime Biases and Late-20th Century Warming," *Science* 309, no. 5740 (September 2, 2005): 1556–59.

27. On the change of first flowering dates of plants, see Daniel Grossman, "Spring Forward," *Scientific American* 290, no. 1 (January 2004): 85–91. A particularly striking recent finding that points to global warming is the decline in the average density of the atmosphere from one hundred to six hundred kilometers above Earth's surface. Although climate theory says that greenhouse gases will warm the lower atmosphere, it also says that the upper atmosphere should simultaneously cool. And as it cools, it will shrink inward toward the planet, which means that at a given altitude, such as one hundred kilometers, the atmosphere should become less dense. In turn, objects like satellites and space debris in low Earth orbit should be slowed less by the atmosphere's friction. Sure enough, when scientists analyzed U.S. Air Force records of the orbits of twenty-seven objects in low Earth orbit since the 1960s, they concluded that the average density of the upper atmosphere had dropped 10 percent in the past thirty-six years. "This is pretty compelling evidence for the effects of carbon dioxide," says Dr. Gerald Keating of NASA's Langley Research Center. "The whole structure of the upper atmosphere will change as this effect becomes stronger and stronger." See J. Emmert, J. Picone, J. Lean, and S. Knowles, "Global Change in the Thermosphere: Compelling Evidence of a Secular Decrease in Density," *Journal of Geophysical Research* 109, no. A02301 (2004). Gerald Keating is quoted in Andrew Revkin, "Pollution Blamed for Thinner Air at Edge of the Atmosphere," *New York Times*, February 10, 2004, national edition, D2.

28. J. Hansen et al., "GISS Surface Temperature Analysis, Global Temperature Trends: 2005 Summation," available at http://data.giss.nasa.gov/gistemp/2005/; and Robert Henson, "The Heat Was On in 2005," *Nature* 438, no. 7071 (December 22–29, 2005): 1062.

29. There is a heated dispute about the use of "proxy" records like tree rings to determine the degree of warming over the past one to two millennia. The skeptical perspective is presented in W. Soon and S. Baliunas, "Proxy Climatic and Environmental Changes of the Past 1000 Years," *Climate Research* 23 (2003): 89–110. Michael Mann and his colleagues respond in "On Past Temperatures and Anomalous Late-20th Century Warmth," *Eos* 84, no. 27 (July 8, 2003): 256–58, and "Response," *Eos* 84, no. 44 (November 4, 2003): 473–74. See also Timothy Osborn and Keith Briffa, "The Spatial Extent of 20th-Century Warmth in the Context of the Past 1200 Years," *Science* 311, no. 5762 (February 10, 2006): 841–44.

30. J. Hansen, "Diffusing the Global Warming Time Bomb," 71.

31. World Meteorological Organization, *WMO Statement on the Status of the Global Climate in 2005* (Geneva: WMO, 2006), available at http://www.wmo.ch/web/wcp/wcdmp/statement/html/WMO998_E.pdf. On the Amazon drought, see Larry Rohter, "A Record Amazon Drought, and Fear of Wider Ills," *New York Times*, December 11, 2005, national edition, 1.

32. However, modeling shows that the droughts and heat waves experienced in the United States, southern Europe, and Southwest Asia from 1998 to 2002 are consis-

tent with greenhouse-gas-induced warming. See Martin Hoerling and Arun Kumar, "The Perfect Ocean for Drought," *Science* 299, no. 5607 (January 2003): 691–94.

33. Associated Press, "UN Says 2003 3rd Hottest Year on Record," *New York Times*, national edition, 17 December 2003.

34. Sallie Baliunas, "Warming up to the Truth: The Real Story about Climate Change," Heritage Lecture No. 758, August 22, 2002 (Washington, DC: The Heritage Foundation), available at http://www.heritage.org/Research/EnergyandEnvironment/HL758.cfm.

35. Intergovernmental Panel on Climate Change, "Summary for Policymakers: A Report of Working Group I of the Intergovernmental Panel on Climate Change," *IPCC Third Assessment Report: Climate Change 2001* (Geneva: IPCC, 2001), 10.

36. BBC News, U.K. edition, "U.S. Science Body Warns on Climate," December 16, 2003, available at http://news.bbc.co.uk/1/hi/sci/tech/3325341.stm.

37. For an assessment of the effects of climate change on global hydrologic cycles, see P. Milly, K. Dunne, and A. Vecchia, "Global Pattern of Trends in Streamflow and Water Availability in a Changing Climate," *Nature* 438, no. 7066 (November 17, 2005): 347–50.

38. This impact is projected to be particularly serious in the western United States (especially Washington, Oregon, and California), the Canadian prairies, and the Rhine in Europe. See Barnett, Adam, and Lettenmaier, "Potential Impacts of a Warming Climate on Water Availability."

39. One of the best assessments of the implications of climate change for large-scale floods is the study by the U.K. Office of Science and Technology of the risks for Britain. See *Foresight: Future Flooding, Executive Summary* (London: Foresight Directorate, Office of Science and Technology, 2004). Global warming can cause sea level to rise by increasing the seawater's volume (the water expands as it's heated) and by increasing its mass (mainly by melting continental ice). Recent research suggests that the observed increase in sea levels over the past century of 1.5 to 2.0 mm a year is principally due to a mass increase. See Laury Miller and Bruce Douglas, "Mass and Volume Contributions to Twentieth-Century Global Sea Level Rise," *Nature* 428, no. 6981 (March 25, 2004): 406–9. Given current rates of warming, very large future increases in sea level (on the order of four to seven meters over several centuries) could occur as a result of melting of the Greenland ice sheet and portions of the Antarctic ice sheet. See Jonathan Overpeck et al., "Paleoclimatic Evidence for Future Ice-Sheet Instability and Rapid Sea-Level Rise," *Science* 311, no. 5768 (March 24, 2006): 1747–50.

40. On trends in hurricane destructiveness, see Kerry Emanuel, "Increasing Destructiveness of Tropical Cyclones over the Past 30 Years," *Nature* 436, no. 7051 (August 4, 2005): 686–88; and P. J. Webster et al., "Changes in Tropical Cyclone Number, Duration, and Intensity in a Warming Environment," *Science* 309, no. 5742 (September 16, 2005): 1844–46.

41. The impacts of climate change on agriculture in the United States are discussed in Joel Smith, *A Synthesis of Potential Climate Change Impacts on the U.S.* (Arlington, VA: Pew Center on Global Climate Change, 2004), 10–11, available at http://www.pewclimate.org/docUploads/Pew%2DSynthesis%2Epdf.

42. On climate change and the decline of coral reefs, see Robert Buddemeier, Joan

Kleypas, and Richard Aronson, *Coral Reefs & Global Climate Change: Potential Contributions of Climate Change to Stresses on Coral Reef Ecosystems* (Arlington, VA: Pew Center on Global Climate Change, 2004), available at http://www.pewclimate.org/docUploads/Coral%5FReefs%2Epdf. See also D. Bellwood, T. Hughes, C. Folke, and M. Nyström, "Confronting the Coral Reef Crisis," *Nature* 429, no. 6994 (June 24, 2004): 827–33.

43. On the health effects of global warming in general, see Jonathan Patz et al., "Impact of Regional Climate Change on Human Health," *Nature* 438, no. 7066 (November 17, 2005): 310–17. The record 2003 heat wave in Europe, which claimed more than twenty thousand lives, was likely caused in part by global warming. See Peter Stott, D. Stone, and M. Allen, "Human Contribution to the European Heatwave of 2003," *Nature* 432, no. 7017 (December 2, 2004): 610–14.

44. The U.S. Central Intelligence Agency estimates that world GDP (purchasing power parity) in 2005 was $60 trillion. See http://www.cia.gov/cia/publications/factbook/rankorder/2001rank.html.

45. John Holdren, "Environmental Change and the Human Condition," *Bulletin of the American Academy of Arts and Sciences* 57, no. 1 (Fall 2003): 27. See also John Holdren, "Memo to the New President: The Energy-Climate Challenge," in Donald Kennedy and John Riggs, eds., *U.S. Policy and the Global Environment: Memos to the President*, A Report of the Environment Policy Forum, July 8–11, 2000, Aspen, Colorado (Washington, DC: Aspen Institute, 2000), 25.

46. The statistics and information in this and the previous paragraph were drawn from Keith Bradsher, "China's Boom Adds to Global Warming Problem," *New York Times*, October 22, 2003, national edition, A1; Jim Yardley, "China's Economic Engine Needs Power (Lots of It)," *New York Times*, Week in Review, March 14, 2004, national edition, 3; Keith Bradsher and David Barboza, "Pollution from Chinese Coal Casts Shadow around Globe," *New York Times*, June 11, 2006, national edition, 1; Jia Hepeng, "High Demand Puts Pressure on Coal Industry," *China Business Weekly*, December 23, 2003, available at http://www.chinadaily.com.cn/en/doc/2003–12/23/content_293519.htm; and Fu Jing, "Coal Output Set to Reach Record High of 2.5 Billion Tons," *China Daily*, March 18, 2006, available at http://www.chinadaily.com.cn/english/doc/2006–03/18/content_544126.htm.

47. Thomas Knutson and his group of researchers at the Geophysical Fluid Dynamics Laboratory at Princeton University have modeled the impact of a quadrupling of atmospheric carbon dioxide. Their report, titled "Climate Impact of Quadrupling Atmospheric CO_2," is available at http://www.gfdl.noaa.gov/~tk/climate_dynamics/climate_impact_webpage.html#title. See also T. L. Delworth et al., "Review of Simulations of Climate Variability with the GFDL R30 Coupled Climate Model," *Climate Dynamics* 19 (2002): 555–74. A similar result is described in G. Bala et al., "Multicentury Changes to the Global Climate and Carbon Cycle: Results for a Coupled Climate and Carbon Cycle Model," *Journal of Climate* 18, no. 21 (2005): 4531–44; the authors of this paper assume that all the carbon dioxide in the planet's fossil fuel resources is released into the atmosphere, boosting atmospheric concentrations to 1,423 ppm.

48. John Holdren, "Environmental Change and the Human Condition," 29.

49. The actual figure depends on the freshness of the snow or other cover on the ice and the state of the ocean surface, all factors that affect surface reflectivity or what scientists call "albedo." Sturm, Perovich, and Serreze, "Meltdown in the North," 66.

50. Ibid., 65.

51. Fred Pearce, "Climate Warms as Siberia Melts," *New Scientist*, August 11, 2005: 12.

52. Overpeck et al., "Arctic System on Trajectory to New, Seasonally Ice-Free State."

53. See the interview with renowned geochemist Wally Broecker, of Columbia University, in Thomas Homer-Dixon, *The Ingenuity Gap: Facing the Economic, Environmental, and Other Challenges of an Increasingly Complex and Unpredictable Future* (New York: Vintage, 2002), 142–46.

54. This kind of "behavioral" equilibrium is distinct from the thermodynamic equilibrium discussed in chapter 2. Although complex systems are never in thermodynamic equilibrium, they can still exhibit stable patterns of behavior (which we call equilibrium) for extended periods.

55. "Large-scale marine ecosystems are dynamically nonlinear, and as such have the capacity for dramatic changes" in response to external influences. See Chih-hao Hsieh et al., "Distinguishing Random Environmental Fluctuations from Ecological Catastrophes for the North Pacific Ocean," *Nature* 435, no. 7040 (May 19, 2005): 336–39. See also Carl Folke et al., "Regime Shifts, Resilience, and Biodiversity in Ecosystem Management," *Annual Review of Ecology, Evolution, and Systematics* 35 (2004): 557–81; and Marten Scheffer et al., "Catastrophic Shifts in Ecosystems," *Nature* 413, no. 6856 (October 11, 2001): 591–96.

56. For a survey of the literature, see J. G. Lochwood, "Abrupt and Sudden Climatic Transitions and Fluctuations: A Review," *International Journal of Climatology* 21 (2001): 1153–79.

57. Richard Alley, Committee on Abrupt Climate Change, National Research Council, *Abrupt Climate Change: Inevitable Surprises* (Washington, DC: National Academy Press, 2002), 1, 25–27, 36. An updated synopsis of the book was republished as "Abrupt Climate Change: Inevitable Surprises," *Population and Development Review* 30, no. 4 (September 2004): 563–68.

58. On the relationship between freshwater discharge in the North Atlantic, disruption of the thermohaline circulation, and changes in oceanic heat transport in the Southern Atlantic Ocean, see R. Knutti et al., "Strong Hemispheric Coupling of Glacial Climate through Freshwater Discharge and Ocean Circulation," *Nature* 430, no. 7002 (August 19, 2004): 851–56.

59. R. Curry and C. Mauritzen, "Dilution of the Northern North Atlantic Ocean in Recent Decades," *Science* 308, no. 5729 (June 2005): 1772–74; R. Curry, R. Dickson, and I. Yashayaev, "A Change in the Freshwater Balance of the Atlantic Ocean over the Past Four Decades," *Nature* 426, no. 6968 (December 18–25, 2003): 826–29; B. Dickson, I. Yashayaev, J. Meincke, B. Turrell, S. Dye, and J. Hoffort, "Rapid Freshening of the Deep North Atlantic Ocean Over the Past Four Decades," *Nature* 416, no. 6883 (April 25, 2002): 832–37; and Robert Gagosian, "Abrupt Climate Change: Should We Be Worried?" (Woods Hole Oceanographic Institution, January 27, 2003), 8.

60. B. Hansen, W. Turrell, and S. Østerhus, "Decreasing Overflow from the Nordic Seas into the Atlantic Ocean Through the Faroe Bank Channel Since 1950," *Nature* 411, no. 6840 (June 21, 2001): 927–30.

61. Harry Bryden, Hannah Longworth, and Stuart Cunningham, "Slowing of the Atlantic Meridional Overturning Circulation at 25° N," *Nature* 438, no. 7068 (December 1, 2005): 655–57. The findings reported in this paper have caused a great deal of debate in the climate-change research community. For a summary, see Quirin Schiermeier, "Climate Change: A Sea Change," *Nature* 439, no. 7074 (January 19, 2006): 256–60.

62. The seminal scientific paper presenting a model of nonlinear transitions of the North Atlantic thermohaline circulation between different equilibriums is Stephan Rahmstorf, "Bifurcations of the Atlantic Thermohaline Circulation in Response to Changes in the Hydrological Cycle," *Nature* 378, no. 6553 (November 9, 1995): 145–49.

63. Robert Gagosian, "A Perspective on Potential Climate Changes" (Woods Hole, MA, 2002).

64. Robert Gagosian, "Abrupt Climate Change: Should We Be Worried?" Address prepared for a panel on abrupt climate change at the World Economic Forum, Davos, Switzerland, January 27, 2003, available at http://www.whoi.edu/administration/president/news_030127.htm.

65. Gagosian, "A Perspective on Potential Climate Changes."

66. Quoted in ibid.

67. Recent research on positive species interactions, or "facilitation," and their implication for ecological theory are discussed in John Bruno, John Stachowicz, and Mark Bertness, "Inclusion of Facilitation into Ecological Theory," *Trends in Ecology and Evolution* 18, no. 3 (March 2003): 119–25.

68. Edward O. Wilson, Foreword, in Stephen Buchmann and Gary Nabhan, *The Forgotten Pollinators* (Washington, DC: Island Press, 1996), xiii–xiv.

69. The high connectivity of mature ecosystems means that damage in different parts of their networks can combine in unexpected ways. In chapter 9 of this book, I discuss the relationship between ecosystem connectivity and resilience in the context of Buzz Holling's panarchy theory.

70. Ecologists have long debated the relationship between ecosystem complexity (including both species diversity and ecosystem connectivity) and ecosystem stability and resilience, and the matter hasn't been entirely settled. See Stuart Pimm, "The Complexity and Stability of Ecosystems," *Nature* 307, no. 5949 (January 16, 1984): 321–26. Some of the most recent and persuasive research supporting a link between diversity and stability is presented by Wolfgang Kiessling in "Long-term Relationships between Ecological Stability and Biodiversity in Phanerozoic Reefs," *Nature* 433, no. 7024 (January 27, 2005): 410–13. See also Thomas Elmqvist et al., "Response Diversity, Ecosystem Change, and Resilience," *Frontiers in Ecology and the Environment* 1, no. 9 (November 2003): 488–94.

71. The new theoretical understanding of scale-free networks has helped ecologists understand how the structure of food webs, especially the number and location of highly connected nodes, influences resilience. See Ricard Solé and José Montoya, "Complexity and Fragility in Ecological Networks," *Proceedings of the Royal Society*

of London 268 (2001): 2039–45; Folke et al, "Regime Shifts"; and Scheffer et al., "Catastrophic Shifts in Ecosystems."

72. Wilson, Foreword.

73. Jay Malcolm et al., *Habitats at Risk: Global Warming and Species Loss in Globally Significant Terrestrial Ecosystems* (Gland, Switzerland: WWF-World Wide Fund for Nature, 2002).

74. Chris Thomas et al., "Extinction Risk from Climate Change," *Nature* 427, no. 6970 (January 8, 2004): 145–48; and Alan Pounds and Robert Puschendorf, "Clouted Futures," *Nature* 427, no. 6970 (January 8, 2004): 107–9.

75. As quoted in "Warming to Extinction," *Environment* 46, no. 2 (March 2004): 8.

76. Warmer surface water in the ocean reduces the amount of mixing of surface water with deeper and cooler water. Because the cooler water is rich in nutrients, less mixing reduces phytoplankton growth in the ocean's upper layers. See Watson W. Gregg et al., "Ocean primary production and climate: Global Decadal Changes," *Geophysical Research Letters* 30, no. 15 (2003): OCE 3–1 to 3–4. Some forms of plankton may also be harmed as higher levels of atmospheric carbon dioxide increase concentrations of carbonic acid in oceans, significantly lowering the oceans' pH. See Richard Feely et al., "Impact of Anthropogenic CO_2 on the $CaCO_3$ System in the Oceans," *Science* 305, no. 5682 (July 16, 2004): 362–66; and Andrew Revkin, "Carbon Dioxide Extends Its Harmful Reach to Oceans," *New York Times*, July 20, 2004, national edition, D3.

77. One of the best treatments of the risks of synergy between climate change and other environmental stresses is the U.S. National Research Council's report on prospects for sustainable development: Board on Sustainable Development, Policy Division, National Research Council, *Our Common Journey: A Transition toward Sustainability* (Washington, DC: National Academy Press, 2001), especially 208–24.

78. James McKinley Jr., "Floodwaters Recede from Haitian City, but Hunger Does Not," *New York Times*, September 25, 2004, national edition, A8.

79. Deborah Sonntag and Lydia Polgreen, "Storm-Battered Haiti's Endless Crises Deepen," *New York Times*, October 16, 2004, national edition, A1.

80. James McKinley Jr., "Weary, Angry Haitians Dig out of Storm," *New York Times*, September 24, 2004, national edition, A3.

81. Emanuel, "Increasing Destructiveness of Tropical Cyclones"; and Webster et al., "Changes in Tropical Cyclone Number, Duration, and Intensity."

CHAPTER EIGHT

1. George Soros, "The Capitalist Threat," *The Atlantic Monthly* 279, no. 2 (February 1997): 45–58.

2. The excerpts of this debate have been taken from Council on Foreign Relations and HBO, "Implications of a Global Economy: A Conversation between George Soros and Paul Krugman," May 1, 1997.

3. For a detailed account of the Long-Term Management Fund crisis and its causes, see Diana Henriques and Joseph Kahn, "Lessons of a Long, Hot Summer," *New York Times*, Sunday Business, December 6, 1998, national edition, 1.

4. Robert Samuelson, "What the Boom Forgot," *The New Republic*, May 3, 2004: 31.

5. George Soros's main books during this period were *The Crisis of Global Capitalism: Open Society Endangered* (New York: PublicAffairs, 1998) and *Open Society: Reforming Global Capitalism* (New York: PublicAffairs, 2000). Soros's superficial understanding of economic theory and his involvement in politics from Eastern Europe to the United States undermined his credibility among economists and mainstream opinion leaders. Especially in the United States, his ideas were marginalized. Robert Solow offered a particularly scathing critique in "The Amateur," *The New Republic*, February 8, 1999: 28–31.

6. An early discussion of sources of disequilibrium is Joseph Schumpeter, "The Instability of Capitalism," *The Economic Journal* 38, no. 151 (September 1928): 361–86.

7. For an excellent summary of the reasons for nonlinear behavior of international capital and exchange rate regimes, see Dan Ciuriak, "Trade and Exchange Rate Regime Coherence: Implications for Integration in the Americas," *The Estey Centre Journal of International Law and Trade Policy* 3, no. 2 (2002): 258–75. A recent analysis of non-equilibrium dynamics in the global economy is George-Marios Angeletos, Christian Hellwig, and Alessandro Pavan, "Information Dynamics and Equilibrium Multiplicity in Global Games of Regime Change," *NBER Working Paper 11017*, National Bureau of Economic Research (December 2004), abstract available at http://www.nber.org/papers/w11017.

8. Michael Bordo, Barry Eichengreen, Daniela Klingebiel, and Maria Martinez-Peria, "Financial Crises: Lessons from the Last 120 Years," *Economic Policy*, April 2001: 53–82.

9. World Bank, "Preventing and Minimizing Crises," chapter 2 in *Finance for Growth: Policy Choices in a Volatile World* (Washington, DC: World Bank, 2001), 75. The entire document is available at http://econ.worldbank.org/prr/FFG/text-24976/. Although there is wide agreement that financial crises have been more frequent in recent decades, experts dispute whether they have become more severe. For instance, Bordo and his colleagues argue that they haven't been more severe. See Bordo et al., "Financial Crises," 72.

10. Some economists blame pegged exchange rates for financial instability; others blame weak banking, accounting, and legal systems or poor fiscal policies; and still others point to the general lack of rules and institutions to help the international financial system cope when a major country defaults on its debts. World Bank, "Preventing and Minimizing Crises," discusses several of these factors. A more accessible treatment of instability's causes is Bill Emmott, "Unstable," chapter 8 in *20:21 Vision: Twentieth-Century Lessons for the Twenty-First Century* (New York: Farrar, Straus, and Giroux, 2003), 208–35. The statistical analysis in Bordo et al., "Financial Crises," highlights the importance of pegged exchange rates. On the risks associated with sovereign default, see Stanley Fischer, "Financial Crises and Reform of the International Financial System," *NBER Working Paper 9297*, National Bureau of Economic Research (October 2002), abstract available at http://www.nber.org/papers/w9297. In his analysis of the precursors to the Great Depression, Harold James provides a historical perspective that emphasizes long-term changes in institutions and popular support for liberalized markets; see *The*

End of Globalization: Lessons from the Great Depression (Cambridge, MA: Harvard University Press, 2001).

11. World Bank, "Preventing and Minimizing Crises," discusses the vulnerabilities of banking systems in emerging markets. Also see Martin Wolf, "Fearful of Finance," chapter 13 in *Why Globalization Works* (New Haven: Yale University Press, 2004), 295–303.

12. The first real examples of international financial panic, according to most economic historians, were the South Sea and Mississippi bubbles of the early eighteenth century. See Charles Kindleberger, *Manias, Panics, and Crashes: A History of Financial Crises* (New York: John Wiley, 1996 [3rd ed.]), 111–12.

13. Jospeh Stiglitz, *Globalization and Its Discontents* (New York: Norton, 2002), 97.

14. Lee Clarke, *Worst Cases: Terror and Catastrophe in the Popular Imagination* (Chicago: University of Chicago Press, 2006), 109.

15. In the technical terms used by social scientists, this situation is "a collective action problem" or "a social dilemma" similar to a prisoner's dilemma. A useful introduction to these issues is Michael Taylor, *The Possibility of Cooperation* (Cambridge: Cambridge University Press, 1995).

16. On herd behavior in exchange-rate regimes, see John Williamson, "Crawling Bands or Monitoring Bands: How to Manage Exchange Rates in a World of Capital Mobility," *International Finance* 1, no. 1 (1998): 59–79.

17. On self-fulfilling pessimistic expectations, see Philippe Martin and Hélène Rey, "Globalization and Emerging Markets: With or without Cash?" *NBER Working Paper 11550*, National Bureau of Economic Research (August 2005); Ciuriak, "Trade and Exchange Rate Regime Coherence"; and Paul Krugman, *The Return of Depression Economics* (New York: Norton, 2000), 109–11. Martin and Rey, along with many other analysts, suggest that while trade liberalization generally has a beneficial effect on emerging market countries, premature financial liberalization can make such countries vulnerable to crashes.

18. Because the country's economic managers were buffered from outside pressure, they were able to keep China's economy healthy by expanding the money supply and spending massively on infrastructure projects like highways and dams. Stiglitz, *Globalization*, 125–26.

19. Krugman, *The Return of Depression Economics*, 145.

20. Stiglitz, *Globalization and Its Discontents*, 67.

21. Ibid., 121.

22. Paul Blustein, *The Chastening: Inside the Crisis that Rocked the Global Financial System and Humbled the IMF* (New York: Public Affairs, 2003), 100–15, 207–21. This book is an excellent comprehensive account of the crisis with a specific focus on the role of the IMF. For contrasting assessments of IMF policies during the East Asia crisis, see Stiglitz, *Globalization*, 77, 104–31, and Wolf, *Why Globalization Works*, 288–95. Kenneth Rogoff defends the IMF in "The IMF Strikes Back," *Foreign Policy* 134 (January–February 2003): 39–46.

23. UNICEF, *The State of the World's Children 2000* (New York: UNICEF, 2000), 22.

24. In 2005, foreign investment (other than in the petroleum and finance sectors) remained at about one-quarter the 1995 level. Asian Development Bank,

"Indonesia," *Asian Development Outlook 2006* (Manila: ADB, 2006), available at http://www.adb.org/Documents/Books/ADO/2006/ino.asp.

25. United Press International, "Unemployment Poses Threat in Indonesia," January 15, 2004.

26. An analysis in 2004 by an independent government agency put underemployment at a staggering 50 percent of the workforce, including many high school and university graduates. See Ridwan Max Sijabat, "Unemployment Splits Top Officials," *Jakarta Post*, April 10, 2004, available at Asian Labour News, http://www.asianlabour.org/archives/001391.php. See United Press International, "Unemployment Poses Threat in Indonesia," for the World Bank official's warning.

27. Stiglitz, *Globalization and Its Discontents*, 122.

28. Emmott, *20:21 Vision*, 23, 223, 228.

29. The World Bank economist Charles Kenny argues that the widening income gap between rich and poor countries isn't as important as generally thought because the broader quality of life in poor countries—as measured by levels of health and education, for example—is rapidly approaching that of rich countries. I take the position here, however, that the income gap does matter because people naturally measure their success or failure by comparing themselves with others using relatively simple metrics such as income or material wealth. Chronic and widening income inequality can therefore engender pervasive grievances that undermine institutional stability, governmental authority, and civil stability. See Kenny, "Why Are We Worried about Income? Nearly Everything That Matters Is Converging," *World Development* 33, no. 1 (January 2005): 1–19.

30. Emmott, *20:21 Vision*, 258.

31. World Bank, *Global Economic Prospects 2005: Trade, Regionalism, and Development* (Washington, DC: World Bank, 2005), 21–22. These World Bank income figures are based on the 1990 value of a U.S. dollar and are adjusted for variations in purchasing power across countries. They include consumption of one's own production as well as income in kind. For a discussion of the methodological, ethical, and philosophical disputes surrounding this kind of poverty measure, see Daniel Altman, "Does a Dollar a Day Keep Poverty Away?" *New York Times*, April 26, 2003, national edition, A19. The economist Xavier Sala-i-Martin, has suggested that the Bank estimates of the number of people living on $1 and $2 a day are far too high, but the World Bank economist Branko Milanovic has in turn shown that Sala-i-Martin's methodology is deeply flawed. See Sala-i-Martin, "The Disturbing 'Rise' of Global Income Inequality," *NBER Working Paper 8904*, National Bureau of Economic Research (May 2002); and Milanovic, *Worlds Apart: Measuring International and Global Inequality* (Princeton: Princeton University Press, 2005), 102, 121.

32. Food and Agriculture Organization, "Table 1. Prevalence of Undernourishment in Developing Countries and Countries in Transition," *The State of Food Insecurity in the World 2004* (Rome: FAO, 2004), 34, available at ftp://ftp.fao.org/docrep/fao/007/y5650e/y5650e00.pdf.

33. UNICEF, *The State of the World's Children 2005: Childhood under Threat* (New York: UNICEF, 2004), 19–22, available at http://www.unicef.org/sowc05/english/sowc05_chapters.pdf.

34. The calculation that produced the statistic in this sentence is derived, in part, from the estimate of the International Labour Organization (ILO) that the world's workforce consists of 2.8 billion people, half of whom are "unable to lift themselves and their families above US$2 a day poverty line." Of these 1.4 billon poorest workers, the ILO continues, "550 million cannot even lift themselves and their families above the extreme US$1 a day poverty threshold." The calculation also incorporates the following reasonable assumptions: each of the world's 1.4 billion poorest workers supports himself/herself plus an additional dependent; the 550 million workers whose families live on $1 a day or less per person earn an average of $1.50 a day (or $0.75 a day for both the worker and his/her dependent) for 365 days a year; and the remaining 850 million workers (1.4 billion minus 550 million), whose families live on income of between $1 and $2 a day per person, earn an average of $3.50 a day (or $1.75 a day per worker and dependent). Using these assumptions, the labor of the poorest 550 million workers would cost $300 billon a year, and the labor of the next poorest 850 million workers would cost about $1.1 trillion a year. The cost of this entire labor force would therefore be $1.4 trillion a year, or $3.8 billion a day; and the billionaires' $2.6 trillion wealth could purchase twenty-three months (or 684 days) of this labor. The figures for the number of billionaires in the world and their total wealth were obtained from Luisa Kroll and Alison Fass, "The World's Billionaires," *Forbes*, March 9, 2006, available at http://www.forbes.com/worldsrichest/. For background statistics on the world's workforce, see International Labour Organization, *World Employment Report 2004–05:Employment, Productivity and Poverty Reduction* (Geneva: International Labour Organization, 2004), 23–25, available at http://www.ilo.org/public/english/employment/strat/wer2004.htm.

35. Stanley Fischer, "Globalization and Its Challenges," paper presented as the Ely Lecture at the American Economic Association meetings in Washington, DC, January 3, 2003, 12.

36. Lant Pritchett, "Divergence, Big Time," *Journal of Economic Perspectives* 11, no. 3 (Summer 1997): 3, 11, 12. According to Branko Milanovic, inequality between average incomes in rich and poor countries has more than doubled since 1820, with only a brief pause in the trend between World War I and World War II. See Milanovic, *Worlds Apart*, 139–40.

37. Danny Quah, "Twin Peaks: Growth and Convergence in Models of Distribution Dynamics," *The Economic Journal* 106, no. 437 (July 1996): 1045–55.

38. Surjit Bhalla, *Imagine There's No Country: Poverty, Inequality, and Growth in the Era of Globalization* (Washington DC: Institute for International Economics, 2002).

39. Glenn Firebaugh, *The New Geography of Global Income Inequality* (Cambridge, MA: Harvard University Press, 2003), 22–30. See also Xavier Sala-i-Martin, "The World Distribution of Income (Estimated from Individual Country Distributions)," *NBER Working Paper 8933*, National Bureau of Economic Research (May 2002). Branko Milanovic critiques Sala-i-Martin's statistical methodology in "The Ricardian Vice: Why Sala-i-Martin's Calculations of World Income Inequality Are Wrong," available at http://papers.ssrn.com/so13/papers.cfm?abstract_id=403020.

40. World Bank, *Global Economic Prospects 2005*, 21–22.

41. On the proportion of the world's population with an income below $1 a day in the early nineteenth century, see Firebaugh, *New Geography*, 13.

42. The trends are not nearly so positive in other parts of the world. In Latin America, between 1990 and 2001, the number of people below the $1-a-day threshold stayed constant at 50 million (although the percentage dropped as the region's population grew), while in sub-Saharan Africa the number expanded by almost 90 million to 313 million, and the percentage increased from 44.6 to 46.4 of the region's population. In the 1990s, poor countries with a total of 12 percent of the world's population had stagnant or falling per capita income. Countries with falling average incomes included six in Latin America and the Caribbean, five in the Arab world, six in East Asia and the Pacific, and twenty in sub-Saharan Africa. Hunger increased in twenty-one countries and under-five mortality increased in fourteen, while life expectancy fell in thirty-four. The United Nations Development Programme notes that such reversals of fortune were rare prior to the 1990s and that these trends point to a "development crisis." See World Bank, *Global Economic Prospects 2005*, 21–22; Table 1, World Commission on the Social Dimension of Globalization, *A Fair Globalization: Creating Opportunities for All* (Geneva: International Labour Office, 2004), 36, available at http://www.ilo.org/public/english/fairglobalization/report/index.htm; and United Nations Development Programme, *Human Development Reported 2003. Millennium Development Goals: A Compact among Nations to End Human Poverty* (New York: Oxford University Press, 2003), especially "Human Development Indicators: Table 12 Economic Performance, 2–3, 278–81, available at http://hdr.undp.org/reports/global/2003/.

43. T. P. Schultz, "Inequality in the Distribution of Personal Income in the World: How It Is Changing and Why." *Journal of Population Economics* 11, no. 3 (1998): 307–44; and Firebaugh, *New Geography*. See also the review of the latter book by John Isbister in *Population and Development Review* 29, no. 4 (December 2003): 731–33.

44. Between 1988 and 1993, inequality increased mainly because average incomes in rural Asia—especially in the still-immense rural populations of China, India, and Bangladesh—grew much more slowly than incomes in rich countries. Inequality increased, too, because of a rapidly widening gap between rural and urban incomes in China. (This bears on China's future political stability, as we'll see in chapter 11.) Between 1993 and 1998, the overall inequality trend reversed itself slightly: while the gap between rural and urban incomes in China continued to grow, the gap between rural incomes in India and China and incomes in rich countries shrank slightly. Milanovic, *Worlds Apart*, 106–16. For an accessible assessment of the importance of Milanovic' research, see Robert Wade, "Winners and Losers," *The Economist*, April 28, 2001, 72–74.

45. Milanovic, *Worlds Apart*, 139–48.

46. One might think that income inequality is the same thing as an income gap and that therefore lower inequality between average incomes in rich and poor countries should mean a smaller gap between these incomes. But the two things are actually not the same, and depending on how we measure inequality, it can fall while the income gap widens. For example, we can use a ratio to measure global economic inequality by dividing the average income of people in rich countries by the average

income of people in poor countries. Inequality between rich and poor then goes down when the number produced by this calculation goes down. Yet under certain conditions, this kind of inequality can fall while the gap between average incomes in rich and poor countries still widens. This will happen, for instance, if the average growth rate of incomes in poor countries is somewhat faster than that in rich countries, but the rich countries have a much higher average income. As I show in the following paragraphs, this is exactly today's situation. For further discussion, Firebaugh, *The New Geography of Global Income Inequality*, 120–21.

In general, people care more about income gaps than income ratios, and they often get angry when they're on the losing side of a widening gap. So if we're interested in things that might erode the world's social and political stability, we should pay attention to the world's income gaps. Unfortunately, one of the most common measures of inequality, the Gini coefficient, is essentially a ratio measure. So it can mislead us into believing that things are getting better, when from the point of view of potential social upheaval, they're actually getting worse. On the derivation of the Gini coefficient, see Milanovic, *Worlds Apart*, 196.

47. Table 1.3, "Long-Term Prospects: Forecast Growth of World GDP Per Capita," World Bank, *Global Economic Prospects 2005*, 17. The figures are as follows: in the 1980s, rich countries experienced 2.5 percent per capita GDP growth, whereas the figure was 0.6 percent in poor countries; in the 1990s, the figures were 1.8 percent and 1.5 percent respectively; and from 2000 to 2006, the Bank estimates per capita income growth of 1.7 percent in rich countries and 3.4 percent in poor countries.

48. After all, the Bank projects such growth only to 2015, and this rate is anomalously high compared with the historical experience: between 1870 and 1990, per capita incomes in rich countries grew on average only about 1.5 percent. See Pritchett, "Divergence, Big Time," 5.

49. There are two reasons why this "optimistic" scenario is highly implausible. First, rich countries will likely have an income growth rate somewhat higher than 1 percent, because policy makers will consider anything lower to be a dangerous political and economic failure (a subject I'll return to later in this chapter). Second, poor countries could easily have an income growth rate below the Bank's 2005 estimate of 3.5 percent. Historical data show that the fastest per capita income growth that a country can sustain for long periods is around 4 percent, which means that the Bank's 3.5 percent prediction is near the upper limit of the plausible range. It seems likely that poor countries won't be able to sustain such a high rate, especially because they're so susceptible to economic instability and often have such difficulty catching up after an economic crisis. On historical growth rates, see ibid., 13.

50. For a particularly influential paper linking globalization and growth, see David Dollar and Aart Kraay, "Trade, Growth, and Poverty," Policy Research Working Paper no. 2199 (Washington, DC: World Bank, 2001). Examples of more general arguments along these lines are Wolf, *Why Globalization Works*; World Bank, *Globalization, Growth, and Poverty: Building an Inclusive World Economy* (Washington, DC, and New York: World Bank and Oxford University Press, 2002), 3–8; and John Micklethwait and Adrian Wooldridge, *A Future Perfect: The Challenge and Hidden Promise of Globalization* (New York: Crown Business, 2000).

51. On the declining trend in the growth of average income, see table 10 of World Commission on the Social Dimension of Globalization, *A Fair Globalization*, 36.

52. Geoffrey Garrett, "Globalization's Missing Middle," *Foreign Affairs* 83, no. 6 (November–December 2004): 84–97.

53. The evidence does not support the assertion that countries with greater openness to trade grow faster. See Francisco Rodriguez and Dani Rodrik, "Trade Policy and Economic Growth: A Skeptic's Guide to the Cross-National Evidence," May 2000, paper available at http://ksghome.harvard.edu/~drodrik/skepti1299.pdf.

54. Richard Easterlin, "Does Economic Growth Improve the Human Lot? Some Empirical Evidence," in Paul David and Melvin Reder, eds., *Nations and Households in Economic Growth* (New York: Academic Press, 1974), 89–125; Richard Easterlin, "Will Raising the Incomes of All Increase the Happiness of All?" *Journal of Economic Behavior and Organization* 27 (1995): 35–47; and Ronald Inglehart and Hans-Dieter Klingemann, "Genes, Culture, Democracy, and Happiness," chapter 7 in Ed Diener and Eunkook Suh, eds., *Culture and Subjective Well-being* (Cambridge, MA: MIT Press, 2000), 165–83. A standard data source is the multi-decade World Values Survey. For the Survey's latest results, see Ronald Inglehart et al., eds., *Human Beliefs and Values: A Cross-Cultural Sourcebook Based on the 1999–2002 Values Surveys* (Mexico: Sigio Veintiuno Editores, 2004). Methodological issues associated with measuring happiness (or what specialists call "subjective well-being") across cultures are discussed in Inglehart and Klingeman, "Genes, Culture, Democracy, and Happiness"; in Ed Diener and Shigehiro Oishi, "Money and Happiness: Income and Subjective Well-being across Nations," chapter 8 in Diener and Suh, eds., *Culture and Subjective Well-being*, 185–218; and in Daniel Kahneman and Alan Krueger, "Developments in the Measurement of Subjective Well-Being," *Journal of Economic Perspectives* 20, no. 1 (Winter 2006): 3–24.

55. David Myers, "The Funds, Friends, and Fate of Happy People," *American Psychologist* 55, no. 1 (January 2000): 61. The statistics are even more startling for Japan: there, real per capita income quintupled between 1958 and 1987—refrigerators, washing machines, and TVs became commonplace, and car ownership rose from 1 percent to 60 percent of the population—but on average people didn't become one iota happier. See Easterlin, "Will Raising the Incomes of All Increase the Happiness of All?" 39–40.

56. The threshold income figures in the previous sentence are stated in 1995 U.S. dollars and adjusted for differences in purchasing power across countries. On the absence of a correlation between incomes above this threshold and happiness, see Inglehart and Klingemann, "Genes, Culture, Democracy, and Happiness," 171.

57. Tibor Scitovsky, "Happiness and Income," chapter 7 in *The Joyless Economy* (New York: Oxford University Press, 1976), 133–45; Doh Shin, "Does Rapid Economic Growth Improve the Human Lot? Some Empirical Evidence," *Social Indicators Research* 8 (1980): 199–221; and Tim Kasser, *The High Price of Materialism* (Cambridge, MA: MIT Press, 2002), especially 77–86.

58. The question of what causes us to seek higher incomes is closely related to the question of what causes us to seek higher consumption. For a superb survey of research relating to high consumption, see Inge Røpke, "The Dynamics of Willingness to Consume," *Ecological Economics* 28 (1999): 399–420.

59. Juliet Schor discusses both these phenomena in *The Overworked American: The Unexpected Decline of Leisure* (New York: Basic, 1992), especially 122–26. In her subsequent book, *The Overspent American: Why We Want What We Don't Need* (New York: HarperPerennial, 1998), Schor offers a close analysis of "competitive consumption." She begins with Thorstein Veblen's famous theory of conspicuous consumption. On the hedonic treadmill, see Martin Seligman, *Authentic Happiness: Using the New Positive Psychology to Realize Your Potential for Lasting Fulfillment* (New York: Free Press, 2002), 49–50. On social comparison, see the classic study by Easterlin, "Will Raising the Incomes of All Increase the Happiness of All?"; and Andrew Clark and Andrew Oswald, "Satisfaction and Comparison Income," *Journal of Public Economics* 61 (1996): 359–81. For a contrary perspective, see Diener and Oishi, "Money and Happiness," 205–7.

60. This kind of argument has deep roots in economics. In the early nineteenth century, the French economist Jean Baptiste Say proposed that the incomes generated in the process of manufacturing goods would create sufficient economic demand for the goods. Say's Law (simplistically stated as "supply creates its own demand") became an axiom of classical economics and still has strong adherents among neoclassical economists today. However, the law has been attacked by a long line of "underconsumption" theorists—including Thomas Malthus, Karl Marx, Thorstein Veblen, John Hobson, and John Maynard Keynes—who have argued that capitalist economies often generate insufficient demand, in part because of technological displacement of labor. These economies are therefore chronically susceptible to overproduction and gluts. See, for instance, Keynes, *The General Theory of Employment, Interest, and Money* (San Diego: Harcourt Brace, 1964 [1953]), 18–34. In popular literature, the need to maintain consumption to compensate for technological displacement of labor was a central theme for Aldous Huxley in his novel *Brave New World* and Kurt Vonnegut Jr. in *Player Piano*. The debate has continued to this day. William Grieder offers an underconsumption argument in *One World, Ready or Not: The Manic Logic of Global Capitalism* (New York: Simon & Schuster, 1997), especially 44–53 and 320–21. He writes, "The present regime is pathological fundamentally because it broadly destroys consumer incomes while it creates the growing surfeit of goods." Grieder focuses on capitalism's tendency to overproduce goods relative to the level of demand, what he calls the "supply problem." In contrast, Jeremy Rifkin emphasizes the technological displacement of labor in *The End of Work: The Decline of the Global Labor Force and the Dawn of the Post-Market Era* (New York: Putnam, 1995). See also Krugman, *The Return of Depression Economics*. My argument in this chapter begins from premises similar to those of the underconsumption theorists but emphasizes that contemporary capitalist economies have at least temporarily solved this problem by using consumerism to sustain high levels of demand, maintain economic growth, and absorb displaced labor.

61. Gary Rivlin, "Who's Afraid of China?" *New York Times*, Sunday Business, December 19, 2004, national edition, 1.

62. This argument implicitly assumes the validity of Say's Law.

63. In ecological terms, this process involves the creation of new economic niches. Arthur shows that competition and cooperation among entities in a coevolutionary environment (for instance, among corporations in a market or among

organisms in an ecosystem) create niches that provide opportunities for the development of new entities. See W. Brian Arthur, "On the Evolution of Complexity," in *Complexity: Metaphors, Models, and Reality*, ed. G. Cowan, D. Pines, and D. Meltzer, Santa Fe Institute Studies in the Sciences of Complexity, Proceedings, Vol. 19 (Reading, MA, 1994), 65–78.

64. For an argument to this effect, see William Nordhaus, "The Sources of the Productivity Rebound and the Manufacturing Employment Puzzle," *NBER Working Paper 11354*, National Bureau of Economic Research (May 2005).

65. A good example of such conviction is Wolf, *Why Globalization Works*, 175–84.

66. Thomas Malthus was one of the first economists to recognize that insatiable desires were key to prosperity. In the preface to his *Principles of Political Economy*, he wrote, "If every person were satisfied with the simplest food, the poorest clothing, and the meanest houses, it is certain that no other sort of food, clothing, and lodging would be in existence." Thomas Malthus, *Principles of Political Economy* (New York: Augustus Kelley, 1964 [1836]), 7.

67. Jack Triplett and Barry Bosworth, "Productivity Measurement Issues in Services Industries: 'Baumol's Disease' Has Been Cured," *FRBNY Economic Policy Review*, September 2003.

68. David Autor, Frank Levy, and Richard Murnane, "The Skill Content of Recent Technological Change: An Empirical Exploration," *Quarterly Journal of Economics* 118 (November 2003): 1279–1333; Norm Leckie, "On Skill Requirements Trends in Canada, 1971–1991," research paper submitted to Human Resources Development Canada and the Canadian Policy Research Network (Ottawa: Human Resource Group, Ekos Research Associates, 1996); and Michael Cox and Richard Alm, Federal Research Bank of Dallas, *A Better Way: Productivity and Reorganization in the American Economy, 2003 Annual Report* (Dallas: Federal Research Bank of Dallas, 2003), 19–21. Cox and Alm write, "Over time, our work moves up a hierarchy of human talents, focusing on new tasks that require higher-order skills, ones that machinery or out-sourcing can't do as well." For a general discussion of these trends, see Frank Levy and Richard Murnane, *The New Division of Labor: How Computers Are Creating the Next Job Market* (New York: Russell Sage, 2004), especially 37–54.

69. Specifically, if the labor market is efficient, and wages are allowed to adjust smoothly to labor supply and demand, the ever-increasing proportion of technologically displaced labor will be employed only at very low wages. If the labor market is inefficient, because of, for instance, minimum wage laws, these circumstances will simply produce greater unemployment.

70. Louis Uchitelle, "Defying Forecast, Job Losses Mount for a 22nd Month," *New York Times*, September 6, 2003, national edition, 1. This growth rate is required not just to absorb labor displaced by technological change but also to absorb the natural increase in the workforce (from population growth) and to employ workers who have lost their jobs when industries move overseas to secure lower labor costs.

71. Lizabeth Cohen, a professor of American studies at Harvard, describes and analyzes the rise of the U.S. consumerist culture in *A Consumers' Republic: The Politics of Mass Consumption in Postwar America* (New York: Knopf, 2003).

72. Juliet Schor develops this argument in detail in *The Overworked American*, espe-

cially 114–22. Schor quotes Charles Kettering, the general director of General Motors' Research Labs, who said in 1929 that the mission of business is "the organized creation of dissatisfaction."

73. Even conservative defenders of modern capitalism acknowledge the vital role of economic growth in preserving social and political stability. For instance, Martin Wolf writes, "When per capita output rises, a society's condition can be described as 'positive sum'—every person in that society can become better off. This outcome makes politics relatively easy to manage. In a static society, however, a 'zero-sum' condition prevails: If anyone is to receive more, someone else must receive less." Wolf, "The Morality of the Market," *Foreign Policy* 138 (September–October 2003): 49.

74. The British author and journalist Jeremy Seabrook writes, "The individual is denuded of everything but appetites, desires, and tastes, wrenched from any context of human obligation or commitment. It is a process of mutilation; and once this has been achieved, we are offered the consolation of reconstituting the abbreviated humanity out of the things and the goods around us, and the fantasies and vapors which they emit." See Seabrook, *What Went Wrong? Working People and the Ideals of the Labour Movement* (London: Victor Gollancz, 1978), 95–96.

75. The argument that Americans, in particular, are addicted to consumption is elaborated in Peter Whybrow, *American Mania: When More Is Not Enough* (New York: Norton, 2005).

76. Røpke proposes that patterns of high consumption persist because (1) competition among corporations lowers prices, renders current products obsolete before they are worn out, and introduces new specialized products; (2) corporations prefer to pass on productivity increases to their employees in the form of higher wages rather than increase leisure time; (3) high consumption and consumption rituals are markers of high social status in our hierarchical societies; and (4) the individualism that comes with modernity encourages people to define their identities through what they own and consume. See Røpke, "The Dynamics of Willingness to Consume."

77. "Today, a young American with at least two years of college can expect to change jobs at least eleven times in the course of working, and change his or her skill base at least three times during those forty years of labor." Richard Sennett, *The Corrosion of Character: The Personal Consequences of Work in the New Capitalism* (New York: Norton, 1998), 22.

78. Ibid., especially 25 and 31; and John Schwartz, "Always at Work and Anxious," *New York Times*, September 5, 2004, national edition, 1.

79. Tim Kasser shows how materialistic values, desires, and pursuits harm people's psychological and social well-being in *The High Price of Materialism*. On the complex relationship between new technologies and consumerism, see Inge Røpke, "New Technology in Everyday Life—Social Processes and Environmental Impact," *Ecological Economics* 38 (2001): 403–22.

80. A contrary perspective, one that emphasizes the social, psychological, and moral benefits of economic growth, is Benjamin Friedman, *The Moral Consequences of Economic Growth* (New York: Knopf, 2005).

81. Greg Critser, *Fat Land: How Americans Became the Fattest People in the World* (Boston: Houghton Mifflin, 2003).

82. Eric Stice, Diane Spangler, and Stewart Agras, "Exposure to Media-Portrayed Thin-Ideal Images Adversely Affects Vulnerable Girls: A Longtitudinal Experiment," *Journal of Social and Clinical Psychology* 20, no. 3 (2001): 270–88.

83. Key studies include Gerald L. Klerman and Myrna M. Weissman, "Increasing Rates of Depression," *Journal of the American Medical Association* 261, no. 15 (April 21, 1989): 2229–36; Cross-National Collaborative Group, "The Changing Rate of Major Depression: Cross-National Comparisons," *Journal of the American Medical Association* 268, no. 21 (December 2, 1992): 3098–3105; Peter M. Lewinsohn et al., "Age-Cohort Changes in the Lifetime Occurrence of Depression and Other Mental Disorders," *Journal of Abnormal Psychology* 102, no. 1 (1993): 110–20; and Ronald Kessler et al., "The Epidemiology of Major Depressive Disorder," *Journal of the American Medical Association* 289, no. 23 (June 18, 2003): 3095–3105.

84. On causes of depression, see Martin Seligman, "Why Is There So Much Depression Today? The Waxing of the Individual and the Waning of the Commons," in Rick E. Ingram, ed., *Contemporary Psychological Approaches to Depression: Theory, Research and Treatment* (New York: Plenum Press, 1990), 1–9; and Daniel Goldman, "A Rising Cost of Modernity: Depression," *New York Times*, December 8, 1992, national edition, C1. On the effects of too much choice, see Sheena Iyengar and Mark Lepper, "When Choice Is Demotivating: Can One Desire Too Much of a Good Thing?" *Journal of Personality and Social Psychology* 79, no. 6 (2000): 995–1006; Barry Schwartz, "Self-Determination: The Tyranny of Freedom," *American Psychologist* 55, no. 1 (2000): 79–88; and. Barry Schwartz, "The Tyranny of Choice," *Scientific American* 290, no. 4 (April 2004): 75.

85. Edward Luttwak presents a strong argument that they are indeed too high in *Turbo-Capitalism: Winners and Losers in the Global Economy* (New York: HarperCollins, 1998).

86. Eric Hobsbawm, *Age of Extremes: The Short Twentieth Century, 1914–1991* (London: Abacus, 1994), 95, 107.

87. The trend in capacity utilization in the U.S. manufacturing sector is indicative. Over the past thirty-five years, U.S. factories have been progressively less busy at the peak of every business cycle: manufacturing capacity utilization has dropped from near 90 percent at the peak of the business cycle in the early 1970s to about 78 percent in early 2005. See "Federal Reserve Statistical Release, G.17 (419) Supplemental Tables, Industrial Production and Capacity Utilization," May 17, 2006, available at http://www.federalreserve.gov/releases/g17/Current/g17_sup.pdf. On chronic surplus productive capacity, see Richard Duncan, *The Dollar Crisis: Causes, Consequences, Cures* (Singapore: John Wiley & Sons [Asia], 2003), 151.

88. This thesis is developed at length in "The Unfinished Recession: A Survey of the World Economy," *The Economist* (September 28, 2002).

89. See Greider, *One World, Ready or Not*, 52–53.

90. For a general discussion and analysis of the propositions in this and the previous sentence, see Oliviero Bernardini and Riccardo Galli, "Dematerialization: Long-Term Trends in the Intensity of Use of Materials and Energy," *Futures* (May 1993): 431–48.

91. Although, as I outlined in chapter 6, such arguments have become a staple of

conservative commentary on environmental issues, they can also be heard on the ideological left. See in particular Ernst von Weizsäcker, Amory Lovins, and L. Hunter Lovins, *Factor Four: Doubling Wealth, Halving Resource Use, The New Report to the Club of Rome* (London: Earthscan, 1997).

92. I discuss the concept of energy intensity in chapter 4. Between 1980 and 2001, energy intensity in OECD countries fell from 12,443 to 9,165 BTUs per dollar of GDP (1995 dollars at purchasing power parity). See Energy Information Administration, "Overview," *World Energy Use and Carbon Dioxide Emissions*, available at http://www.eia.doe.gov/emeu/cabs/carbonemiss/chapter1.html.

93. "Sui Genocide," *The Economist* (December 19, 1998): 130–31.

94. Peter Huber, *Hard Green: Saving the Environment from the Environmentalists: A Conservative Manifesto* (New York: Basic, 1999), 81.

95. In their standard textbook, the economist Rudiger Dornbusch and his colleagues argue that a discussion of resource constraints is more appropriate for "a course in astrophysics, or perhaps theology, than for a course in economics," because "technical progress permits us to produce more using fewer resources." And writing in *The Atlantic Monthly* about the oil industry, and by extension about resource-extraction industries more generally, Jonathan Rauch proposes that "[in] every sense except the one that is most literal and least important, the planet's resource base is growing larger, not smaller. Every day the planet becomes less an object and more an idea." See Rudiger Dornbusch, Stanley Fischer, and Richard Startz, *Macroeconomics* (Boston: McGraw-Hill, 1998), 74–75; and Jonathan Rauch, "The New Old Economy: Oil, Computers, and the Reinvention of the Earth," *The Atlantic Monthly* 287, no. 1 (January 2001): 49.

96. There is, of course, one big exception to this trend: the energy-intensity of energy production itself. The EROI of petroleum production (discussed in chapters 2 and 4) has been increasing for decades in the United States and almost certainly elsewhere too. See Cutler Cleveland, "An Exploration of Alternative Measures of Natural Resource Scarcity: The Case of Petroleum Resources in the U.S.," *Ecological Economics* 7 (1993): 123–57.

97. "Claims that a substantial decoupling of economic production from material inputs has occurred or is feasible should be viewed for what they are: assertions that currently have little convincing empirical support." Cutler Cleveland and Matthias Ruth, "Indicators of Dematerialization and the Materials Intensity of Use: A Critical Review with Suggestions for Future Research," *Journal of Industrial Ecology* 2, no. 3 (Summer 1998): 40. In correspondence with the author (July 6, 2005), Cleveland reiterated his concerns about the subjective and impressionistic methodologies used by many researchers investigating dematerialization. "Decoupling must be substantiated with appropriate time-series econometrics," he wrote. "The dematerialization literature has almost none of this." See also Iddo Wernick, Robert Herman, Shekhar Govind, and Jesse Ausubel, "Materialization and Dematerialization: Measures and Trends," *Daedalus* 125, no. 3 (Summer 1996): 171–98.

98. Between 1975 and 1994, the total material requirement for the U.S. economy, which includes both direct and indirect material flows, changed from 21.4 to 21.9 million metric tons. Albert Adriaaanse et al., *Resource Flows: The Material Basis of Industrial Economies* (Washington, DC: World Resources Institute, U.S.A.;

Wuppertal Institute, Germany; Netherlands Ministry of Housing, Spatial Planning, and Environment; National Institute for Environmental Studies, Japan; 1997). See especially appendix, 33–64. See also Emily Matthews, *The Weight of Nations: Material Outflows from Industrial Economies* (Washington, DC: World Resources Institute, 2000), especially figure 3, "Trends in Total Domestic Output, 1975–1996," 15.

99. There is, indeed, strong evidence that improvements in resource-use efficiency, by lowering resource prices, sometimes actually boost resource consumption—a phenomenon sometimes called the "Jevons Paradox," after the nineteenth-century English economist William Stanley Jevons. See Horace Herring, "Does Energy Efficiency Save Energy? The Debate and Its Consequence," *Applied Energy* 63 (1999) 209–26; and Stephen Bunker, "Raw Material and the Global Economy: Oversights and Distortions in Industrial Ecology," *Society & Natural Resources* 9 (1996): 419–29. Also, on the tendency of new technologies to raise consumption, see Røpke, "New Technology in Everyday Life."

100. Iddo Wernick, "Dematerialization and Secondary Materials Recovery in the U.S.," *Journal of the Minerals, Metals, and Materials Society* 46, no. 4 (April 1994): 39–42.

101. Statistics on U.S. energy consumption are derived from the U.S. Energy Information Administration, "Table 1.1 Energy Overview," available at http://www.eia.doe.gov/emeu/mer/pdf/pages/sec1_3.pdf, and on U.S. carbon dioxide output from "Table 12.1 Emissions of Greenhouse Gases, 1980–2003," available at http://www.eia.doe.gov/emeu/aer/txt/ptb1201.html, and "Table ES1. Summary of Estimated U.S. Emissions of Greenhouse Gases, 1990 and 1996–2004," available at http://www.eia.doe.gov/oiaf/1605/ggrpt/summary/pdf/execsum_tables.pdf.

102. Ernst von Weizsäcker, Amory Lovins, and Hunter Lovins are an exception. They argue that efficiency improvements of 4 to 5 percent a year are possible. See Weizsäcker, Lovins, and Lovins, *Factor Four: Doubling Wealth, Halving Resource Use, The New Report to the Club of Rome* (London: Earthscan, 1997), 142.

103. See Bunker, "Raw Material and the Global Economy." The economist Dan Ciuriak writes that the "shift towards services and the intellectual content of GDP" in the U.S. and other industrialized countries "was to some extent made possible by the transfer of the weightier parts of GDP to developing countries. Globally, GDP continued to have a higher material content than its declining weight in the U.S. would indicate. . . . In this sense, China and other developing countries to some extent made possible the much-vaunted 'knowledge economy' of the U.S. and other economies in the industrialized world. . . ." See Ciuriak, "Resource Implications of China's Development Path: The Weight of GDP Revisited," *American Journal of Chinese Studies* 12, no. 1 (Spring 2005): 25–44.

104. "Globalization thus helps to sustain the illusion among the rich and powerful that the limits to material growth have been permanently abolished and eliminates any reason to question the prevailing expansionist myth." William Rees, "Economics and Sustainability: Conflict or Convergence?" paper prepared for StatsCan Economic Conference, Ottawa, June 5, 2001; published as "Globalization and Sustainability: Conflict or Convergence?" *Bulletin of Science, Technology and Society* 22, no. 4 (August 2002): 249–68.

105. Eric Williams, Robert Ayers, and Miriam Heller, "The 1.7 Kilogram Microchip: Energy and Material Use in the Production of Semiconductor Devices," *Environmental Science and Technology* 36, no. 24 (December 15, 2002): 5504–10. The authors write, "The relative use of secondary materials is much higher for the microchip than for traditional goods."

106. The political scientist Benedict Anderson coined the term "imagined communities" in his famous book *Imagined Communities: Reflections on the Origins and Spread of Nationalism* (London: Verso, 1983).

107. World Commission on the Social Dimension of Globalization, 4.

108. Wade, "Winners and Losers," 74.

CHAPTER NINE

1. The common assertion that Ptolemy used eighty epicycles in his system is incorrect and appears to be derived from a mistaken count by Nicolas Copernicus in his effort to refine the Ptolemaic system. See Arthur Koestler, *The Sleepwalkers: A History of Man's Changing Vision of the Universe* (London: Penguin Arkana, 1989 [1959]), 195.

2. Copernicus did not, as is commonly thought, refute the Ptolemaic system or claim Earth orbits the sun; rather, he did his best to preserve the system's core and in so doing may have actually increased the number of epicycles. See ibid., 195–97.

3. Ibid., 433–34.

4. I develop this idea in chapter 3, "The Big I," in Thomas Homer-Dixon, *The Ingenuity Gap: Facing the Economic, Environmental, and Other Challenges of an Increasingly Complex and Unpredictable Future* (New York: Vintage, 2002), 71–98.

5. Essentially, each of these epicycles was what philosophers of science would call an "auxiliary hypothesis"—an ad hoc adjustment to the theory that dealt with an annoying bit of contradictory evidence. But the cost of such adjustments is always greater complexity in the overall theory, and the Santucci armillary sphere shows that this complexity can become extreme.

6. For discussion of these and related issues, see Imre Lakatos and Alan Musgrave, eds., *Criticism and the Growth of Knowledge* (Cambridge: Cambridge University Press, 1970), especially Lakatos, "Falsification and the Methodology of Scientific Research Programmes," 91–196.

7. Thomas S. Kuhn, *The Structure of Scientific Revolutions* (Chicago: University of Chicago Press, 1970 [1962]).

8. I discuss the power of this cognitive shift in *The Ingenuity Gap*, 92–94.

9. Individual temperament—that basic distinction, mentioned in chapter 1, between optimism and pessimism—matters a lot here. Psychological research shows that we have, on average, an optimistic bias when it comes to assessing possible threats in our environment and our ability to respond to those threats. Most of us tend to underestimate the difficulties facing us, and we tend to overestimate our ability to respond to those difficulties. The anthropologist Lionel Tiger writes that "anticipating optimistic outcomes of undecided situations is as much part of human

nature, of the human biology, as are the shape of the body, the growth of children, and the zest of sexual pleasure" and that there is a "neurophysiology for a sense of the benignity of the future." See Lionel Tiger, *Optimism: The Biology of Hope* (New York: Simon and Schuster, 1979), 15, 51. The psychologists Charles Carver and Michael Scheier argue that optimists "tend to take a position of confidence (even if progress is presently difficult or slow)"; people with confidence, in turn, will "continue to display efforts and engagement, even as the cues imply increasingly less basis for confidence." See Carver and Scheier, "Optimism, Pessimism, and Self-Regulation," in Edward C. Chang, ed., *Optimism and Pessimism: Implications for Theory, Research and Practice* (Washington, DC: American Psychological Association, 2001), 41, 46. And the researchers Lauren Alloy and Lyn Abramson have shown that nondepressives overestimate their personal efficacy. "Nondepressed people succumb to cognitive illusions that enable them to see both themselves and their environment with a rosy glow." Alloy and Abramson, "Judgment of Contingency in Depressed and Nondepressed Students: Sadder but Wiser?" *Journal of Experimental Psychology: General* 108, no. 4 (1979): 441–85.

10. Paradoxically, the same people who have great confidence in science as a tool to solve the world's problems often disparage the scientific evidence about those problems.

11. A wide range of technological, economic, and social trends—some of which I've described in this book—are making global problems harder, perhaps too hard, given human beings' cognitive and social characteristics. These trends include the rising complexity of technologies and institutions, which makes them opaque to operators and managers; tighter coupling and greater specialization and speed of technological and social networks, which make them more vulnerable to unexpected negative synergies; increasing scarcity of high-quality energy, which limits societal capacity to boost solution complexity; widespread disruption of Earth's environmental systems, which boosts the risk of ecosystem flips; and increasingly severe imbalances in global society, such as wide gaps in fertility and wealth between rich and poor regions, which increases the likelihood of major social and political instability. For a full discussion, see Homer-Dixon, *The Ingenuity Gap*.

12. Koestler, *The Sleepwalkers*, 79.

13. Richard Posner, "The 9/11 Report: A Dissent," *New York Times Book Review*, August 29, 2004, 1.

14. Robert Ornstein and Paul Ehrlich, *New World, New Mind: Moving toward Conscious Evolution* (New York: Touchstone, 1989).

15. Herbert Simon, *Reason in Human Affairs* (Stanford, CA: Stanford University Press, 1983), 20–23, 79–83.

16. Ross Gelbspan, *The Heat Is On: The High Stakes Battle over Earth's Threatened Climate* (Reading, MA: Addison-Wesley, 1997); and Gelbspan, *Boiling Point: How Politicians, Big Oil and Coal, Journalists, and Activists Are Fueling the Climate Crisis— And What We Can Do to Avert Disaster* (New York: Basic, 2004).

17. The most accessible rendering of the argument that follows can be found in "Structure," chapter 3 of Joshua Cohen and Joel Rogers, *On Democracy: Towards a Transformation of American Society* (New York: Penguin, 1983), 47–87. See also Adam Przeworksi, "Proletariat into a Class: The Process of Class Formation

from Karl Kautsky's *The Class Struggle* to Recent Controversies," *Politics and Society* 7 (1977): 343–401; Przeworksi, "Material Bases of Consent: Economics and Politics in a Hegemonic System," *Political Power and Social Theory*, 1 (1980): 21–66; and Przeworksi and Michael Wallerstein, "The Structure of Class Conflict in Democratic Capitalist Societies," *American Political Science Review* 76, no. 2 (June 1982): 215–38.

18. Branko Milanovic provides a particularly compelling analysis of such resistance in today's global economy. Elites that he collectively labels a "global plutocracy"—and that operate through international financial institutions they've designed and largely control—use their power to prevent any significant change in the global distribution of wealth. See Milanovic, *Worlds Apart: Measuring International and Global Inequality* (Princeton: Princeton University Press, 2005), 149–52. Capitalism's ideological hegemony hasn't always been so pervasive and unchallenged, of course, especially during the first half of the past century.

19. "Both an unemployed worker and a millionaire owner of a major television station enjoy the same formal right of free speech," the political philosophers Joshua Cohen and Joel Rogers note, "but their powers to express and give substance to that right are radically different." Cohen and Rogers, *On Democracy*, 50.

20. Indeed, today's capitalist discourse perpetuates itself partly because it orders reality—specifying roles, expectations, and norms—and in so doing facilitates communication between people. In terms that economists use, this discourse has positive "network externalities" because the incentive for any one person to adopt the discourse increases as other people adopt it.

21. "The interests of capitalists appear as general interests of the society as a whole, the interests of everyone else appear as merely particular, or 'special.'" Cohen and Rogers, *On Democracy*, 53. In making this point, the authors note their indebtedness to Karl Marx and Antonio Gramsci. See, in particular, Karl Marx and Frederick Engels, *The German Ideology*, in Marx and Engels, *Collected Works*, vol. 5 (New York: International Publishers, 1976); and Antonio Gramsci, *Selections from the Prison Notebooks*, ed. and trans. Quintin Hoare and Geoffrey Nowell Smith (New York: International Publishers, 1971). See also Charles Lindblom, "The Privileged Position of Business," chapter 13 in *Politics and Markets: The World's Political-Economic Systems* (New York: Basic Books, 1977), 170–88.

22. I discuss the fallibility of experts in chapter 4, "Glimpsing the Abyss," and chapter 11, "White-Hot Landscapes" of *The Ingenuity Gap*, 149–69 and 270–309.

23. Yehezkel Dror, *The Capacity to Govern: A Report to the Club of Rome* (London: Frank Cass, 2001), 40–41.

24. Joseph Tainter, *The Collapse of Complex Societies* (Cambridge: Cambridge University Press, 1988).

25. Ibid., 195.

26. Ibid., 91.

27. Ibid., 195.

28. The best single short summary of panarchy theory is C. S. Holling, "Understanding the Complexity of Economic, Ecological, and Social Systems," *Ecosystems* 4, no. 5 (2001): 390–405. A complete treatment is offered in Lance

Gunderson and C. S. Holling, eds., *Panarchy: Understanding Transformations in Human and Natural Systems* (Washington, DC: Island Press, 2002). Also see Lance Gunderson, C. S. Holling, and Stephen Light, eds., *Barriers and Bridges to the Renewal of Ecosystems and Institutions* (New York: Columbia University Press, 1995); and Fikret Berkes, Johan Colding, and Carl Folke, eds., *Navigating Social-Ecological Systems: Building Resilience for Complexity and Change* (Cambridge: Cambridge University Press, 2003).

29. Holling, "Understanding," 394. In thermodynamic terms, Holling's "potential"corresponds to physicists' concept of "exergy," which is an energy form's capacity to do work. See the discussion of energy quality in chapter 2 and the associated endnotes.

30. Ibid.

31. Ibid., 395.

32. The image of the adaptive cycle in the accompanying figure is a modification of that used in articles by Holling and other panarchy theorists. It ensures that the characteristics of the cycle particularly relevant to the present discussion can be easily seen and understood. Specifically, in this figure's back loop, potential, connectedness, and resilience are shown to collapse simultaneously almost to zero. Then potential and resilience recover first, before connectedness. In contrast, in the schemas of panarchy theorists, potential generally collapses first. Then, while potential partially recovers, connectedness and resilience decline. Once connectedness reaches zero, potential collapses a second time, while resilience falls rapidly to zero. At this point both potential and connectivity begin to grow again in the adaptive cycle's front loop.

33. Holling, "Understanding,," 399.

34. This idea reappears in many domains of thought. For instance, "The Chinese have traditionally interpreted their past as a series of dynastic cycles in which successive dynasties repeat a boringly repetitious story: a heroic founding, a period of great power, then a long decline, and finally total collapse." See John Fairbank, Edwin Reischauer, and Albert Craig, *East Asia: Tradition and Transformation*, rev. ed. (Boston: Houghton Mifflin, 1989), 70. In economics, the idea of regular cycles occurs in neo-Schumpeterian analyses of innovation and in Kondratieff long-wave theories of technological change. See for instance, R. U. Ayres, *Technological Transformations and Long Waves* (Laxenburg, Austria: International Institute for Applied Systems Analysis, 1989); and Arnulf Grubler and Nebojsa Nakicenovic, "Long Waves, Technology Diffusion, and Substitution," International Institute for Applied Systems Analysis, *Review* 14, no. 2 (Spring 1991): 313–42. A comprehensive summary of grand theories of societal change is Johan Galtung and Sohail Inayatullah, eds., *Macrohistory and Macrohistorians: Perspectives on Individual, Social, and Civilizational Change* (Westport, CT: Praeger, 1997).

35. Among other things, panarchy theory shows us that different kinds of complexity in different contexts have different effects. Tainter's theory in contrast doesn't differentiate between kinds of complexity. It treats all complexity as basically the same—as a homogeneous input into a society's problem-solving processes, an input that eventually does more harm than good. But while some common features of complex systems, like tight connectivity and scale-free structure (discussed in chapter 5 of this book), can make systems vulnerable to certain kinds of damage,

other features, like species diversity in an ecosystem (chapter 7), increase system resilience. Holling's theory is rich enough to cope with these differences.

36. "Periods of success carry the seeds of subsequent downfall, because they allow stresses and rigidities to accumulate." Holling, "Understanding," 399.

CHAPTER TEN

1. A comprehensive statement of this research is Sander van der Leeuw, F. Favory and J. L. Fiches, eds., *Archéologie et systèmes socio-environnementaux: études multi-scalaires sur la vallée du Rhône dans le programme ARCHAEOMEDES* (Valbonne: CNRS, Monographies du CRA, 2003).

2. Jean-Pierre Adam, "Civil Engineering," chapter 10 in *Roman Building: Materials and Techniques*, trans. Anthony Mathews (London: Routledge, 2001), 239–47. On Roman water engineering and aqueducts in general, see Deane Blackman and A. Trevor Hodge, *Frontinus' Legacy: Essays on Frontinus' De aquis urbis Romae* (Ann Arbor: University of Michigan Press, 2001).

3. The classicist A. Trevor Hodge writes, "[So] tiny was the gradient to be established, and hence so vital the need for precision, that in these circumstances one would be tempted . . . to describe the feat as not almost but wholly impossible, except for the fact that the Romans did it. . . . The degree of fine accuracy required for surveying in these conditions . . . is almost unbelievable, but it was achieved and the aqueduct did flow." Hodge, *Roman Aqueducts and Water Supply* (London: Duckworth, 1992), 184–91.

4. Jean-Louis Guendon and Jean Vaudour, "Concrétions et fonctionnement de l'aqueduc: étude morpho-stratigraphique," chapter 11 in Guilhem Fabre, Jean-Luc Fiches, and Jean-Louis Paillet, eds., *L'Aqueduc de Nîmes et Le Pont du Gard: Archéologie Géosystème Histoire*, 2d ed. rev. and augmented (Paris: CNRS Édition, 2000), 233–48.

5. Over the aqueduct's active lifetime, some eighty thousand metric tons of this material built up along its fifty-kilometer length.

6. The rock is so fine that locals often used it as a building material after the aqueduct fell into disuse.

7. Claire Rodier, Christian Joseph, and Jean-Claude Gilly, "Étude à la microsonde de la géochimie de concrétions internes dans l'aqueduc à Bezouce"; Jean-Luc Fiches, Michiel Gazenbeek, Jean-Louis Paillet, "Prospections et fouilles: archéologie d'un aqueduct,": and Jean-Luc Fiches and Jean-Louis Paillet, "De la mise en eau au démantèlement: essai de périodisation," chapters 13, 16, and 19 in Fabre et al., eds., *L'Aqueduc de Nîmes et Le Pont du Gard*, 263–71, 315–55, 407–22.

8. "Many traits of modern society were in effect already part of Roman society some two millennia ago: the rapid colonization of most or all of the known world, an elaborate military and civil organization that managed to control the empire, an urban base, as well as major investments in infrastructure such as roads, aqueducts, ports and large-scale semi-industrial agriculture organized in [large landed estates]." Sander van der Leeuw and Bert de Vries, "Empire: The Romans in the

Mediterranean," chapter 7 in Bert de Vries and Johan Goudsblom, eds., *Mappae Mundi: Humans and Their Habitats in a Long-Term Socio-Ecological Perspective: Myths, Maps and Models* (Amsterdam: RIVM, Amsterdam University Press, 2002), 210.

9. Sander van der Leeuw, "Land Use, Settlement Pattern and Degradation in the Ancient Rhône Valley," chapter 6 in Sander van der Leeuw, ed., *The Archaeomedes Project: Understanding the Natural and Anthropogenic Causes of Land Degradation and Desertification in the Mediterranean Basin: Research Results* (Luxembourg: Office for Official Publications of the European Commission, 1998), 176.

10. Sander van der Leeuw and Bert de Vries, "Empire," 210.

11. For two centuries, the empire's currency consisted mainly of two coins—the denarius, made of silver, and the aureus, made of gold. The denarius was the standard coin; it was used to pay soldiers and was also the common unit of public and private account. As such, it was often profitable for authorities to debase the denarius, while the aureus was never significantly debased. A. H. M. Jones, *The Roman Economy: Studies in Ancient Economic and Administrative History*, P. A. Brunt ed. (Totawa, NJ: Rowman and Littlefield, 1974), 191–95. On the use of gold coin for transporting value over distances, see Richard Duncan-Jones, *Structure and Scale in the Roman Economy* (Cambridge: Cambridge University Press, 1990), 45; and on the conversion of agricultural produce into coin, see Keith Hopkins, "Taxes and Trade in the Roman Empire, 200 B.C.–A.D. 200," *Journal of Roman Studies* 70 (1980): 101–25, especially 102.

12. James C. Scott discusses the importance of cadastral maps to the implementation of effective taxation and the rise of the modern state in *Seeing like a State: How Certain Schemes to Improve the Human Condition Have Failed* (New Haven: Yale University Press, 1998).

13. Van der Leeuw and de Vries, "Empire," 234–36.

14. Nearly a thousand years before Rome became an empire, cities had started appearing along the east coast of the Mediterranean. By 500 BCE, they'd spread to the coasts of the western Mediterranean, and in Spain and Gaul they'd migrated inland, creating a surprisingly dense network of roads.

15. Caesar remarked, for instance, that in Belgium "there is nothing to control—no cities, no forts, no installations." Quoted in N. Roymans, *Tribal Societies in Northern Gaul: An Anthropological Perspective* (Amsterdam: University of Amsterdam Press, 1990), cited in Van der Leeuw and de Vries, "Empire," 217. See also Edward Luttwak, *The Grand Strategy of the Roman Empire: From the First Century A.D. to the Third* (Baltimore: Johns Hopkins University Press, 1976), 45.

16. Jones notes that under the empire tax farming was largely abolished and that "city governments acted as agents for the collection of [property and poll taxes]." Jones, *The Roman Economy*, 27.

17. For a detailed description of the empire's elaborate and costly postal system, see A. H. M. Jones, *The Later Roman Empire, 284–602: A Social, Economic, and Administrative Survey*, Vol. 2 (Baltimore: Johns Hopkins University Press, 1964), 830–34.

18. The extent and importance of trade in the Roman empire is a subject of considerable controversy among scholars. The historian Aurelio Bernardi writes that Rome's economic development had "created an integrated system with intensive

trade relations over an immense area." He goes on to say that "[with] the emergence of organized urban centers the traditional agrarian economy, poor and simple, rapidly developed into an economy which, although still mainly based on agriculture, was vitalized by intense commercial exchange." However, in his classic study of the Roman economy, A. H. M. Jones argued that because of limited demand and the cost of overland transportation, production for almost all goods was localized, and trade within the empire was limited to high-value products, especially luxury items. Long-distance trade in grain was "commercially profitable only when the [grain] was grown in areas close to a port or inland waterway, and the market was a large town which lay on the sea or on a navigable river." Wine and oil, he acknowledged, "were probably more important objects of trade," along with papyrus, iron, copper, high-quality textiles, superior glassware, jewelry, and exotic spices. The historian Moses Finley has also argued that the Roman economy was primitive and characterized by localized production and limited surpluses and consumption. Although the Jones/Finley view has long prevailed among scholars of antiquity, archaeologists have recently found evidence of the large-scale production and long-distance transport of a wide range of goods during Roman times and of urban demand's role as a stimulus to trade. In an analysis of early Roman Gaul, the professor of ancient history Greg Woolf takes an intermediate position. While the scale of Gaul's production was often large, and some of its products were traded over long distances, most trade remained within the region. See Aurelio Bernardi, "The Economic Problems of the Roman Empire at the Time of Its Decline," in Carlo Cipolla, ed., *The Economic Decline of Empires* (London: Methuen, 1970), 16–83, especially 30–31; Jones, *The Later Roman Empire*, Vol. 2, chapter 21, "Industry, Trade and Transport," 824–72, especially 824–27 and 844–50, and chapter 25, "The Decline of the Empire," especially 1039–40; and M. I. Finley, *The Ancient Economy* (Berkeley: University of California Press, 1973). Archaeological findings are discussed in Duncan-Jones, *Structure and Scale;* Michael Fulford, "Economic Interdependence among Urban Communities of the Roman Mediterranean," *World Archaeology* 19 (1987): 58–75, which provides evidence of widespread seaborne grain shipments during the empire; and Hopkins, "Taxes and Trade." See also Woolf, "Regional Productions in Early Roman Gaul," chapter 3 in David Mattingly and John Salmon eds., *Economies beyond Agriculture in the Classical World* (London: Routledge, 2001), 49–65.

19. Van der Leeuw and de Vries, "Empire," 236; and van der Leeuw, "Land Use, Settlement Pattern and Degradation in the Ancient Rhône Valley," 210–11.

20. Hopkins, "Taxes and Trade," 103. Van der Leeuw and de Vries, "Empire," 224. Once again, for inland areas, the exception appears to have been grain: because of the prohibitive cost of overland transport, emergency imports of grain were feasible only between regions that could be accessed easily by water, and local famines were common. See Jones, *The Later Roman Empire*, Vol. 2, 844–45; and Chris Wickham, *Early Medieval Italy: Central Power and Local Society, 400–1000* (Ann Arbor: University of Michigan Press, 1989), 14.

21. Jason Dowdle, "Road Networks and Exchange Systems in the Aeduan Civitas, 300 B.C.–A.D.300," in C. Crumley and W. Marquardt, eds., *Regional Dynamics:*

Burgundian Landscapes in Historical Perspective (San Diego: Academic Press, 1987), 265–94.

22. T. F. H. Allen, Joseph Tainter, and Thomas Hoekstra, *Supply-Side Sustainability* (New York: Columbia University Press, 2003), 148.

23. Van der Leeuw and de Vries, "Empire,"p. 238.

24. The rents paid to the aristocracy, Jones writes, "must have been considerable, a factor in the economy of the empire comparable with the imperial revenue." Jones, *The Roman Economy*, 126. On the range of taxes in the empire through the third century, see Bernardi, "Economic Problems," 38.

25. Van der Leeuw and de Vries, "Empire," 242.

26. Joseph Tainter, *The Collapse of Complex Societies* (Cambridge: Cambridge University Press, 1988).

27. Luttwak, *Grand Strategy*, especially chapter 2, "From the Flavians to the Severi: 'Scientific' Frontiers and Preclusive Defense from Vespasian to Marcus Aurelius," 51–126.

28. On the general dilemma of dealing with barbarian threat through expansion, see Bennet Bronson, "The Role of Barbarians in the Fall of States," chapter 8 in Norman Yoffee and George Cowgill, eds., *The Collapse of Ancient States and Civilizations* (Tucson: University of Arizona Press, 1988), 196–218, especially 216.

29. In fact, these distant zones were net beneficiaries of a large-scale redistribution of wealth within the empire. While most revenue was raised in productive regions around the Mediterranean, a large portion was spent on the armies stationed along the northern frontier. Jones, *The Roman Economy*, 127. On the diminishing returns to territorial expansion, see Tainter, *Collapse*, 148–49; and Joseph Tainter, "Problem Solving: Complexity, History, Sustainability," *Population and Environment* 22, no. 1 (September 2000): 19–20.

30. Jones, *The Roman Economy*, 135.

31. Luttwak, *Grand Strategy*, 87, 113–14, 117.

32. "The new strategy of perimeter defense inaugurated by the Flavians required an investment of colossal proportions over the course of three centuries." Ibid., 61.

33. Calculated on the basis of 1.5 kilograms of grain per soldier per day. See Martin Van Creveld, *Technology and War: From 2000 BC to the Present* (New York: Free Press, 1991), 107.

34. Writes Jones, "The number of man-hours of labor required to feed, clothe and arm each man and to transport the food, clothing and arms from where they were produced to the frontier was, in view of the primitive method of production and transport, immense." Jones, *The Roman Economy*, 135; on army pay, see 192 and 194.

35. J. Donald Hughes thoroughly describes the damage Romans caused to their environment in *Pan's Travail: Environmental Problems of the Ancient Greeks and Romans* (Baltimore: Johns Hopkins University Press, 1994).

36. "As the number of people involved in a system grows linearly," note van der Leeuw and de Vries, "the number of messages required to keep all of them in touch grows exponentially, and so does the time involved to reach everyone concerned." And "as the number of people involved grows, so does their diversity, and with it the time it takes to 'negotiate' collaboration between them." Van der Leeuw and de Vries, "Empire," 242.

37. Local producers became "more and more dependent on the economic conditions in faraway regions over which they had no control whatsoever." Ibid., 246.

38. Jones discusses the increasing burden on the empire of unproductive consumers, including senators and other aristocrats, the civil service, the poor, and (later) the Church's clergy. "The basic economic weakness of the empire was that too few producers supported too many idle mouths." See Jones, *The Later Roman Empire*, Vol. 2, 1045–47; also, Bernardi, "Economic Problems," 33; and on the dole, see Greg Aldrete and David J. Mattingly, "Feeding the City: The Organization, Operation, and Scale of the Supply System for Rome," in D. S. Potter and D. J. Mattingly, eds., *Life, Death, and Entertainment in the Roman Empire* (Ann Arbor: University of Michigan Press, 1999), 177–78.

39. "A man feels poor and shabby unless the walls are alight with large and precious circular mirrors, unless there are slabs of Alexandrine marble picked out with overlay of Numidian; unless the marbles themselves are treated everywhere with elaborate surface-color designs as variegated as fresco; unless the ceiling is masked with glass; unless the basins into which we plunge our sapless, over-sweated bodies have kerbs of the Thasian marble that once was only a rarity to be stared at in some temple; unless the nozzles from which the water spouts are silver." Lucius Annaeus Seneca, c. 4 BCE–65 CE, *Seneca's Letters to Lucilius*. E. Phillips Barker, trans. (Oxford: Clarendon Press, 1932), 43–44.

40. Bernardi, "Economic Problems," 30.

41. Tainter, *Collapse*, 129.

42. "Energy subsidies are resources outside a society's normal territory or sphere of influence, or resources that are suddenly usable with new technologies." Joseph Tainter, "Evolutionary Consequences of War," in G. Ausenda, ed., *Effects of War on Society* (San Marino: Center for Interdisciplinary Research on Social Stress, 1992), 103–30. See also Tainter, *Collapse*, 124.

43. Joseph Tainter, "Societal Metabolism in the Roman Empire," in *Advances in Energy Studies, 3rd Biennial International Workshop: Reconsidering the Importance of Energy*, eds. Sergio Ulgiati, Mark T. Brown, Mario Giampietro, Robert A. Herendeen, and Kozo Mayumi (Padua, Italy: SG Editoriali, 2003), 125–32.

44. "To the ancients to create a new tax or increase the rate of an old one was a major operation, only to be undertaken in desperate circumstances. Many taxes went on unchanged for centuries." Jones, *The Roman Economy*, 189, 193. See also A. H. M. Jones. *The Later Roman Empire, 284–602: A Social, Economic, and Administrative Survey*, vol. 1 (Baltimore: Johns Hopkins University Press, 1964), 9–11.

45. Tainter, *Collapse*, 135; and Tainter, "Societal Metabolism."

46. On the "technical retardation" of Roman society because of the absence of economic incentives to innovate, see F. W. Walbank, *The Awful Revolution: The Decline of the Roman Empire in the West* (Toronto: University of Toronto Press, 1969), 45–47.

47. A good summary of this research can be found in van der Leeuw, "Land Use, Settlement Pattern and Degradation in the Ancient Rhône Valley," 212–17.

48. Sander van der Leeuw, "Desertification, Land Degradation and Land Abandonment in the Rhône Valley, France," in Graeme Barker and David

Gilbertson, eds., *The Archaeology of Drylands: Living at the Margin* (London: Routledge, 2000), 342–43.

49. "The drainage network was conceived as a multi-level system to regulate the water balance over the whole plane, collecting it in individual plots and bringing it down to the Rhône." Van der Leeuw, "Land Use, Settlement Pattern and Degradation in the Ancient Rhône Valley," 217.

50. Van der Leeuw and de Vries, "Empire," 246.

51. Ibid., 245.

52. Alexander Demandt, *Der Fall Roms: Die Auflösung des Römischen Reiches im Urteil der Nachwelt* (Munich: Beck, 1984), 695. Some explanations of Rome's fall are worth special note. The historian Ramsay MacMullen argues that endemic corruption and proliferating patron-client relations eroded norms of public honor and responsibility to the commonweal, and in so doing undermined the broader conception of the empire's purpose. F. W. Walbank stresses the inability of Roman society and economy to innovate, which polarized the society between antagonistic upper and lower classes. And Michael Grant highlights three sets of "disunities" that split Roman society apart: conflict between the military and the state (exacerbated by the absence of a succession mechanism for the emperor) and between the people and the military; alienation of the productive classes from the state, especially because of excessive taxation; and the increasing corruption, inefficiency, and rigidity of an expanding bureaucracy. For the most part, contemporary scholars regard Edward Gibbon's famous treatment of Rome's decline as of merely historiographic interest; it does not provide a clear or coherent thesis on the causes of Rome's fall. See MacMullen, *Corruption and the Decline of the Roman Empire* (New Haven: Yale University Press, 1988); Walbank, *The Awful Revolution*; Grant, *The Fall of the Roman Empire: A Reappraisal* (Radnor, PA: The Annenberg School of Communication Press, 1976); and David Jordan, "Gibbon and the Fall of Rome," chapter 7 in *Gibbon and His Roman Empire* (Urbana, IL: University of Illinois Press, 1971), 213–30.

53. The story in the eastern Roman empire was markedly different, involving a directed and systematic simplification of institutions, including the government and the military. See Allen, Tainter, and Hoekstra, *Supply-Side Sustainability*, 122–36.

54. Luttwak, *Grand Strategy*, 128; Tainter, *Collapse*, 140.

55. "The government's fiscal obligations may well have doubled, and yet this was a government that even before the crisis had been strapped for funds." Tainter, *Collapse*, 139.

56. Jones, *The Roman Economy*, 197–99; and Bernardi, "Economic Problems," 43.

57. Allen, Tainter, and Hoekstra, *Supply-Side Sustainability*, 113.

58. Chris Wickham, "The Other Transition: From the Ancient World to Feudalism," *Past and Present* 103 (May 1984): 3–36, especially 13.

59. Jones, *The Roman Economy*, 84, 86.

60. Luttwak, *Grand Strategy*, 130. Tainter and his colleagues concur: "The government taxed its citizens more heavily, conscripted their labor, regulated their lives, and dictated their occupations. It was an omnipresent, coercive organization that subdued individual interests and amassed all resources toward one overarching goal: sustaining the empire." Allen, Tainter, and Hoekstra, *Supply-Side Sustainability*, 112.

61. Ibid., 177; Tainter, *Collapse*, 141.

62. Bernardi, "Economic Problems," 23–26.

63. Jones writes that "[e]ven the privileged categories of landowners paid more than twice as much in the sixth century A.D. as the provincials had in the first century B.C., and the ordinary landowner well over three times as much." Jones, *The Roman Economy*, 83; and Bernardi, "Economic Problems," 55–56.

64. "A late Roman peasant, for example, might have paid from one-fourth to one-third of gross yields in taxes if he owned his own land. If he rented land, taxes and rent together took from one-half to two-thirds of a crop. Thus, Roman peasants were perpetually impoverished, often undernourished, and sometimes forced to sell children into slavery when they could not be fed." Joseph Tainter, subsection 6, "Energy Flow in the Aftermath of Collapse," in "Sociopolitical Collapse, Energy and," *Encyclopedia of Energy*, Cutler Cleveland, ed. (San Diego: Academic Press/Elsevier Science, 2004), 529–43. On damage to cropland, see Hughes, *Pan's Travail*, 191.

65. Jones, *The Later Roman Empire*, Vol. 2, 1040–44.

66. Bernardi calls this period a "military monarchy." See "Economic Problems," 40, 42.

67. On the increasing concentration of land, see Bernardi, "Economic Problems," 44–52; and on tax evasion see Bernardi, "Economic Problems," 57–65, and Alexander Motyl, *Imperial Ends: The Decay, Collapse, and Revival of Empires* (New York: Columbia University Press, 2001), 57.

68. Allen, Tainter, and Hoekstra, *Supply-Side Sustainability*, 118.

69. Bernardi, "Economic Problems," 69–73. Tainter argues that when the marginal returns to rising complexity in a social system turn negative, "decomposition" is a rational response: some units within the system—for instance regions, cities, villages, and even individual landowners—may decide that it makes sense to separate from the system and go their own way, reducing complexity in the process. Tainter, *Collapse*, 120–22.

70. Tainter, "Problem Solving," 23.

71. Cutler Cleveland, "An Exploration of Alternative Measures of Natural Resource Scarcity: The Case of Petroleum Resources in the U.S.," *Ecological Economics* 7 (1993): 123–57.

72. The U.S. Central Intelligence Agency estimates that world GDP (purchasing power parity) in 2005 was $60 trillion. See http://www.cia.gov/cia/publications/factbook/rankorder/2001rank.html.

73. Tainter and his coauthors write that compared with high-EROI societies, low-EROI systems "may capture even more energy, but, because they must capture it from more extensive sources, organization is required to aggregate resources. The Roman empire exemplifies this. Late in its history, the empire greatly expanded its organizational control to amass the dispersed resources needed to survive." Joseph Tainter et al., "Resource Transitions and Energy Gain: Contexts of Organization," *Conservation Ecology* 7, no. 3 (2003), available at www.consecol.org/vol7/iss3/art4.

74. The proposition that there is a positive correlation between the complexity of a problem, the complexity of actors generating solutions to the problem, and the complexity of the solutions themselves is derived from a principle that theorists call the "law of requisite variety." According to this law, a successful adaptive system

must have a repertoire of behaviors at least as wide as the range of behaviors expressed by its surrounding environment. To widen its repertoire of behaviors, in turn, an adaptive system must increase its internal complexity. See Yaneer Bar-Yam, "Multiscale Variety in Complex Systems," *Complexity* 9, no. 4 (2004): 37–45; and Elinor Ostrom, "Designing Complexity to Govern Complexity," in *Property Rights and the Environment*, eds. Susan Hanna and Mohan Munasinghe (Washington, DC: Beijer International Institute of Ecological Economics and the World Bank, 1995).

75. In the following paragraphs, I use "Holland" to refer to the Netherlands. Although such usage is commonplace, technically "Holland" refers solely to the central-western provinces of the Netherlands, North and South Holland.

76. At almost four hundred people per square kilometer, the Netherlands ranks fourth in population density, if we don't count city-states and islands. Using 2005 population estimates, and including in national land area inland water bodies (lakes, reservoirs, and rivers), the world's most densely populated countries are Bangladesh (1,002 people per square kilometer), Taiwan (636), and South Korea (491).

77. The Dutch system of pumping stations is described by Jared Diamond in *Collapse: How Societies Choose to Fail or Succeed* (New York: Viking, 2005), 519–20.

78. The Dutch have a term—*maakbaar*—that doesn't translate well into English, but means something like "moldable" or "creatable." Things that were maakbaar can be molded or shaped at will, and until the past fifteen years or so the Dutch commonly accepted that Holland's citizens and Dutch society were maakbaar. More recently, with Holland's chronic and sometimes violent friction between native Dutch and new (often Muslim) immigrants, the term has been used with a certain degree of skepticism or even irony. Still, compared with people in many other societies, the Dutch are far more receptive to intrusive social policies designed to improve or defend the common good.

79. The historian William McNeill notes that the Dutch are "among the world's largest per capita importers of timber, most of it tropical hardwoods from Southeast Asia," and that they maintain their livestock largely with imported fodder. "They can survive handsomely and harmoniously in part because the deforestation, soil erosion, and degradation associated with cutting timber, growing cocoa, and soya happen in Indonesia, West Africa, and Brazilian Amazonia, not in the Netherlands." McNeill, "Diamond in the Rough: Is There a Genuine Environmental Threat to Security?" *International Security* 30, no. 1 (Summer 2005): 190. On the "Netherlands fallacy," see also Paul Ehrlich and Anne Ehrlich, *One with Nineveh: Politics, Consumption, and the Human Future* (Washington, DC: Island, 2004), 100.

80. There is a direct link between energy costs and the level of international trade. Indeed, the surge in global trade in the 1960s and 1990s can be attributed as much to low energy prices as to tariff cuts. Research by economists at the World Bank and elsewhere indicates that a 25 percent increase in fuel prices leads to a 10 percent increase in freight rates, which in turn can depress international trade by about 5 percent. Even with the modest energy-price increases we've seen up to 2006, manufacturers are already rethinking their globalized production models. Some North American manufacturers are considering bringing factories closer to consumers—transferring them back from China to Mexico, for instance—

because of rising long-distance transportation costs. See Jeff Rubin and Benjamin Tal, "Soaring Oil Prices Will Make the World Rounder," *CIBC World Markets: Occasional Report #55* (October 19, 2005), available at http://research.cibcwm.com/economic_public/download/occ_55.pdf. On likely future restrictions of travel because of energy costs and environmental considerations, see Andrew Curry, et al., *Intelligent Infrastructure Futures: The Scenarios—Towards 2055* (London: Foresight Programme, Office of Science and Technology, 2005).

81. Thomas Friedman argues that the world is evolving toward a largely frictionless global economy, in which anyone anywhere can compete against anyone else. He is right that electronic barriers to entry to international commerce may continue to decline as information technology improves. Tasks that can be digitized and broken into components may continue to be distributed widely around the planet. But a truly "flat" world economy requires trade of huge quantities of raw materials and manufactured goods—and therefore abundant low-cost energy. See Friedman, *The World Is Flat: A Brief History of the Twenty-First Century* (New York: Farrar, Straus & Giroux, 2005).

82. Holling calls a system's capacity for regeneration its "memory."

83. C. S. Holling, "From Complex Regions to Complex Worlds," *Ecology and Society* 9, No. 1 (2004), available at http://www.ecologyandsociety.org/vo19/iss1/art11/print.pdf.

84. Allen, Tainter, and Hoekstra, *Supply-Side Sustainability*, 154; and Van der Leeuw and de Vries, "Empire," 248. In the third century, Saint Cyprian, bishop of Carthage, wrote, "[The] age is now senile . . . the World itself . . . testifies to its own decline by giving manifold concrete evidence of the process of decay. There is a diminution of the winter rains that give nourishment to the seeds in the earth, and in the summer heats that ripen the harvests. . . . The mountains, disembowelled and worn out, yield a lower output of marble; the mines, exhausted, furnish a smaller stock of the precious metals: the veins are impoverished, and they shrink daily. There is a decrease and deficiency of farmers in the field, of sailors on the sea, of soldiers in the barracks, of honesty in the marketplace, of justice in court, of concord in friendship, of skill in technique, of strictness in morals. . . . [This] loss of strength and loss of stature must end, at last, in annihilation." Quoted in Joseph Tainter, "Post-Collapse Societies," *Companion Encyclopedia of Archaeology*, Graeme Barker, ed. (London: Routledge, 1999), 1022.

85. Niall Ferguson develops a scenario somewhat similar to that outlined in this paragraph in "A World without Power," *Foreign Policy* 143 (July–August 2004): 32–39. See also Robert Harvey, *Global Disorder: How to Avoid a Fourth World War* (New York: Carroll & Graf, 2003); and Robert Cooper, *The Breaking of Nations: Order and Chaos in the Twenty-First Century* (New York: Grove, 2003).

86. Following the 2005 urban riots in France, the Oxford historian Timothy Garton Ash wrote, "On all reasonable assumptions Europe's population of immigrant descent and Muslim culture will grow significantly over the next decade. . . . If we cannot make even those who have lived in Europe since birth feel at home here, there will be hell to pay. Six thousand burning cars will seem like nothing more than an hors-d'oeuvre. Garton Ash, "A Fear Not Only for France," *Guardian Weekly*, November 18–24, 2005, 6.

87. The literature on the sources of civil violence is huge and encompasses scholarly work on causes of revolution, rural rebellion in poor countries, and social protest movements. A superb summary of recent theory and research is Jack Goldstone, "Toward a Fourth Generation of Revolutionary Theory," *Annual Review of Political Science, Vol. 4, 2001*, Nelson Polsby, ed. (Palo Alto, CA: Annual Reviews, 2001), 139–87. See also Doug McAdam, Sidney Tarrow, and Charles Tilly, *Dynamics of Contention* (Cambridge: Cambridge University Press, 2001); Doug McAdam, John McCarthy, and Mayer Zald, eds., *Comparative Perspectives on Social Movements: Political Opportunities, Mobilizing Structures, and Cultural Framings* (Cambridge: Cambridge University Press, 1996); and Charles King, "Review Article: The Micropolitics of Social Violence," *World Politics* 56, no. 3 (April 2004): 43–55.

88. On the role of "protest identities," see Goldstone, "Toward a Fourth Generation of Revolutionary Theory," 153–54.

89. On framing, see Doug McAdam, Sidney Tarrow, and Charles Tilly, "Toward an Integrated Perspective on Social Movements," in Mark Irving Lichbach and Alan Zuckerman, eds., *Comparative Politics: Rationality, Culture, and Structure* (Cambridge: Cambridge University Press, 1997), 142–73, especially 157–58.

90. An excellent discussion of the psychology of extremism, including the relationship between poverty and terrorism and how feelings of humiliation contribute to terrorism, is Jessica Stern, *Terror in the Name of God: Why Religious Militants Kill* (New York: Ecco, HarperCollins, 2003), especially 32–62 and 281–86.

91. "Globalization must, by changing the reference point upward," writes Branko Milanovic of the World Bank, "make people in poor countries feel more deprived." Milanovic, *Worlds Apart: Measuring International and Global Inequality* (Princeton: Princeton University Press, 2005), 155–56.

92. The middle class's interests are usually aligned closely enough with those of the rich that it opposes efforts by the poor to expropriate the wealth of the rich; at the same time, though, it's suspicious enough of the power of the rich that it opposes overt efforts by the rich to exploit the poor.

93. Milanovic uses 1998 purchasing power parity figures for per capita incomes. Milanovic, *Worlds Apart*, 130–31.

94. "While in the year 1960," Milanovic writes, "there were forty-one rich countries—nineteen of them non-Western—in 2000, there were only thirty-one rich countries, and only nine of them were non-Western. None of the African countries (except for Mauritius) and none of the Latin American and Caribbean countries (except for the Bahamas) were left among the rich. Latin America and the Caribbean, probably for the first time in 200 years, had no country that was richer than the poorest West European country." Milanovic, *Worlds Apart*, 62–65.

95. See James Fearon and David Laitin, "Ethnicity, Insurgency, and Civil War," *American Political Science Review* 97, no. 1 (February 2003): 75–90; Paul Collier et al., *Breaking the Conflict Trap: Civil War and Development Policy* (Washington, DC: World Bank and Oxford University Press, 2003); and "The Failed States Index," *Foreign Policy* 149 (July–August 2005): 56–65.

96. Jessica Stern has introduced the concept of vicarious humiliation. See James Bennet, "Blowing Up in the West," *New York Times*, Week in Review, Sunday, July 17, 2005, national edition, 1.

97. Robert Leiken, "Europe's Angry Muslims," *Foreign Affairs* 84, no. 4 (July–August 2005): 120–36; and R. Scott Appleby and Martin Marty, "Fundamentalism," *Foreign Policy* 128 (January–February 2002): 16–22.

98. Goldstone highlights the role of elite disaffection as a cause of major civil violence. He also notes that a governing regime's authority and legitimacy derives largely from perceptions of the regime's effectiveness and justness. "States and rulers that are perceived as ineffective may still gain elite support for reform and restructuring if they are perceived as just. States that are considered unjust may be tolerated as long as they are perceived to be effective in pursuing economic and nationalist goals, or just too effective to challenge. However, states that appear both ineffective and unjust will forfeit the elite and popular support they need to survive." Goldstone, "Toward a Fourth Generation of Revolutionary Theory," 146–48.

99. Jessica Tuchman Mathews, "Power Shift," *Foreign Affairs* 76, no. 1 (January–February 1997): 50–66.

100. National security experts often refer to this as the problem of "asymmetric threat" because small groups have power wholly disproportionate to their resources and size.

101. "Hezbollah and the West African Diamond Trade," *Middle East Intelligence Bulletin* 6, no. 6–7 (June–July 2004), available at http://www.meib.org/articles/0407_12.htm; and Douglas Farah and Richard Shultz, "Al Qaeda's Growing Sanctuary," *Washington Post*, July 14, 2004, A19.

102. In 2004, Graham Allison of the Kennedy School of Government at Harvard University wrote, "In my own considered judgment, on the current path, a nuclear terrorist attack on America in the decade ahead is more likely than not." See Alison, *Nuclear Terrorism: The Ultimate Preventable Catastrophe* (New York: Times Books, 2004), 15. At the end of 2003, according to David Albright, president of the Institute for Science and International Security in Washington, D.C., the world stockpile of HEU was 1,895 metric tons, of which about 1,100 metric tons was in Russia. See also Alexander Glaser and Frank von Hippel, "Thwarting Nuclear Terrorism," *Scientific American* 294, no. 2 (February 2006): 56–63; and Matthew Bunn and Anthony Wier, *Securing the Bomb: An Agenda for Action* (Cambridge, Massachusetts: Project on Security the Atom, Belfer Center, Harvard University, May 2004).

103. David Sanger, "U.S. Rebukes Pakistanis for Lab's Aid to Pyongyang," *New York Times*, April 1, 2003, national edition, B15.

104. Jingdong Tian et al., "Accurate Multiplex Gene Synthesis from Programmable DNA Microchips," *Nature* 432, no. 7020 (December 23, 2004): 1050–54. See also, Andrew Pollack, "Scientists Create Live Polio Virus," *New York Times*, July 12, 2002, national edition, A1.

105. A good survey of the issue is provided by Mark Williams in "The Knowledge," *Technology Review* 109, no. 1 (March–April 2006): 44–53.

106. Bennet Bronson argues that barbarians—which he defines as members "of a political unit that is in direct contact with a state but is not itself a state"—have specific advantages in a contest with complex states. When it comes to warfare, they have far lower resource requirements, greater tactical flexibility, and a favourable cost-benefit ratio. Also, states tend to concentrate wealth in major centers that are attractive targets for plunder. See Bronson, "The Role of Barbarians in the Fall of States."

107. For much of the Middle Ages, locals attributed the construction of the Pont du Gard to supernatural beings.

108. Greg Easterbrook makes such an argument in "The End of War?" *The New Republic* 232, no. 4715 (May 30, 2005): 18–21. Data on the decline in armed conflicts are available in Monty Marshall and Ted Robert Gurr, *Peace and Conflict 2005: A Global Survey of Armed Conflicts, Self-Determination Movements, and Democracy* (College Park, MD: Center for International Development & Conflict Management, 2005); and Human Security Centre, *Human Security Report 2005: War and Peace in the 21ˢᵗ Century* (New York: Oxford University Press, 2005).

109. Ferguson, "A World without Power." See also, Michael Mandelbaum, *The Case for Goliath: How America Acts As the World's Government in the Twenty-First Century* (New York: PublicAffairs, 2006). The stability of the international financial system, in particular, appears to depend in significant measure on perceived international political stability provided by a dominant power. See William Brown, Richard Burdekin, and Marc Weidenmier, "Volatility in an Era of Reduced Uncertainty: Lessons from Pax Britannica," *Working Paper 11319*, National Bureau of Economic Research (May 2005).

110. Smil points out that in the 1850s the U.S. was almost entirely rural, powered by wood and a marginal actor in global affairs. "A century later, after more than tripling its per capita consumption of useful energy and becoming the world's largest producer and consumer of fossil fuels, it was both an economic and military superpower." Vaclav Smil, *Energy in World History* (Boulder, CO: Westview, 1994), 237.

111. Definitive discussions of the steadily increasing risks of conflict over nonrenewable resources, especially oil, are Michael Klare, *Blood and Oil: The Dangers and Consequences of America's Growing Dependency on Imported Petroleum* (New York: Metropolitan, 2004); and Klare, *Resource Wars: The New Landscape of Global Conflict* (New York: Owl, 2002). Klare writes that "the American military is increasingly being converted into a global oil-protection service." See Klare, "Oil Wars," available at http://www.commondreams.org/views04/1008–23.htm.

112. Perhaps the best description of this challenge is provided by John Robb. See Robb, "Security: Power to the People," *Fast Company* 103 (March 2006): 120, available at http://www.fastcompany.com/magazine/103/essay-security.html.

113. These groups engage in what specialists have come to call "netwar." The protagonists "are likely to consist of dispersed small groups who communicate, coordinate, and conduct their campaigns in an internetted manner, without a precise central command." See David Ronfeldt et al., *The Zapatista Social Netwar in Mexico* (Santa Monica: RAND, 1998), 9; and John Arquilla and David Ronfeldt, *The Advent of Netwar* (Santa Monica: RAND, 1996).

114. These features were evident in the Iraqi insurgency that developed in 2004 and 2005. Wrote the RAND Corporation analyst Bruce Hoffman in 2004, "The insurgency in Iraq is taking place in an ambiguous and constantly shifting environment, with constellations of cells and individuals gravitating toward one another—to carry out armed attacks, exchange intelligence, trade weapons, and engage in joint training—and then dispersing, sometimes never to operate together again." And in late 2005, U.S. intelligence sources determined that over one hundred insurgent groups were operating in Iraq. Based on this analysis,

Dexter Filkins of *New York Times* noted that the insurgency "is horizontal as opposed to hierarchical, and ad hoc as opposed to unified. [This] central characteristic, similar to that of terrorist organizations in Europe and Asia, is what is making the Iraqi insurgency so difficult to destroy. Attack any single part of it, and the rest carries on largely untouched. It cannot be decapitated, because the insurgency, for the most part, has no head." Hoffman, "Plan of Attack," *The Atlantic Monthly* (July–August 2004); and Filkins, "Profusion of Rebel Groups Helps Them Survive in Iraq," *New York Times*, December 2, 2005, national edition, A1.

115. "Al Qaeda is no longer a global disciplined and centralized network. It has become a grassroots phenomenon. It is a flag, a loosely connected body of home-grown terror groups and freelance terrorists, inspired but not beholden by bin Laden, each going their own way without central command, unaffiliated with any group." Rik Coolsaet, *Al Qaeda: The Myth. The Root Causes of International Terrorism and How to Tackle Them*, trans. Erika Peeters (Gent, Belgium: Academia Press, 2005), vii. On the roots of anti-Western ideologies in general, see Ian Buruma and Avishai Margalit, *Occidentalism: The West in the Eyes of Its Enemies* (New York: Penguin, 2004).

116. John Robb, a security analyst with a background in counterterrorism and complex systems analysis, has developed the concept of global guerrilla warfare at length. See his blog at http://globalguerrillas.typepad.com/globalguerrillas/.

117. Analysts distinguish between guerrilla war and insurgency. Both involve hit-and-run attacks by dispersed and usually small units taking advantage of local terrain or cover (in Iraq, mainly urban cover) and the passive support of at least a portion of the local population. Both seek the long-term exhaustion of the conventional forces they're opposing—in other words, they win by not losing. But guerrilla war is usually thought to exhibit a high degree of centralized operational control, whereas insurgency can be far more loosely coordinated.

CHAPTER ELEVEN

1. Francis Fukuyama, *The End of History and the Last Man* (New York: Free Press, 1992). Fukuyama defined history narrowly as a contest of ideologies of governance. For an excellent response, see Roger Kimball, "Francis Fukuyama and the End of History," *The New Criterion* 10, no. 6 (February 1992), available at http://www.newcriterion.com/archive/10/feb92/fukuyama.htm#.

2. For an insightful discussion of the difficulties that complexity and low predictability pose for the management of social systems, see Jake Chapman, *System Failure: Why Governments Must Learn to Think Differently* (London: Demos, 1992), especially chapter 2, "Current Policy Making," 18–24.

3. The dean of the School of Forestry and Environmental Studies at Yale University, James Gustave Speth, elaborates on these transitions in "Part Four: The Transition to Sustainability," in *Red Sky at Morning: America and the Crisis of the Global Environment* (New Haven: Yale University Press, 2004), 149–202.

4. The effects of today's greenhouse gas emissions—especially on glaciers, polar ice, and sea levels—won't fully work their way through Earth's complex natural

systems for up to a millennium. This long lag time is what scientists call the "inertia" of the climate systems. See James Hansen, "A Slippery Slope: How Much Global Warming Constitutes 'Dangerous Anthropogenic Interference'?" editorial essay, *Climate Change* 68, no. 3 (February 2005): 269–79.

5. "The idea that we can address climate-change matters successfully at the expense of economic growth is not only unrealistic but also unacceptable," said Prime Minister John Howard of Australia in early 2006 at a summit meeting of countries that are major emitters of carbon dioxide and nonsignatories of the Kyoto Protocol. Agence France-Presse, "Polluters Vow No Sacrifice of Growth in Climate Fight," January 12, 2006.

6. T. F. H. Allen, Joseph Tainter, and Thomas Hoekstra, *Supply-Side Sustainability* (New York: Columbia University Press, 2003), 150.

7. Of course, depending on the criteria used, different analysts will come up with different lists of countries. An interesting example of an effort that focuses on indicators of instability is "The Failed States Index," *Foreign Policy* 149 (July–August 2005): 56–65. Pakistan and Saudi Arabia are included in this list but China is not.

8. For a complete discussion of Saudi reserves, see Matthew Simmons, *Twilight in the Desert: The Coming Saudi Oil Shock and the World Economy* (Hoboken, NJ: John Wiley, 2005), especially 265–79.

9. "A production decline of 30 to 50 percent in a period of five years or less in any or all of Saudi Arabia's key production fields is not out of the question." Ibid., 358–59.

10. Population Division of the Department of Economic and Social Affairs of the United Nations Secretariat, *World Population Prospects: The 2004 Revision* and *World Urbanization Prospects: The 2003 Revision*, available at http://esa.un.org/unpp.

11. Richard Cincotta, Robert Engelman, and Daniele Anastasion, "Appendix 4: Country Data Table," in *The Security Demographic: Population and Civil Conflict after the Cold War* (Washington, DC: Population Action International, 2003), 100.

12. Gerald Posner, "The Kingdom and the Power," *New York Times*, August 2, 2005, national edition, A23.

13. Jad Mouawad, "Saudi Arabia Looks Past Oil; Enriched by Record Prices, the Nation Seeks to Diversify," *New York Times*, December 13, 2005, national edition, B1.

14. In late February 2006, Al Qaeda terrorists attacked the Saudi Abqaiq oil-processing facility. According to a statement released by "al Qaeda in the Arab peninsula," the bombing was intended to stop "crusaders and Jews" from stealing Muslim wealth. On the general threat to Saudi oil facilities, see "What If? Terrorists are now targeting Saudi Arabia's oil infrastructure. How bad could things get?" *The Economist* (May 29, 2004): 67–70.

15. Robert Baer, *Sleeping with the Devil: How Washington Sold Our Soul for Saudi Crude* (New York: Crown, 2003).

16. Population Division, *World Population Prospects: The 2004 Revision*, available at http://esa.un.org/unpp. A somewhat more optimistic report on Pakistan's fertility decline, which sees a 2020 population in the neighbourhood of 200 million, is Griffith Feeney and Iqbal Alam, "New Estimates and Projections of Population Growth in Pakistan," *Population and Development Review* 29, no. 3 (September 2003): 483–92.

17. Even with high inputs of technology, irrigation, and fertilizer, Pakistan's available agricultural land can sustainably support barely one-tenth of today's population. See F. H. Beinroth, Hari Eswaran, and Paul Reich, "Land Quality and Food Security in Asia," paper presented at the Second International Conference on Land Degradation (Khon Kean, Thailand, 2001), published on CD-ROM by the Department of Land Development, Bangkok, Thailand.

18. Somini Sengupta. "Pakistan Is Booming Since 9/11, at Least for the Well-Off," *New York Times*, March 23, 2005, national edition, A 3; "3.72 Million Unemployed in Labour Force in Pakistan," *Daily Times* (Lahore), June 12, 2004.

19. See Ahmad Faruqui, "The Political Economy of Militarism," chapter 3 in Faruqui, *Rethinking the National Security of Pakistan: The Price of Strategic Myopia* (Burlington, VT: Ashgate, 2003), 19–39.

20. Husain Haqqani, "The Role of Islam in Pakistan's Future," and C. Raja Mohan, "What If Pakistan Fails? India Isn't Worried . . . Yet," *Washington Quarterly* 28, no. 1 (Winter 2004–2005): 85–96 and 117–28.

21. David Rohde, "Turning Away from the U.S., Pakistan's Elite Gravitate Toward Islamic Religious Parties," *New York Times*, October 13, 2002, national edition, 6.

22. David Rohde, "Anti-American Sentiment Intensifies in Pakistan as U.S. Confronts Iraq," *New York Times*, December 22, 2002, national edition, 17; and David Rohde and Somini Sengupta, "Qaeda on the Run? Raids Seem to Belie Pakistan's Word," *New York Times*, August 5, 2005, national edition, A8.

23. For an account of how the metallurgist and spy A. Q. Khan built Pakistan's first uranium weapons, see William Langewiesche, "The Wrath of Khan," *The Atlantic Monthly* 296, no. 4 (November 2005): 62–85.

24. "[Evidence suggests] that growth may have caused distributional conflicts in Pakistan, and their eventual eruption into violence. According to this evidence, each decade of high growth has been followed by one marked by widespread violence, political instability, and low-growth." Tarique Niazi, "Economic Growth and Social Violence in Pakistan," *International Journal of Contemporary Sociology* 38, no. 2 (October 2001): 171–92.

25. Robert Kaplan writes, "Pulsing with consumer and martial energy, and boasting a peasantry that, unlike others in history, is overwhelmingly literate, China constitutes the principal conventional threat to America' liberal imperium." Kaplan, "How We Would Fight China," *The Atlantic Monthly* 295, no. 5 (June 2005): 49–64.

26. Minxin Pei provides an overview of these stresses, with a particular focus on skyrocketing official and corporate corruption, in "The Dark Side of China's Rise," *Foreign Policy* 153 (March–April 2006): 32–40.

27. In fact, capital extracted from farmers (estimated at some $80 billion a year) has provided essential funds—channeled by state banks—for urban infrastructure and industrial, commercial, and residential development.

28. The ranks of the poorest people, those making less than $75 a year, actually grew in 2004. Joseph Kahn and Jim Yardley, "Amid China's Boom, No Helping Hand for Young Qingming," *New York Times*, August 1, 2004, national edition, 1.

29. Jim Yardley, "Rural Exodus for Work Fractures Chinese Family," *New York Times*, December 21, 2004, national edition, A1.

30. Sai Liang and Zhongdong Ma, "China's Floating Population: New Evidence from the 2000 Census," *Population and Development Review* 30, no. 3 (September 2004): 467–88; and Jim Yardley, "In a Tidal Wave, China's Masses Pour from Farm to City," *New York Times*, September 12, 2004, national edition, 6.

31. For a survey of China's environmental problems, see Jianguo Liu and Jared Diamond, "China's Environment in a Globalizing World," *Nature* 435, no. 7046 (June 30, 2005): 1179–86.

32. Careful estimates put the economic cost of environmental damage in China as high as 18 percent of the country's GDP. See Mao Yu-Shi, "The Economic Cost of Environmental Degradation in China: A Summary," *Project on Environmental Scarcities, State Capacity, and Civil Violence* (Toronto: Trudeau Centre for Peace and Conflict Studies, 1998), available at http://www.library.utoronto.ca/pcs/state/chinaeco/summary.htm.

33. Edward O. Wilson, "The Bottleneck," *Scientific American* 286, no. 2 (February 2002): 89.

34. "The Irresistible Desert," *Environment* 46, no. 5 (June 2004): 4; and Geoffrey York, "Northern China Choked by Massive Dust Storm," *Globe and Mail* (Toronto), April 19, 2006, A13.

35. Jim Yardley, "China Races to Reverse Its Falling Production of Grain," *New York Times*, May 2, 2004, national edition, 6.

36. On China's energy predicament, see Bernard Cole, *"Oil for the Lamps of China"— Beijing's 21ˢᵗ-Century Search for Energy*, McNair Paper No. 67, Institute for National Strategic Studies (Washington, DC: National Defense University, 2003).

37. The rural protests clearly alarmed the Beijing regime, and it has moved to eliminate agricultural taxes and increase farm subsidies, but the practical effect on rural conditions has so far been minimal. See Joseph Kahn, "China to Cut Taxes on Farmers and Raise Their Subsidies," *New York Times*, February 3, 2005, national edition, A3. On the escalating rate of protest, see Joseph Kahn, "Pace and Scope of Protest in China Accelerated in '05," *New York Times*, national edition, January 20, 2006, A10; Howard French, "Land of 74,000 Protests (But Little Is Ever Fixed)," *New York Times*, August 24, 2005, national edition, A4; and Howard French, "Riots in Shanghai Suburb as Pollution Protest Heats Up," *New York Times*, July 19, 2005, national edition, A3.

38. In just the past ten years, some 70 million farmers have had their land seized. See Jim Yardley, "Farmers Being Moved Aside by China Real Estate Boom," *New York Times*, December 8, 2004, national edition, A1.

39. Joseph Kahn, "Rioting in China over Label on College Diplomas," *New York Times*, June 22, 2006, national edition, A1.

40. As extensions of the state, Chinese banks often provide loans for political or social reasons—for example, to keep inefficient but worker-heavy factories afloat or to fund the pet building projects of favored officials and businessmen—paying little attention to economic fundamentals. This easy money, especially between 2001 and 2004, created a frenzy of investment, and some parts of the economy have developed the kind of massive overcapacity characteristic of a speculative bubble: thousands of redundant factories litter the countryside, and underoccupied office buildings pepper the skylines of cities like Shanghai. (Shanghai now has some

four thousand skyscrapers, more than New York City.) Because of high savings rates, Chinese consumers don't generate enough domestic demand to put this excess capacity to use. But the country's leaders have found another source of demand to fill the gap—the American consumer. On the Chinese real estate boom, see David Barboza, "China Builds Its Dreams and Some Fear a Bubble," *New York Times*, October 18, 2005, national edition, A1.

41. Drawing on home equity, U.S. households spent on an annual basis half a trillion dollars more than they earned in the third quarter of 2005. Gretchen Morgenson, "After the Debt Feast Comes the Heartburn," *New York Times*, November 27, 2005, national edition, section 3, 1.

42. "Never in the history of modern economics has a large industrial country run persistent current account deficits of the magnitude posted by the U.S. since 2000." Sebastian Edwards, "Is the U.S. Current Account Deficit Sustainable? And If Not, How Costly Is Adjustment Likely To Be?" *NBER Working Paper* 11541, National Bureau of Economic Research (August 2005), abstract available at http://papers.nber.org/papers/W11541. The *Financial Times* columnist Martin Wolf provides a marvelously lucid overview of the dangers to the U.S. and world economies in "Super Power on Borrowed Money: Can It Last?" The Robert J. Peolosky, Jr. Distinguished Lecture Speaker Series, Elliott School of International Affairs, George Washington University, April 5, 2005, available at http://www.gwu.edu/~elliott/news/transcripts/wolf.html.

43. In January 2004, the International Monetary Fund issued a blunt warning about the dangers of these deficits and "ballooning" liabilities to foreigners. Martin Mühleisen and Christopher Towe, eds., *U.S. Fiscal Policies and Priorities for Long-Run Sustainability*, Occasional Paper 227 (Washington, DC: IMF, 2004). The sentiment was echoed by the former U.S. Treasury Secretary Lawrence Summers in "America Overdrawn," *Foreign Policy* 143 (July–August 2004): 47–49.

44. On the dangers for the Chinese economy of a U.S. downturn, see Gerard Baker, "The Deficit Debacle," *Foreign Policy* 147 (March–April 2005): 42–47.

45. Niall Ferguson, "A World without Power," *Foreign Policy* 143 (July–August 2004): 90.

46. Chinese resentment of Japan is especially strong especially among relatively educated, Internet-savvy urban Chinese from their teens to their thirties. It erupted in April 2005 in surprisingly vicious attacks against Japanese consulates and the local offices of Japanese businesses. See Howard French, "By Playing at 'Rage,' China Dramatizes Its Rise," *New York Times*, April 21, 2005, national edition, A4; and Jim Yardley, "In Soccer Loss, a Glimpse of China's Rising Ire at Japan," *New York Times*, August 9, 2004, national edition, A3. On anti-Chinese sentiments in Japan, see Norimitsu Onishi and Howard French, "Ill Will Rising between China and Japanese," *New York Times*, August 3, 2005, national edition, A1. For an argument, contrary to the one I make here, that the Chinese regime's political insecurity tends to produce compromise in the country's foreign policy rather than nationalist hostility, see M. Taylor Fravel, "Regime Insecurity and International Cooperation: Explaining China's Compromises in Territorial Disputes," *International Security* 30, no. 2 (Fall 2005): 46–83.

47. For instance, China and Japan are competing for energy exports from Russia and angrily disputing ownership of potentially large offshore oil and gas reserves that

straddle their shared frontier in the East China Sea. James Brooke, "Drawing the Line on Energy: China and Japan Wrangle over Oil and Gas Projects in Disputed Waters," *New York Times*, March 29, 2005, national edition, C1; and Norimitsu Onishi and Howard French, "Japan's Rivalry with China Roils a Crowded Sea," *New York Times*, September 11, 2005, national edition, 1. Japan's energy situation is, in fact, even more critical than China's. Despite some of the world's most aggressive energy-conservation efforts since the 1973 oil shock, the country's overall energy consumption has increased nearly 50 percent. Japan now imports 80 percent of its primary energy and virtually all its oil. See Communications Office, Agency for Natural Resources and Energy, Ministry of Economy, Trade, and Industry, *Energy in Japan, 2005* (Tokyo, 2005), especially figure 1, p. 3.

48. Wilson, "The Bottleneck," *Scientific American* 286, no. 2 (February 2002): 89.

49. Radical Islamic groups have found these communities fertile places to recruit jihadis to pursue holy war, either in the Middle East, especially in Iraq, or within Europe itself. Often members aren't recruited directly; instead, they seek out like-minded extremists after being initially radicalized by Internet materials and chat groups. Leaders tend to be middle class and educated, often with technical degrees in subjects like engineering, while the grunts are economically unsuccessful residents of the segregated enclaves. These groups—dispersed across Europe—share few elements of ethnicity, grievance, or ideology. Instead, they're rooted in distinctly local grievances and social networks, although members may be motivated by a myth of global Muslim solidarity in the face of Western encroachment and oppression. See Timothy Savage, "Europe and Islam: Crescent Waxing, Cultures Clashing," *Washington Quarterly* 27, no. 3 (Summer 2004): 25–50; Rik Coolsaet, "Between al-Andalus and a Failing Integration: Europe's Pursuit of a Long-Term Counterterrorism Strategy in the Post-al-Qaeda Era," *Egmont Paper No. 5* (Brussels: Royal Institute for International Relations, 2005); Olivier Roy, *Globalised Islam: The Search for a New Ummah* (London: Hurst, 2004); Craig Smith, "Feeling of Being the Outsiders," *New York Times*, November 9, 2005, national edition, A1; Hassan Fattah, "Anger Burns on the Fringe of Britain's Muslims," *New York Times*, July 16, 2005, national edition, A1; Craig Smith and Don van Natta Jr., "Officials Fear Iraq's Lure for Muslims in Europe," *New York Times*, October 23, 2004, national edition, A1.

50. Association for the Study of Peak Oil and Gas, "Country Re-Assessment: United Kingdom," *ASPO Newsletter* 63 (March 2006): 3–5.

51. European Commission, Directorate-General for Energy and Transport, *Report on the Green Paper on Energy: Four Years of European Initiatives* (Luxembourg: Office for Official Publications of the European Commission, 2005).

52. Food and Agriculture Organization, *The State of Food and Agriculture 2005* (Rome: FAO, 2005), table 2, "Origin of Agricultural Imports by Region (percent)," 23.

53. Some types of personality are particularly likely to exploit moments of contingency to produce harm and to translate fear, confusion, and anger into violent action. Especially dangerous are people with narcissistic personality disorder. "Unable to face his own inadequacies," says the American political psychologist Jerrold Post, "the individual with this personality style needs a target to blame and attack for his own inner weakness and inadequacies." The disorder is common

among terrorists and suicide bombers, and it is also, Post says, the "dominant mechanism of the destructive charismatic, such as Hitler, who projects the devalued part of himself onto the interpersonal environment and then attacks and scapegoats the enemy without." Any crisis gives pathological narcissists a wonderful opportunity to rise to prominence because people in general are disoriented, scared, and looking for answers. Once in leadership positions of any kind, they're adept at pointing fingers and firing up old hatreds. And because they take offense easily and lack empathy, they have few qualms about inflicting horrible suffering on others. See Jerrold Post, "Terrorist Psycho-Logic: Terrorist Behavior As a Product of Psychological Forces," in Walter Reich, ed., *Origins of Terrorism: Psychologies, Ideologies, Theologies, States of Mind* (Washington, DC: Woodrow Wilson Centre Press, 1998), 27–28.

54. The term "fundamentalism" first appeared among American Protestant evangelicals in the first decades of the twentieth century. Today the term generally refers to all forms of absolutist religious doctrine and practice that are reactions against modernity, although this usage's pejorative connotation makes it contentious in some quarters. Moreover, the term is often misapplied. R. Scott Appleby and Martin Marty helpfully clarify the concept in "Fundamentalism," *Foreign Policy* 128 (January–February 2002): 16–22. See also Marty and Appleby, *The Glory and the Power: The Fundamentalist Challenge to the Modern World* (Boston: Beacon Press, 1992); and the five volumes of *The Fundamentalism Project*, eds. Marty and Appleby, published by the University of Chicago Press from 1993 to 2004.

55. Yehezkel Dror, *The Capacity to Govern: A Report to the Club of Rome* (London: Frank Cass, 2001), 46; and Langdon Gilkey, "The Flight from Reason: The Religious Right," in Paul Gross, Norman Levitt, and Martin Lewis, eds., *The Flight from Science and Reason* (New York: New York Academy of Sciences, 1996), 523–25.

56. Sam Harris provides a sweeping indictment of religion's role in the abridgement of reason in *The End of Faith: Religion, Terror, and the Future of Reason* (New York: Norton, 2004). Astonishingly, a religiously motivated disdain for reason, evidence, and the scientific method has insinuated its way into the highest decision-making levels of the United States. See Ron Suskind, "Without a Doubt," *New York Times Magazine*, October 17, 2004, 44–106.

57. Yeats—poet, mystic, and Anglo-Irish nationalist—was an aristocrat at heart and no fan of popular democracy. He loathed the mob and social disorder, and his attraction to authority led to his notorious, if brief, dalliance with Irish fascists in the early 1930s (but he later sided with the Spanish Republic against Franco). So people have debated what he meant by "the best" and "the worst" in this famous line from "The Second Coming." Critics suggest that for Yeats the best were members of society's elite—educated, refined, and naturally possessing the character and moral authority to govern—while the worst were members of the general public, especially the middle class, who were preoccupied with mundane pursuits and were often ruled by emotion. Others argue that Yeats was deeply disturbed by the Russian Revolution, so the best might have been Kerensky and his followers, while the worst were Lenin and the Bolsheviks. Either way, it's unfair to shackle such powerful writing to Yeats's peculiar preoccupations and

sentiments. Frank Tuohy provides a marvelous account of the man, his personality, politics, and writing in *Yeats* (New York: Macmillan, 1976).

58. In panarchy theory, Buzz Holling calls this kind of phenomenon, in which a system becomes locked into a degenerative cycle that doesn't allow for creative reorganization and regrowth, a "poverty trap." He writes, "This condition can then propagate downward through levels of the panarchy, collapsing levels as it goes. An ecological example is the productive savanna that, through human overuse and misuse, flips into an irreversible, eroding state, beginning with sparse vegetation. Thereafter, subsequent drought precipitates further erosion, and economic disincentives maintain sheep production. The same persistent collapse might also occur in a society traumatized by social disruption or conflict, so that its cultural cohesion and adaptive abilities are lost. In such a situation, the individual members of the society would be able to depend only on themselves and perhaps their immediate family members." C. S. Holling, "Understanding the Complexity of Economic, Ecological, and Social Systems," *Ecosystems* 4, no. 5 (2001): 400.

59. Richard Norgaard and Paul Baer, "Seeing the Whole Picture," Chapter 10 in Ervin Laszlo and Peter Seidel, eds., *Global Survival and Its Implications for Thinking and Acting* (New York: Select, 2006), 139–57.

60. As the historian Eric Hobsbawm wrote of the disorientation that the Great Depression caused among the West's elites, "It was precisely the absence of any solutions within the framework of the old liberal economy that made the predicament of the economic decision makers so dramatic." Hobsbawm, *Age of Extremes: The Short Twentieth Century, 1914–1991* (London: Abacus, 1994), 94.

61. I discuss the problem of unknown unknowns in Thomas Homer-Dixon, "Unknown Unknowns," chapter 7 of *The Ingenuity Gap: Facing the Economic, Environmental, and Other Challenges of an Increasingly Complex and Unpredictable Future* (New York: Vintage, 2002), 171–87.

62. Members of the Resilience Alliance offer the following definition: "Resilience is the capacity of a system to absorb disturbance and reorganize while undergoing change so as to still retain essentially the same function, structure, identity, and feedbacks." See Brian Walker, C. S. Holling, Stephen Carpenter, and Ann Kinzig, "Resilience, Adaptability and Transformability in Social-Ecological Systems," *Ecology and Society* 9, no. 2 (2004): 5. Available at http://www.ecologyandsociety.org/vol9/iss2/art5/print.pdf.

63. Rabun Taylor, *Roman Builders: A Study in Architectural Process* (Cambridge: Cambridge University Press, 2003), 49.

64. The British energy analyst and nuclear physicist Walt Patterson provides a detailed plan for highly decentralized electricity production in *Transforming Electricity* (London: Earthscan, 1999).

65. A tax on speculative transactions in international financial markets is one method often proposed to loosen coupling in these markets and lower the risk of cascading failures. Such a tax was originally proposed in the 1970s by the economist James Tobin, and for this reason it is often referred to as the Tobin tax. For a full account, see Mahbub ul Haq, Inge Kaul, and Isabelle Grunberg, *The Tobin Tax: Coping with Financial Volatility* (New York: Oxford University Press, 1996). Contagion in international financial markets might also be reduced by establishing an institution

roughly equivalent to a global central bank that could act as a lender of last resort in the event of financial panic. For an argument along these lines, see Guillermo Calvo, "Crises in Emerging Market Economies: A Global Perspective," Working paper 11305, National Bureau of Economic Research, April 2005.

66. It's commonly assumed among economists, business analysts, and systems managers that just-in-time production increases vulnerabily to cascading failures. But in some circumstances just-in-time production can actually boost resilience by stimulating continuous learning of better problem-solving and crisis-management techniques among firms in the production chain. See Toshihiro Nishiguchi and Alexandre Beaudet, "Case Study: The Toyota Group and the Aisin Fire," *Sloan Management Review* 40, no. 1 (Fall 1998): 49–59. More information on improving the resilience of industrial enterprises is available from the Center for Resilience at the Ohio State University. See http://resilience.osu.edu/.

67. The Bank of New York—a main clearinghouse for U.S. Treasury securities—was much less resilient. To the extent that its managers had planned for terrorist attacks, they appear to have thought purely in terms of an IRA-style attack against a single building, so the Bank's main and backup offices were surprisingly near each other in lower Manhattan, and both were inside the zone cordoned off after the attack.

68. The San Francisco Fire Department considers these cisterns the final line of defense for a "last-resort, worst-case, drop-dead scenario." They are part of a remarkably elaborate and unique Auxiliary Water Supply System that was built in the quake and fire's aftermath. For more information, see Steve Van Dyke, "San Francisco Fire Department Water Supply System," available at http://www.sfmuseum.org/quake/awss2.html.

69. The general principle articulated in this sentence holds, even though we've seen in chapters 5, 7, and 9 that the relationship between system connectivity and complexity, on the one hand, and system stability and resilience, on the other, is not straightforward. Connectivity's effect on resilience depends not just on its degree but also on its nature—specifically on how strong the connections are, whether they create feedback loops, what kinds of synergies they induce among disparate network components, and whether the topography of the network is scale-free or random. Regarding connection strength, research generally shows that networks with large numbers of relatively weak ties tend to be more resilient than those with a disproportionate number of strong ties. See the classic sociological studies by Mark Granovetter, "The Strength of Weak Ties," *The American Journal of Sociology* 78, no. 6 (May 1973): 1360–80; and "The Strength of Weak Ties: A Network Theory Revisited," *Sociological Theory* 1 (1983): 201–33. The argument is extended in Deborah Wallace and Rodrick Wallace, "Life and Death in Upper Manhattan and the Bronx: Toward an Evolutionary Perspective on Catastrophic Social Change," *Environment and Planning A* 32 (2000): 1245–66.

70. This is only partly a prescription for more development aid to poor countries. More aid is indeed needed, especially from the world's wealthiest countries, like the United States, and especially to achieve pragmatic local-level goals like reforestation, better use of scarce water, and treatment of disease. But it's also a prescription for muscular intervention—as we've seen (somewhat halfheartedly) in

Afghanistan—to stabilize failed states and provide basic security. And perhaps most important, it's a prescription for an international structure of incentives that helps poor countries prosper and draws their elites into the task of nation building. For example, a sharp reduction in Western domestic agricultural subsidies would raise prices for southern agricultural products, both internationally and domestically, and would help poor countries build the agricultural capital that's essential for greater investment elsewhere in the economy.

71. The details in this and the following paragraphs are drawn largely from Kerry Odell and Marc Weidenmier, "Real Shock, Monetary Aftershock: The 1906 San Francisco Earthquake and the Panic of 1907," *Journal of Economic History* 64, no.4 (December 2004): 1002–27.

72. William Greider, *Secrets of the Temple: How the Federal Reserve Runs the Country* (New York: Simon and Schuster, 1987), 273–74.

73. Accounts of the meeting are available in Robert West, *Banking Reform and the Federal Reserve: 1863–1923* (London: Cornell University Press, 1947), 71–72; Thibaut de Saint Phalle, *The Federal Reserve: An International Mystery* (New York: Praeger, 1985), 49–50; and William Barton McCash and June Hall McCash, *The Jekyll Island Club: A Southern Haven for America's Millionaires* (London: University of Georgia Press, 1989), 124–27. Frank Vanderlip provides a participant's account in *From Farm Boy to Financier* (New York: Appleton-Century, 1935), 210–19. On the history of the Fed, see "The Federal Reserve and the Money Trust," chapter 4 of Jerry Markham, *A Financial History of the United States, Volume II: From J. P. Morgan to the Institutional Investor (1900–1970)* (New York: Shape, 2002), especially 42–46.

74. C. S. Holling, "Understanding the Complexity of Economic, Ecological, and Social Systems," *Ecosystems* 4, no. 5 (2001): 399.

75. Economists heatedly debate whether recessions produce, on balance, economic benefit or harm. Some, mainly on the ideological left, believe that they generally cause harm and so assume that it's the central purpose of economic policy makers and institutions to prevent any significant downturn in economic output. Others, often on the right and influenced by the twentieth-century Austrian school of economics that included Ludwig von Mises and Friedrich Hayek, believe that moderate recessions are a useful purgative of excessive, inefficient, and unprofitable investment. The various arguments are nicely summarized for lay readers in "The Unfinished Recession: A Survey of the World Economy," *The Economist* (September 28, 2002), especially 22–26.

76. "The opening up of new markets, foreign or domestic, and the organizational development from the craft shop and factory to such concerns as U. S. Steel illustrate the same process of industrial mutation—if I may use that biological term—that incessantly revolutionizes the economic structure from within, incessantly destroying the old one, incessantly creating a new one. This process of Creative Destruction is the essential fact about capitalism. It is what capitalism consists in and what every capitalist concern has got to live in." Joseph Schumpeter, *Capitalism, Socialism, and Democracy* (London: Unwin Paperbacks, 1943), 83.

77. Recessions have become less severe also because of firm-level technological and organizational changes that have permitted, for instance, more efficient inventory management. But because these changes often tighten coupling among firms,

they may exacerbate the economy's overall vulnerability to cascading failure. On changes in the risk and severity of recessions, see Steven Weber, "The End of the Business Cycle?" *Foreign Affairs* 76, no. 4 (1997): 65–82; and David Leonhardt, "The FedEx Economy: Have Recessions Absolutely, Positively Become Less Painful?" *New York Times*, October 8, 2005, national edition, B1.

78. This argument is developed at length in "The Unfinished Recession," *The Economist*.

79. "One way to achieve this," write Resilience Alliance members Brian Walker and Buzz Holling, "is through a series of deliberate, small shocks . . . to prevent later, catastrophic shifts. That is, the creative destruction is transferred to a smaller and faster scale where learning and change can occur without destroying the larger organization." Walker and Holling, "Resilience, Adaptive Capacity and Transformative Capacity," Working Paper of the Resilience Alliance, prepared for a workshop at the Jekyll Island Club Hotel, April 9 and 10, 2003, 8.

80. Research in fields as diverse as archaeology and organizational theory suggests that periods of instability or discontinuous change—which I would call break-down—are key to the successful adaptation of human social systems. For example, see James McGlade and Sander Van der Leeuw, "Introduction: Archaeology and Non-Linear Dynamics—New Approaches to Long-Term Change," in Van der Leeuw and McGlade, eds., *Time, Process, and Structured Transformation in Archaeology* (London: Routledge, 1997), 1–31, especially 9–13; and Michael Tushman and Elaine Romanelli, "Organizational Evolution: A Metamorphosis Model of Convergence and Reorientation," in L. L. Cummings and Barry Staw, eds., *Research in Organizational Behavior, Volume 7* (Greenwich, CT: JAI Press, 1985), 171–222.

81. Two useful discussions of scenario-based forecasting are John Ratcliffe, "Scenario Planning: An Evaluation of Practice," *Futures Research Quarterly* 19, no. 4 (Winter 2003): 5–25; and Hugues de Jouvenel, *An Invitation to Foresight*, trans. Helen Fish (Paris: Futuribles Perspectives, 2004).

82. Howard Rheingold analyses how new information technologies allow rapid social mobilization in Rheingold, *Smart Mobs: The Next Social Revolution* (Cambridge, MA: Perseus, 2002).

83. The Israeli political scientist Yehezkel Dror makes a similar point: "Grave socio-political and cultural crises may be conducive, and perhaps essential, to radical innovations in politics and governance. But there is no assurance that they will result in beneficial institutions and regimes, as illustrated by the victory of Nazism in the Weimar Republic. To increase the chances that crises will result in desirable innovations, promising ideas have to be prepared well in advance. . . . *Government redesigns meeting real needs should therefore be prepared, even when they seem not to be feasible, so as to be ready when radical governance restructuring becomes feasible—such as result from crisis or enlightened democratic rulership."* (emphasis in original). Dror, *The Capacity to Govern: A Report to the Club of Rome* (London: Frank Cass, 2001), 220.

84. The collective action problem facing nonextremists described in this paragraph is a strategic dilemma commonly called, by game theorists, a "stag hunt." Everyone would be better off if they cooperated to stop the actions of extrem-ists, but the price of going alone is very high, and when one is faced with

uncertainty about whether other nonextremists will act, it makes sense to wait on the sidelines. If others do act and are successful, a person waiting on the sidelines can reap the gains without risking anything (game theorists call this "free riding"). Community building through planning for breakdown would reduce the uncertainty of nonextremists about the actions of other nonextremists and increase the reputational cost of free riding, thus helping to solve their collective action problem. On the stag hunt, see Brian Skyrms, *The Stag Hunt and the Evolution of Social Structure* (Cambridge: Cambridge University Press, 2004); and on applications of game theory to problems of cooperation, see Michael Taylor, *The Possibility of Cooperation* (Cambridge: Cambridge University Press, 1987).

85. "Do not try to plan the details," Holling recommends. Instead we should "invent, experiment, and build." Also, because we often learn most from failure, we should design our experiments to be "safe-fail." If they don't work, the costs to people and the environment should be low. C. S. Holling, "From Complex Regions to Complex Worlds," *Ecology and Society* 9, no. 1 (2004), available at http://www.ecologyandsociety.org/vol9/iss1/art11/print.pdf.

86. An essential element of any type of reformed capitalism will be the redistribution of economic risk and insecurity so that it is far more evenly shared across social and economic classes. This will likely involve some kind of broadly based system of insurance against economic catastrophe available to both individuals and groups. See Jacob Hacker, "Insurance Policy," *The New Republic* 233, no. 1 (July 4, 2005): 18–21. For a vigorous defense of economic growth as a source of social tolerance and generosity that provides the basis for political liberty, see Benjamin Friedman, *The Moral Consequences of Economic Growth* (New York: Knopf, 2005).

87. Herman Daly and John Cobb Jr. attempt to imagine a very different economic future in their classic book *For the Common Good: Redirecting the Economy Toward Community, the Environment, and a Sustainable Future* (Boston: Beacon Press, 1989). See also, Herman Daly, "Economics in a Full World," *Scientific American* 293, no. 3 (September, 2005): 100–107.

88. Some of these challenges to conventional economics are discussed in Kenneth Arrow et al., *Journal of Economic Perspectives* 18, no. 3 (Summer 2004): 147–72.

89. In Hegelian terms, it's useful to think of the dominant rationalization as a "thesis" that becomes more elaborate and rigid as it evolves upward along the adaptive cycle's front loop, or growth phase. The cycle's collapse is brought about, in part, by the emergence of an "antithesis," while the recombination and regeneration in the backloop is a process of finding a new "synthesis."

90. Many of the ideas in this paragraph, especially regarding how greater diversity of a system's elements can increase creativity, are developed further in Stuart Kauffman, *At Home in the Universe: The Search for the Laws of Self-organization and Complexity* (New York: Oxford University Press, 1995). See particularly his comments on autocatalytic sets on pp. 59–66 and on systems that reside at the end of chaos on pp. 86–92.

91. Research shows that such "shadow networks" are often key to successful social adaptation. See Lance Gunderson, "Resilience, Flexibility and Adaptive Management—Antidotes for Spurious Certitude?" *Ecology and Society* 3, no. 1

(1999), available at http://www.ecologyandsociety.org/vo13/iss1/art7/. See also Per Olsson et al., "Shooting the Rapids—Navigating Transitions to Adaptive Ecosystem Governance," *Ecology and Society* (in review).

92. This proposal has some faint similarity to suggestions made by other authors, especially the English novelist and social critic H. G. Wells. In the 1920s, Wells suggested that world revolution leading to a liberal world state could be achieved through the coordinated action, simultaneously across many societies, of men of high intelligence and professional accomplishment, drawn especially from business, scientific, academic, and managerial elites. In today's light, Wells's idea seems quaint and even bizarre. In contrast to the present proposal, for instance, it was based on the assumption that the process of governance is mainly a matter of technocratic decision making. It was also profoundly antidemocratic, because only highly educated people could contribute to the revolutionary effort. See H. G. Wells, *The Open Conspiracy: Blue Prints for a World Revolution* (London: Gollancz, 1928); and Warren Wager, *H. G. Wells and the World State* (New Haven: Yale University Press, 1961).

CHAPTER TWELVE

1. Neil MacFarquhar, "Target: Saudi Oil Industry," *International Herald Tribune*, May 31, 2004, 1; "4 U.S. Soldiers Die in Land-Mine Blast in Afghanistan," *International Herald Tribune*, May 31, 2004, 2; Salman Masood, "Pro-Taliban Sunni Cleric Is Killed in Pakistan," *International Herald Tribune*, May 31, 2004, 2; Tim Weiner, "Flood Challenges Haiti's Small Resources," *International Herald Tribune*, May 31, 2004, 6; and Tim Weiner and Lydia Polgreen, "Grief as Haitians and Dominicans Tally Flood Toll," *New York Times*, May 28, 2004, online archive.

2. "Lebanese Soldiers Kill 3 Fuel-Price Protesters," *International Herald Tribune*, May 28, 2004, 5; Reuters, "5 Killed in Beirut in Protests over Rapidly Rising Fuel Prices," *New York Times*, May 28, 2004, online archive; and Adnan El-Ghoul, "Calm Returns to Beirut As Army Redeploys," *Daily Star* (Beirut), May 29, 2004, 1.

3. An excellent history of the precursors to this conflict is Samir Khalaf, *Civil and Uncivil Violence in Lebanon: A History of the Internationalization of Communal Conflict* (New York: Columbia University Press, 2002).

4. A recent comprehensive treatment of the history and ideas of the Axial Age is Karen Armstrong, *The Great Transformation: The Beginning of Our Religious Tradition* (New York: Knopf, 2006).

5. Karl Jaspers, *Vom Ursprung und Ziel der Geschichte*, 1st ed. (München: Piper Verlag, 1949).

6. In Roman times, the town of Baalbek was called Heliopolis, the "City of the Sun."

7. Rabun Taylor, *Roman Builders: A Study in Architectural Process* (Cambridge: Cambridge University Press, 2003), 120–25.

8. Andrew Collins provides one of the more credible and well-researched arguments to this effect in "Baalbek: Lebanon's Sacred Fortress," *New Dawn Magazine* 43 (July–August 1997), available at http://www.newdawnmagazine.com/articles/Baalbek%20-%20Lebanons%20Sacred%20Fortress1.html.

9. Friedrich Ragette, *Baalbek* (London: Chatto and Windus, 1980). On the likely mechanism used to transport these megaliths, see Jean-Pierre Adam, *Roman Building: Materials and Techniques*, trans. Anthony Mathews (London: Routledge, 2001), 28–29. On Roman stone transport in general, see Colin Adams, "Who Bore the Burden: The Organization of Stone Transport in Roman Egypt," chapter 8 in David Mattingly and John Salmon, eds., *Economies beyond Agriculture in the Classical World* (London: Routledge, 2001), 171–92.

10. On the importance of value change, see James Gustave Speth, *Red Sky at Morning: America and the Crisis of the Global Environment* (New Haven: Yale University Press, 2004), especially 191–96. See also Paul Ehrlich and Anne Ehrlich, *One with Nineveh: Politics, Consumption, and the Human Future* (Washington, DC: Island, 2004), 270–82.

11. Herman Daly and John Cobb Jr., *For the Common Good: Redirecting the Economy toward Community, the Environment, and a Sustainable Future* (Boston: Beacon Press, 1989).

12. My position is therefore radically at odds with that adopted by Benjamin Friedman in *The Moral Consequences of Economic Growth* (New York: Knopf, 2005).

13. On the role of technology in reducing human impact on nature, see Jesse Ausubel, "Maglevs and the Vision of St. Hubert or the Great Restoration of Nature: Why and How," in W. Steffen Jaeger and D. Carson, eds., *Challenges of a Changing Earth* (Heidelberg: Springer, 2002), 175–82.

14. On the importance of expanded democracy to break through the resistance of today's "global plutocracy" to change, see Branko Milanovic, *Worlds Apart: Measuring International and Global Inequality* (Princeton: Princeton University Press, 2005), 149–52.

15. Philip Selznick, *The Moral Commonwealth: Social Theory and the Promise of Community* (Berkeley: University of California Press, 1992).

16. "The fact is that the people on this planet live under one global atmosphere, on the shores of one global ocean, our countries linked by flows of people, money, goods, ideas, images, diseases, drugs, weapons, and, perhaps ultimately, nuclear explosives. We cannot keep one end of the boat afloat while the other end sinks." John Holdren, "Environmental Change and the Human Condition," *Bulletin of the American Academy of Arts and Sciences* 57, no. 1 (Fall 2003): 31.

17. Peter Singer, *One World: The Ethics of Globalization* (New Haven: Yale University Press, 2d ed., 2004); and Alexander Wendt, "Why a World State Is Inevitable," *European Journal of International Relations* 9, no. 4 (December 2003): 491–542. Drawing on the thinking of the political scientist Daniel Deudney, Wendt stresses the role of the increasing destructive power of weapons as a key motivation for the creation of a world state.

18. According to folklore, women increase their fertility when they touch the stone.

ILLUSTRATION

CREDITS

_____

PROLOGUE: FIRESTORM

Page 4
Title: The Roman Forum
Photo Credit: Thomas Homer-Dixon

CHAPTER ONE: TECTONIC STRESSES

Page 10
Title: September 11, 2001, won't be the last time we walk out of our cities.
Photo credit: Andrea Mohin/*New York Times*

Page 15
Title: Terrorists need less than a ten-thousandth of the world's highly enriched uranium (HEU) to build a crude atomic bomb.
Artists: Karen Frecker and Todd Barsanti

Page 19
Title: "Let's change 'brink of chaos' to 'Everything is wonderful.'"
Artist: David Sipress, *New Yorker* cartoon, October 18, 2004

Page 27
Title: The traveler confronts path dependency.
Artist: Todd Barsanti

CHAPTER TWO: A KEYSTONE IN TIME

Page 32
Title: The keystone in the Colosseum
Photo Credit: Laura Bondy

Page 33
Title: The design of the voussoir arch gives it strength.
Artist: Karen Frecker

Page 35
Title: The Colosseum and its foundation originally contained about a million metric tons of rock, concrete, and brick.
Photo credit: Thomas Homer-Dixon

Page 39
Title: The light won't last forever in a closed system.
Artist: Todd Barsanti

Page 45
Title: A-frame cranes lifted the rock for Level 1 of the Colosseum.
Source: Rabun Taylor, *Roman Builders: A Study in Architectural Process* (Cambridge: Cambridge University Press, 2003), 153. Reprinted with Permission of Cambridge University Press.

Page 46
Title: Cranes were rolled up the sloped seating area to construct Level 4.
Source: Rabun Taylor, *Roman Builders: A Study in Architectural Process* (Cambridge: Cambridge University Press, 2003), 171. Reprinted with Permission of Cambridge University Press.

Page 50
Title: Ancient Rome and its surrounding regions
Artist: Todd Barsanti

CHAPTER THREE: WE ARE LIKE RUNNING WATER

Page 59
Title: The city of Rome's population rose and fell dramatically between 200 BCE and 600 CE.
Source: Roberto Luciani, *Roma Sotterranae* (Rome: Fratelli Palombi, 1984), 9.

Page 66
Title: Migrants trek across Africa in overloaded trucks, hoping to get to Europe.
Photo credit: Toru Morimoto

Title: But many pay a terrible price.
Photo credit: Javier Bauluz

Page 71
Title: Forty-three percent of the urban population of poor countries lives in slums.
Photo credit: Thomas Homer-Dixon

CHAPTER FOUR: SO LONG, CHEAP SLAVES

Page 78
Title: Drilling for gas at the Sikanni Chief River in northern B.C. in 1974
Photo credit: Thomas Homer-Dixon

Page 79
Title: The author as a roughneck on the floor of the Sikanni Chief rig
Photo credit: Thomas Homer-Dixon

Page 86
Title: Decades can pass between oil discovery and production peaks.
Artists: Karen Frecker and Todd Barsanti

Page 90
Title: Global oil discovery peaked in the early 1960s.
Source: Harry Longwell, "The Future of the Oil and Gas Industry: Past Approaches, New Challenges,"*World Energy* 5, no. 3 (2002): 100–4, and Colin Campbell, personal correspondence.

Page 91
Title: Most giant oil fields were found decades ago.
Source: Matthew Simmons, *The World's Giant Oilfields* (Houston: Simmons and Company International, 2001).

CHAPTER FIVE: EARTHQUAKE

Page 108
Title: Smoke from the October 2003 wildfires in Southern California, as seen from space
Photo credit: NASA, Earth Observatory

Page 114
Title: Tight coupling at high speed contributes to multi-car accidents.
Photo credit: AP/Worldwide Photos

Page 117
Title: The world's networks can be divided into two general categories.
Source: Albert-László Barabási, *Linked: The New Science of Networks* (Cambridge, Massachusetts: Perseus, 2002), 71. Reprinted with permission.

Page 126
Title: Modern empires have collapsed more quickly than early-modern empires.
Source: Rein Taagepera, "Expansion and Contraction Patterns of Large Polities," *International Studies Quarterly* 41 (1997): 475–504.

CHAPTER SIX: FLESH OF THE LAND

Page 138
Title: As the soil washes away, only the rocks stay behind.
Photo credit: Thomas Homer-Dixon

Page 144
Title: Each year, Indonesia loses an area of forest the size of Connecticut.
Photo credit: Togu Manurung

Page 145
Title: The annual global catch of cod, hake, and haddock has declined one-third from its peak.
Source: FAO FISHSTAT, "Capture Production, 1950–2001," Database.

CHAPTER SEVEN: CLOSING THE WINDOWS

Page 154
Title: Kilimanjaro at dawn in April 1983
Photo credit: Thomas Homer-Dixon

Page 156
Title: In Alaska, melting permafrost has caused telephone poles to tip.
Photo credit: United States Geological Survey

Page 163
Title: Since 1970, global temperature and the atmosphere's level of carbon dioxide have increased together.
Sources: Temperature data courtesy of the Climatic Research Unit, University of East Anglia and Hadley Centre, the Met Office, U.K. CO_2 data obtained from C. D. Keeling and T. P. Whorf, "Atmospheric CO_2 Records from Sites in the SIO Air Sampling Network," in *Trends: A Compendium of Data on Global Change* (Carbon Dioxide Information Analysis Center, Oak Ridge National Laboratory, U.S. Department of Energy, Oak Ridge, Tenn., U.S.A. 2003), and D. M. Etheridge, L. P. Steele, R. L. Langenfelds, R. J. Francey, J. M. Barnola and V. I. Morgan, "Historical CO_2 records from the Law Dome DE08, DE08–2, and DSS ice cores," in *Trends* (US DOE, Oak Ridge, 1998).

CHAPTER EIGHT: NO EQUILIBRIUM

Page 202
Title: "No creature adapts so ingeniously as the human being."
Artist: Joel Pett, *Lexington Herald-Leader*

CHAPTER NINE: CYCLES WITHIN CYCLES

Page 208
Title: Antonio Santucci's armillary sphere in the Museum of the History of Science, Florence, Italy
Photo credit: Franca Principe, IMSS–Florence

Page 224
Title: The gate to the Temple of Bacchus in Baalbek
Illustration credit: (1996.1.81) David Roberts, "Doorway, Baalbec," 1842–1845, tinted lithograph, 23 1/2 x 17 inches. Collection of the Nasher Museum of Art at Duke University, Museum purchase.

Page 229
Title: The adaptive cycle can be represented in three dimensions.
Artist: Todd Barsanti

CHAPTER TEN: DISINTEGRATION

Page 236
Title: The Pont du Gard, just northeast of Nîmes, France
Photo credit: Thomas Homer-Dixon

Page 238
Title: Limestone deposits of the Nîmes aqueduct provide a record of its past.
Photo credit: Thomas Homer-Dixon

Page 245
Title: The Rhône valley
Artist: Todd Barsanti

CHAPTER ELEVEN: CATAGENESIS

Page 285
Title: The 1906 San Francisco Fire
Source: J. D. Givens, *San Francisco in Ruins: A Pictorial History of Eight Score Photo-views of the Earthquake Effects, Flames Havoc, Ruins Everywhere, Relief Camps* (San Francisco: Leon Osteyee, 1906).

CHAPTER TWELVE: BAALBEK: THE LAST ROCK

Page 303
Title: The last standing columns of the Temple of Jupiter, Baalbek, Lebanon
Photo credit: Thomas Homer-Dixon

Page 307
Title: The Hajar el Hibla rests in a quarry near the Temple of Jupiter.
Photo credit: Thomas Homer-Dixon

NOTES

Page 342
Title: Frequency distributions of random and scale-free networks
Artist: Karen Frecker

ACKNOWLEDGMENTS

D URING THE FIVE YEARS I've been working on *The Upside of Down*, my wife, Sarah Wolfe, has been a real source of strength. Sarah has given me love, intellectual inspiration, creative insight, and constant encouragement. I doubt the book would exist if not for her, and if it did exist it would surely be much less than the book it is now. I'll always be especially thankful for several vital interventions she made—one with only five minutes to spare before a critical deadline—that solved problems that had long stumped me. And one event that Sarah and I have shared during this time informs every sentence I've written here: our son, Benjamin Homer-Dixon, was born in early 2005. When Benjamin is my age, humankind will be in the midst of the great challenges I've outlined in this book. I've never felt such an intimate and passionate connection to our collective future.

Bruce Westwood, my literary agent, gently urged me to begin a new book, and his wisdom, kindness, and friendship have been great solace at difficult moments. In Canada, I've been blessed to have had Louise Dennys and Michael Schellenberg as my editors at Knopf Canada. Michael's marvelously thoughtful editing has made my arguments and prose far more readable and convincing. Louise has never wavered in her confidence in this project, and she has been supportive throughout. And in the years I've worked with her—now nearly a decade on two books—I've learned more about the craft of writing than I could possibly recount. In the United States, my editor at Island Press, Jonathan

Cobb, has been with me every step of the way. Early on he helped me distill my argument's essential points. He has since read every chapter innumerable times and spent many hours discussing them with me on the phone, often paragraph by paragraph. His unflagging generosity, wry sense of humor, and astonishingly insightful mind have always been available when I needed them, and I needed them often.

Several people helped me with specific sections of the book. Ian Graham, a physicist, senior technology consultant to a major bank, and a dear friend for over four decades walked me through the physics of energy, and then read and reread—and in some cases rewrote—the passage on the thermodynamics of empire as it evolved through many drafts. Rabun Taylor of Harvard University discussed with me the architecture and engineering of the Colosseum and provided valuable leads that helped me estimate more accurately the energy required for the building's construction. Branko Milanovic of the World Bank helped me understand the tangled problem of global income inequality, and James Risbey of the Australian Commonwealth Scientific and Research Organization reviewed my chapter on climate change. Joseph Tainter gave me a day of his time in Albuquerque to discuss his theory of societal collapse. Buzz Holling met me several times in Florida and Ontario to talk about panarchy theory. I spent an afternoon with Sander van der Leeuw in Paris to discuss his research on ancient societies as complex adaptive systems. Tainter, Holling, and van der Leeuw have all read and commented on the sections of this book that deal with their research, although they are in no way responsible for my interpretations of their work.

I've been fortunate to have had an extraordinary staff at the University of Toronto. Ashllie Claassen, my assistant at the Trudeau Centre for Peace and Conflict Studies, has stepped into the breach an uncountable number of times to keep the Centre running and meet the vital needs of our students while I've been writing at home or away on research trips. Over the past five years, six of the Centre's students— Vanessa Corlazzoli, Karen Frecker, Pierre Gemson, Jenna Mequid, Edan Rotenberg, and Jenna Slotin—have worked as researchers on a broad range of topics associated with this book, always with intelligence, care, and enthusiasm. My heartfelt thanks go to these remarkable young people who are already changing the world for the better. Karen deserves a very special thanks, of course, not only for her truly heroic work on the Colosseum calculations but even more for her generosity and patience and her commitment to seeing through to a successful conclusion what had become an immensely difficult project.

Many, many others have helped in vital ways. I especially wish to thank John Bongaarts for data on population growth and the demographic transition; Alastair Cairns for his report on the impact of the 9/11 attacks on the New York financial system; Colin Campbell for data and much advice on the coming peak of conventional oil production; Richard Cincotta for insight into the relationship between demographic trends and social conflict; Dan Ciuriak for thoughtful comments on international economics; Culter Cleveland for advice on how to understand EROI statistics and calculations; Josh Cohen for reviewing my passage on capitalist hegemony; William Conway for information on total human biomass relative to other species; Dan Deudney for references to the writings of H. G. Wells; Yehezkel Dror for his panoramic view of the evolution of institutions of governance; Hari Eswaran for data on Pakistan's ability to feed itself; Henry Farrell for information on the causes of Rome's collapse; Julio Friedmann for help understanding coal gasification and carbon sequestration; David Galbraith for leads to research on the pollinator crisis; Jack Goldstone for sage thoughts on the nature of social breakdown in the face of converging stresses; John Holdren for the idea of stages of denial; Terry Joyce and Ruth Curry for insight into the operation of the North Atlantic thermohaline circulation; Amory Lovins for his view of the stability of the North American electrical grid; Fiona Mackintosh for insights on spirituality and leadership; Michael Marien for his encyclopedic knowledge on everything from human adaptation to infoglut; John McRuer for his helpful ideas about central dependence and economic inequality; Andrew Miller for his comments on the state of mind of New Yorkers in the aftermath of 9/11; Sandra Postel, Brian Halweil, and Michael Renner for data on water scarcity and Chinese food production; Steve Purdey for feedback on my argument about the growth imperative; Mike Rice for information on the network of cisterns that help protect San Francisco from catastrophic fire; Paul Ritvo for insights on the psychology of growth, addiction, and denial; Vaclav Smil for help understanding the global nitrogen cycle and Roman energy requirements; Ian Sterling for his research on global warming's threat to polar bears; Marc Tessier-Lavigne for information on the technical requirements to manufacture and manipulate viruses for bioterrorism; Ad van Dommelen for a translation and cultural interpretation of the Dutch concept of "maakbaar"; Kim Vicente for pointing me to Arthur Koestler's writings on the Copernican revolution; Brian Walker for his review of my passage on panarchy theory; and, last but certainly not least, Mary Langford for being such a loving caregiver for Ben while Sarah and I worked.

Karen Frecker prepared many of the initial drawings and graphs for the chapters on Rome and energy, while Todd Barsanti provided the rest and standardized the design, format, and quality of all photographs and illustrations. Todd's advice as a professional illustrator and designer and his aesthetic sense have helped make the book's illustrations come alive. Others who deserve great thanks for their help with illustrations are Steve Reiter at the United States Geological Survey Information Services, Maureen Morin and Terry Jones at the University of Toronto Information Services, and Kathleen Freer and Jenilynn Johnson of the Leo Epp Company, property managers of what was once the Call Building in San Francisco, who provided the extraordinary photograph of the great 1906 fire from their company archives.

From the earliest days of this project, Wren Wirth's enthusiasm, warmth, and ceaseless support have inspired me more than she could possibly know. And I greatly appreciate the financial support provided by the Winslow Foundation to cover some of the costs of my research and travel.

And finally, my sister-in-law, Laura Bondy, deserves particular credit and thanks. When I visited the Colosseum in Rome in May 2003 and spied the keystone that led me to look at the Roman empire's carved rock in a whole new way, I came away with only a poor photograph of the keystone. A year later, on a visit to Italy to study Renaissance art, Laura made a special trip to the Colosseum on my behalf, located the keystone, and took the splendidly clear photo printed at the beginning of chapter 2. Thank you, Laura, for getting this photo of what is, to me, the most important rock in the world, and for never saying how truly bizarre the request must have seemed.

INDEX

Thomas Homer-Dixon is a Professor of Political Science at the University of Toronto, where he teaches in the Trudeau Centre for Peace and Conflict Studies. His previous book, *The Ingenuity Gap*, won the Governor General's Literary Award for Non-Fiction. For more information about his work, visit www.homerdixon.com.

A NOTE ABOUT THE TYPES

The Upside of Down has been set in Janson, a misnamed typeface designed in or about 1690 by Nicholas Kis, a Hungarian in Amsterdam. In 1919 the original matrices became the property of the Stempel Foundry in Frankfurt, Germany. Janson is an old-style book face of excellent clarity and sharpness, featuring concave and splayed serifs, and a marked contrast between thick and thin strokes.

BOOK DESIGN BY CS RICHARDSON